BIODIVERSITY IN LAND–INLAND WATER ECOTONES

MAN AND THE BIOSPHERE SERIES

Series Editor:

JOHN JEFFERS
The University, Newcastle-upon-Tyne, NE1 7RU, United Kingdom and
Glenside, Oxenholme, Cumbria LA9 7RF, United Kingdom

Editorial Advisory Board

EDOUARD G. BONKOUNGOU
Centre de Recherche en Biologie et Ecologie Tropicale,
B.P. 7047, Ouagadougou, Burkina Faso

GONZALO HALFFTER
Instituto de Ecologia, A.C., Apartado Postal 63,
91000 Xalapa, Veracruz, Mexico

OTTO LANGE
Lehrstuhl für Botanik II, Universität Würzburg, Mittlerer Dallenbergweg 64,
D-8700 Würzburg, Germany

LI WENHUA
Commission for Integrated Survey of Natural Resources,
Chinese Academy of Sciences, P.O. Box 787,
Beijing, People's Republic of China

GILBERT LONG
Centre d'Ecologie Fonctionelle et Evolutive Louis Emberger (CNRS-CEFE),
Route de Mende, B.P. 5051, 34033 Montpellier Cedex, France

IAN NOBLE
Research School of Biological Sciences, Australian National University,
P.O. Box 475, Canberra City A.C.T. 2601, Australia

P.S. RAMAKRISHNAN
School of Environmental Sciences, Jawaharlal Nehru University,
New Delhi 110067, India

VLADIMIR SOKOLOV
Institute of Evolutionary Morphology and Animal Ecology,
Academy of Sciences, 33 Leninsky Prospect,
117071 Moscow, Russian Federation

ANNE WHYTE
Environment and Natural Resources Division, International Development Research
Centre (IDRC),
P.O. Box 8500, Ottawa K1G 2119, Canada

Ex officio:

MALCOLM HADLEY
Division of Ecological Sciences, UNESCO, 7, Place de Fontenoy,
75352 Paris 07SP, France

MAN AND THE BIOSPHERE SERIES

Series Editor J.N.R. Jeffers

VOLUME 18

BIODIVERSITY IN LAND–INLAND WATER ECOTONES

Edited by
Jean-Bernard Lachavanne and Raphaëlle Juge

University of Geneva
Switzerland

PUBLISHED BY

PARIS

AND

The Parthenon Publishing Group
International Publishers in Science, Technology & Education

Published in 1997 by the United Nations Educational, Scientific and Cultural Organization,
7 Place de Fontenoy, 75700 Paris, France—UNESCO ISBN 92-3-103161-9
and
The Parthenon Publishing Group Inc.
One Blue Hill Plaza
PO Box 1564, Pearl River,
New York 10965, USA—ISBN 1-85070-735-9
and
The Parthenon Publishing Group Limited
Casterton Hall, Carnforth,
Lancs LA6 2LA, UK—ISBN 1-85070-735-9

© Copyright **UNESCO 1997**

The designations employed and the presentation of the material throughout this publication do not imply the expression of any opinion whatsoever on the part of the publishers and the UNESCO Secretariat concerning the legal status of any country or territory, city or area or of its authorities, the delimitations of its frontiers or boundaries.

The authors are responsible for the choice and the presentation of the facts contained in this book and for the opinions expressed therein, which are not necessarily those of UNESCO and do not commit the organization.

Mention of a trademark or a proprietary product does not constitute a guarantee or a warranty of the product and does not imply its approval to the exclusion of other products that also may be suitable.

No part of this publication may be reproduced,
in any form, without permission from the
publishers except for the quotation of
brief passages for the purpose of review.

British Library Cataloguing in Publication Data

Biodiversity in land–inland water ecotones. – (Man and the Biosphere ; v. 18)
1. Biological diversity 2. Ecotones 3. Aquatic ecology 4. Freshwater ecology
I. Lachavanne, J. II. Juge, R.
333.9'52
ISBN 1-85070-735-9

Library of Congress Cataloging-in-Publication Data

Biodiversity in land–inland water ecotones / edited by Jean-Bernard Lachavanne and Raphaëlle Juge
    p.   cm. — (Man and the Biosphere series ; v. 18)
  Papers from a special session held at the 25th Congress of the International Association of Theoretical and Applied Limnology, held in Barcelona, Aug. 21–27, 1992; with invited papers.
  Includes bibliographical references and index.
  ISBN 1-85070-735-9
  1. Land-water ecotones—Congresses. 2. Biological diversity—Congresses. 3. Freshwater ecology—Congresses. 4. Riparian ecology—Congresses. I. Lachavanne, J. B. II. Juge, Raphaëlle. III. International Association of Theoretical and Applied Limnology. Congress (25th : 1992 : Barcelona, Spain) IV. Series.
QH541.15.E27B56   1996
574.5'2632—dc20   96-33108
                                                                                                                                    CIP

Typeset by Martin Lister Publishing Services, Carnforth, UK
Printed and bound in the UK by Butler and Tanner Ltd., Frome and London

PREFACE

UNESCO's Man and the Biosphere Programme

Improved scientific understanding of natural and social processes associated with the environment is the focus of UNESCO's Man and the Biosphere Programme. It works towards this aim through the provision of information useful to resource managers and administrators and by promoting the conservation of genetic diversity. Much work is devoted to problem-solving ventures that bring together scientists, policy-makers and local people. In parallel with this, the programme seeks to develop scientific networks and strengthen regional co-operation.

Launched in the early 1970s, the programme is nationally based but internationally co-ordinated. MAB emphasizes problem solving research, training and demonstration. Often interdisciplinary teams are assembled to analyse interactions between ecological and social systems. A systems approach is used to understand relationships between development and the environment.

MAB is a decentralized programme with field projects and training activities in all regions of the world. Universities, academies of sciences, national research laboratories and other research and development institutions are all involved and brought together under the auspices of more than 100 MAB National Committees. Co-operation with a range of international government and non-governmental organizations is characteristic of the MAB Programme.

Man and the Biosphere Book Series

The Man and the Biosphere Series was launched to help communicate results from the MAB Programme. Primarily aimed at upper level university students, scientists and resource managers, the books are not normally suitable as undergraduate textbooks but rather seek to provide additional resource material. Some books are based on primary case studies that involve original data collection. Others provide a synthesis of global and regional research across several sites or countries. In all cases, a state of the art assessment of knowledge and methodological approaches is sought. In some cases, books are commissioned. The series editor is John Jeffers, formerly Director of the Institute of Terrestrial Ecology in the United Kingdom, who has been associated with MAB since its inception.

Biodiversity in Land–Inland Water Ecotones

Philosophers, writers and scientists, from cell biologists to ecologists, have long recognized the special nature of boundaries and interface areas of all kinds. Among ecologists in particular, there has been an upsurge in interest in the sensitive boundary areas of interaction between ecosystems, which they call 'ecotones'. The term was first used in 1905 by Frank Clements, one of the fathers of ecology in North America, to denote the junction between two plant communities where processes of competition or exchange might readily be observed. It has now been extended to refer to interfaces, edges, transition zones or boundaries between adjacent ecosystems.

Within the MAB Programme, concern with ecotones has focused particularly on land–water interfaces, in coastal regions but more particularly between land and inland waters of various kinds. The present volume, *Biodiversity in Land–Inland Water Ecotones*, constitutes one of several synthesis activities within a collaborative research project on such ecotones. The objectives and content of this project, 'The role of land–inland water ecotones in landscape management and restoration', were worked out by an international workshop held at Sopron, Hungary, in May 1988. Proposals on comparative research activities were put forward in the form of nineteen working hypotheses, set out on MAB Digest 4, with further elaboration in the contributions to an earlier volume (Number 4) in the present series (*Ecology and Management of Aquatic–Terrestrial Ecotones*). Several of these hypotheses are concerned with biological diversity, including the following: 'Land–inland water ecotones are often characterized by higher biological diversity than adjacent patches, consequently the frequency of ecotones across a landscape directly affects biodiversity in a predictable manner'.

The initial sponsors of the Ecotones project were two of UNESCO's intergovernmental programmes (MAB and the International Hydrological Programme, IHP), in co-operation with non-governmental scientific bodies such as the International Association of Theoretical and Applied Limnology (SIL). The project was organized in the form of a major network of more than fifty field projects involving forty countries and two more specific sub-networks, one dealing with the dynamics of aquatic populations and land–inland water ecotones (thirty-six activities shared among sixteen countries) and the other with 'vertical ecotones', the relations between surface water and groundwater. Since the ecotones project was launched, more than a dozen technical workshops and seminars have been convened, on methodological aspects (United States and France), technical issues (Australia, Austria, Poland, Spain and Sweden), and regional perspectives (Hong Kong, India and Uruguay). Synthesis activities within the project have

Preface

focused on such topics as surface water–groundwater ecotones, fish populations, and biodiversity, the subject of the present volume.

The volume is based on contributions presented at a special session on biodiversity at the 25th Congress of SIL, held in Barcelona in August 1992. In its fourteen chapters, the book seeks to provide an overview of current knowledge on biodiversity in land–inland water ecotones, to highlight the role of biodiversity in the functioning of transition systems, to distil recommendations and insights useful for management, planning and educational purposes, and to identify the main scientific issues to be addressed in future research. Of the fourteen chapters, the more generic and wide-ranging contributions examine such topics as the reasons for studying biodiversity in land–inland water ecotones, the dynamics of diversity and its expression over gradients and boundaries, habitat complexity in land–inland water ecotones, ecotonal biodiversity and sustainability in unique tropical landscapes, and the scientific basis for conserving biodiversity along river margins. More finely focused chapters take up the role and function of diversity in particular trophic and taxonomic groups, including micro-organisms, plants, invertebrates, amphibians, fish, birds, and mammals. A final chapter presents a synthesis of land–inland water ecotones as transitional systems of high diversity.

In thanking volume editors Jean-Bernard Lachavanne and Raphaëlle Juge for planning this volume and seeing it through to print, UNESCO would also like to take this opportunity of paying tribute to Frederic Fournier, for his long-term association with the MAB Programme, stretching back to the '1968 Biosphere Conference' and before, and in particular for his role in nurturing and servicing the 'Ecotones' project since its inception in the late 1980s. UNESCO hopes that the present volume will prove a useful contribution not only to the outputs of the Ecotones project, but also to the preoccupations of several other programmes of scientific co-operation, such as the *Diversitas* initiative to promote and catalyse scientific knowledge about biodiversity. I would also like to thank Malcolm Hadley for his critical appraisal of the papers. Thanks are also due to C. Lazzero and M. Al-Khudri for their editorial work.

> *The limits that we think we discern in nature are in reality within ourselves... Every time we define a space we discover that it is a space in transition.*
>
> Molina
>
> *To love anything is to love its boundaries; thus children will always play on the edge of anything. They build castles on the edge of the sea, and can only be restrained by public proclamation or private violence from walking on the edge of the grass. For when we have come to the end of a thing we have come to the beginning of it.*
>
> G.K. Chesterton

MAN AND THE BIOSPHERE SERIES

1. The Control of Eutrophication of Lakes and Reservoirs, S.-O. Ryding; W. Rast (eds.), 1989.
2. An Amazonian Rain Forest. The Structure and Function of a Nutrient Stressed Ecosystem and the Impact of Slash-and-Burn Agriculture. C.F. Jordan, 1989.
3. Exploiting the Tropical Rain Forest: An Account of Pulpwood Logging in Papua New Guinea. D. Lamb, 1990.
4. The Ecology and Management of Aquatic–Terrestrial Ecotones. R.J. Naiman; H. Décamps (eds.), 1990.
5. Sustainable Development and Environmental Management of Small Islands. W. Beller; P. d'Ayala; P. Hein (eds.), 1990.
6. Rain Forest Regeneration and Management. A. Gómez-Pompa; T.C. Whitmore; M. Hadley (eds.), 1991.
7. Reproductive Ecology of Tropical Forest Plants. K. Bawa; M. Hadley (eds.), 1990.
8. Biohistory: The Interplay between Human Society and the Biosphere – Past and Present. S. Boyden, 1992.
9. Sustainable Investment and Resource Use: Equity, Environmental Integrity and Economic Efficiency. M.D. Young, 1992.
10. Shifting Agriculture and Sustainable Development: An Interdisciplinary Study from North-Eastern India. P.S. Ramakrishnan, 1992.
11. Decision Support Systems for the Management of Grazing Lands: Emerging Issues. J.W. Stuth; B.G. Lyons (eds.), 1993.
12. The World's Savannas: Economic Driving Forces, Ecological Constraints and Policy Options for Sustainable Land Use. M.D. Young; O.T. Solbrig (eds.), 1993.
13. Tropical Forests, People and Food: Biocultural Interactions and Applications to Development. C.M. Hladik; A. Hladik; O.F. Linares; H. Pagezy; A. Semple; M. Hadley (eds.), 1993.
14. Mountain Research in Europe: An Overview of MAB Research from the Pyrenees to Siberia. M.F. Price, 1995.
15. Brazilian Perspectives on Sustainable Development of the Amazon Region. M. Clüsener-Godt; I. Sachs (eds.), 1995.
16. The Ecology of the Chernobyl Catastrophe: Scientific Outlines of an International Programme of Collaborative Research. V.K. Savchenko, 1995.
17. Ecology of Tropical Forest Tree Seedlings. M.D. Swaine (ed.), 1996.

CONTENTS

PREFACE	v
LIST OF CONTRIBUTORS	xvii
1. WHY STUDY BIODIVERSITY IN LAND–INLAND WATER ECOTONES?	1
Jean-Bernard Lachavanne	
The erosion of biodiversity – a major problem facing humanity	2
A worldwide concern	2
The challenges to humanity	3
A difficult challenge to take up	6
About land–inland water ecotones	8
What is a land–inland water ecotone?	8
Ecotone – an ambiguous concept	11
Toward a hierarchical application of the ecotone concept	15
Importance of land–inland water ecotones for the conservation of biodiversity	19
Biodiversity and land–inland water ecotones	19
Threats to the biodiversity of land–inland water ecotones	22
The land–inland water ecotone: a life-sized laboratory for studying the patterns and regulating factors of biodiversity	25
Land–inland water ecotones have to receive more attention in the future	25
Key biodiversity-related questions	27
Objectives of this book	30
Acknowledgements	33
References	33

2. **DYNAMICS OF DIVERSITY AND ITS EXPRESSION OVER GRADIENTS AND BOUNDARIES**
   Ramon Margalef
Introduction	47
A dynamic model	48
The play between two feedback loops	49
Time-dependence of diversity, as expressed in succession	50
Diversity and space	51
Boundaries	52
Question of names	54
Diversity and biodiversity, the examples of streams and plankton	54
References	58

3. **HABITAT COMPLEXITY IN LAND-INLAND WATER ECOTONES**
   Ewa Pieczyńska and Maciej Zalewski
Introduction	61
Size and physical characteristics of streamside and lakeside ecotones	61
Biological aspects of ecotone heterogeneity	65
Ecotones and feedback processes regulating ecosystem dynamics	68
Human influences, management, restoration	71
Anthropogenic changes to ecotones	71
Filtering function of ecotones – management	72
Conclusions	74
References	75

4. **MICROBIAL DIVERSITY AND FUNCTIONS IN LAND-INLAND WATER ECOTONES**
   Michel Aragno and Blanka Ulehlova
Land–inland water ecotones generate unique microbial habitats	81
Sampling, biomass and biodiversity evaluations	83
Sampling	83
Biomass measurement	83
Microbial biodiversity	84
Bacterial functions associated with land–inland water ecotone microenvironments	86
Primary production: phototrophic bacteria	87
Chemotrophic metabolism	87
Other functions to be considered	92

Relationship of bacteria with other organisms in the biocoenoses	94
Considerations and hypotheses on the microbial ecology of some microecotones (gradients, transitions, interfaces) existing in land–inland water ecotones	95
Carbon flux and related functions associated with litter decomposition in submersed conditions	96
Rhizospheric environment of aquatic plants	98
The possible consequences of a changing water table	102
The role of microorganisms as depolluting agents	103
Conclusions	104
Acknowledgements	105
References	106

5. **PATTERNS AND REGULATION OF PLANT DIVERSITY IN LACUSTRINE ECOTONES**
Raphaëlle Juge and Jean-Bernard Lachavanne

Introduction	109
Vegetation: a key to ecotone complexity	110
Patterns of plant diversity in the circumlacustrine ecotone	111
Criteria for determination of plant diversity	111
Types of diversity	112
Zonation and patchiness of vegetation in ecotones	115
Regulation of plant diversity in ecotones	119
Modes of regulation of plant diversity related specifically to land–water ecotones	120
Roles of plant diversity	126
Conclusions	127
Acknowledgements	129
References	129

6. **INVERTEBRATE BIODIVERSITY IN LAND–INLAND WATER ECOTONAL HABITATS**
Jean Giudicelli and Michel Bournaud

Introduction	143
Evaluation of biodiversity and representativity of results	143
Sampling macroinvertebrates in ecotonal habitats	143
Evaluation of biodiversity	144
Illustration of ecotonal biodiversity in land–stream boundaries	144
Role of terrestrial–aquatic boundary characteristics in influencing biodiversity	149
(1) The intra-ecotonal complexity and heterogeneity	149

(2) The use of terrestrial–aquatic ecotonal systems by invertebrates — 150
Taxonomic richness and specificity in the ecotonal communities — 151
Conclusions — 155
References — 157

7. AMPHIBIAN DIVERSITY AND LAND–WATER ECOTONES
Pierre Joly and Alain Morand

Introduction — 161
Typology of the ecotones used by amphibians — 162
Reasons for the presence of amphibians in ecotonal habitats — 163
   Temperature — 163
   Larval diet and ecotones — 164
   Ecotones and the availability of refuges against predation — 165
Ecological constraints on the use of ecotonal habitats as breeding sites — 165
   Fragmentation and connectivity — 165
   Disturbance through hydrological processes — 166
Production and fluxes — 167
Adaptation of amphibians to the ecological constraints of littoral ecotones — 168
   Adaptation to the seasonality of feeding resources — 168
   Adaptation to the unpredictability and transience of suitable habitats — 169
   Adaptation to foraging in ecotones — 170
   Adaptations of the amphibian larvae to escape fish predation — 171
   Demographic adaptation to the fragmentation of ecotones — 172
Impact of human activities and guidelines for ecotone management — 173
   Fragmentation and isolation of ecotonal habitats — 173
   Quality of ecotonal habitats — 174
   Guidelines for the restoration of ecotonal habitats suitable for amphibian breeding — 174
Conclusions — 175
Acknowledgements — 176
References — 176

8. FISH DIVERSITY AND ECOTONAL HABITAT
Maciej Zalewski

Introduction — 183
Increase of fish diversity with shift from abiotic to biotic mode of ecosystem regulation — 184
The change of biodiversity as an effect of the ecosystem eutrophication — 187

Shoreline ecotones – fish as regulators of community dynamics and patterns of energy flow		189
The role of the land–water ecotones at the river in agricultural catchments		193
Stocking as a method of compensation for loss of ecotonal habitats		194
Conclusions		195
Acknowledgements		199
References		199
9.	**BIRD DIVERSITY IN ECOTONAL HABITATS**	205
	Kazimierz A. Dobrowolski	
	River ecotones	209
	Lake ecotones	217
	References	220
10.	**MAMMAL DIVERSITY IN INLAND WATER ECOTONE HABITATS**	
	Rüdiger Schröpfer	
	Introduction	223
	The bank as a typical ecotone	223
	The diversity of phenotype of semiaquatic mammals	223
	Diversity of form	224
	Referring to ecotones in general	224
	Diversity of size	224
	Diversity of strategies	225
	The ecotone situation for mammals	228
	Acknowledgements	229
	References	230
11.	**ECOTONAL BIODIVERSITY AND SUSTAINABILITY IN UNIQUE TROPICAL LANDSCAPES**	
	Heath J. Carney	
	Introduction	233
	Survey, sampling and monitoring of tropical ecotones	235
	Ecotonal biodiversity and sustainability	242
	Conclusions	246
	Acknowledgements	246
	References	247

12. SCIENTIFIC BASIS FOR CONSERVING DIVERSITY ALONG RIVER MARGINS
Geoff Petts

Introduction	249
The character of river margins	249
Rationale for managing river margins	253
The case for restoration	254
The scientific basis of restoring river margin ecosystems	256
An integrated approach	256
The fluvial hydrosystem perspective	258
Principles for ecologically-sensitive river management	260
The river margin as an ecotone	261
Management perspective	265
Acknowledgements	266
References	266

13. BIODIVERSITY: A REVIEW OF THE SCIENTIFIC ISSUES
Stephen C. Stearns

Introduction	269
What politicians need from biodiversity research	270
Where basic research is needed	272
Connecting genetic variation and ecological processes: within species and among species within communities	272
Why are some communities and ecosystems more diverse than others?	273
How many species are there and what can they tell us?	274
Conclusions	274
References	275

14. LAND–INLAND WATER ECOTONES AS TRANSITIONAL SYSTEMS OF PARTICULARLY HIGH BIODIVERSITY: TOWARDS A SYNTHESIS
Jean-Bernard Lachavanne and Raphaëlle Juge

Introduction	277
The ecotone concept as applied to the transitional zones between terrestrial and aquatic ecosystems: the necessity of an agreement	277
Biodiversity analysed in an ecotonal context	279
Ecotones between terrestrial and aquatic ecosystems are generally characterized by high biodiversity	281
Biodiversity in land–inland water ecotones also depends on characteristics of related ecosystems	282

Hydrology, a prominent regulatory factor of biodiversity in land–inland water ecotones	283
Higher plants as a major determinant factor for biodiversity	284
Biodiversity in land–inland water ecotones at risk	285
Managing ecotonal zones for preserving and restoring biodiversity: an urgent need	286
Conclusions	290
References	291
INDEX	299

LIST OF CONTRIBUTORS

Michel Aragno
Laboratoire de Microbiologie
Institut de Botanique
Université de Neuchâtel
CH-2007 Neuchâtel
Switzerland

Michel Bournaud
URA CNRS 1451
Ecologie des eaux douces et des
grands fleuves
Université Claude Bernard-Lyon 1
43, Bd. du 11 Novembre 1918
69622 Villeurbanne Cédex
France

Heath J. Carney
Institute of Ecology
University of California
Davis, California 95616
USA

Kazimierz A. Dobrowolski
Institute of Ecology
Polish Academy of Sciences
05-092 Dziekanow Lesny
near Warsaw
P.O. Lomianki
Poland

Jean Giudicelli
Laboratoire de biologie animale
(Ecologie)
Université d'Aix-Marseille
Av. Escadrille Normandie-Niemen
Case 331
13397 Marseille Cédex 13
France

Pierre Joly
URA CNRS 1451
Ecologie des eaux douces et des
grands fleuves
Université Claude Bernard-Lyon 1
43, Bd. du 11 Novembre 1918
69622 Villeurbanne Cédex
France

Raphaëlle Juge
Laboratoire d'écologie et de biologie
aquatique
Université de Genève
18, ch. des Clochettes
1206 Genève
Switzerland

Jean-Bernard Lachavanne
Laboratoire d'écologie et de biologie
aquatique
Université de Genève
18, ch. des Clochettes
1206 Genève
Switzerland

Ramon Margalef
Department d'Ecologia
Universitat de Barcelona
Avenida Diagonal 645
08028 Barcelona
Spain

Alain Morand
URA CNRS 1451
Ecologie des eaux douces et des
grands fleuves
Université Claude Bernard-Lyon 1
43, Bd. du 11 Novembre 1918
69622 Villeurbanne Cédex
France

Geoffrey E. Petts
Environmental Research and
Management
University of Birmingham
Edgbaston
Birmingham, B15 2TT
United Kingdom

Ewa Pieczyńska
Department of Hydrobiology
University of Warsaw
Nowy Swiat 67
00-046 Warszawa
Poland

Rüdiger Schröpfer
Fachbereich Biologie/Chemie
Universität Osnabrück
Postfach 4469
4500 Osnabrück
Germany

Stephen C. Stearns
Zoology Institute
University of Basle
Rheinsprung 9
CH-4051 Basle
Switzerland

Blanka Ulehlova
Institute of Systematic and
Ecological Biology
Nerudova 354
66701 Zidlochovice
Brno
Czech Republic

Maciej Zalewski
Institute of Environmental Biology
Department of Applied Ecology
University of Lodz
ul, Banacha 16
90-237 Lodz
Poland

CHAPTER 1

WHY STUDY BIODIVERSITY IN LAND–INLAND WATER ECOTONES?

Jean-Bernard Lachavanne

Biodiversity refers to variety within the living world. It is defined as the property of groups or classes of living entities to be varied. Biodiversity manifests itself at all hierarchically related levels of biological organization, from genes through cells, tissues, organs, individuals, populations, species, communities and life forms to ecosystems (Solbrig, 1991a; Hansen *et al.*, 1992), in both their structural and functional dimensions (ecological roles). The relative contribution of these levels of integration to biological diversity of the biosphere is described in the authoritative work published by Groombridge (1992).

Biodiversity is at the same time the most integrated expression of ecological systems, whatever the level of the biosystem under consideration. It reflects the importance of the living world's constantly renewed potential to adapt to the changing complexity of the life conditions and to the variety of habitats which these dynamic conditions generate. It is the result of the history of organism–habitat and organism–organism interactions involving processes of a varied nature (physical, chemical, biochemical, genetic, physiological, ecological) which operate at different places and at varied time scales (geological, secular, annual, seasonal, daily) and spatial scales (from the biosphere as a whole to an individual's particular habitat).

The biodiversity observed today is the result of a long evolution. Life began as single-cell beings, then diversified in ever more complex forms maintaining unstable and subtle equilibria among themselves and with the abiotic and biotic environmental conditions. Thus biodiversity reflects the variety of the complexities of life since its origin on earth.

The present picture of biodiversity in temperate countries is the reflection of the far-reaching changes induced, firstly, by the quaternary glacier cycle and, secondly, by the development of human populations and their activities (Barbault, 1994). Thus the biodiversity measured today is the result of the combined influence of autogenic and allogenic factors (disturbances, disasters) which have been operative during the last 10,000 years and which are still at work. It therefore constitutes a kind of memory of the abiotic and biotic events which are responsible for shaping the ecological systems. Moreover, biodiversity informs us directly about the degree of organization of an ecological system and indirectly about other important characteristics such as stability, resistance, resilience, etc.

Three major reasons justify the focus on biodiversity in general and on land–inland water ecotones in particular:

(1) It is accepted that it is in the best interests of humanity for diversity at all levels (genetic, species, landscape) to be maintainable. Now biodiversity is seriously endangered by increasing human pressure on natural ecosystems. There is an urgent need for understanding in order to manage natural resources sustainably.

(2) Land–inland water ecotones are generally characterized by high biological diversity and at the same time by heavy human pressure owing to their great attractiveness (direct access to water resources, navigable waterways, fishing, leisure areas, etc.).

(3) It is widely acknowledged that studies of ecotones are very promising and of both theoretical and practical value in developing ecological theories useful for understanding biodiversity regulating factors.

Why study biodiversity? Why study biodiversity in land–inland water ecotones? These questions will be answered by the different sections into which this introductory chapter has been divided.

THE EROSION OF BIODIVERSITY – A MAJOR PROBLEM FACING HUMANITY

A worldwide concern

The loss of biological diversity is a natural process which takes many forms but at its most fundamental and irreversible outcome it involves the extinction of species. However, it is beyond question that extinctions caused directly or indirectly by man are occurring at a rate which far exceeds any reasonable estimates of background extinction rates, and which, to the extent that it is correlated with habitat perturbation (the habitat being made unsuitable for the species due, for example, to clear-felling of forests or severe water pollution, or through the habitat becoming fragmented) must be increasing (Groombridge, 1992). This phenomenon affects above all highly developed organisms whose generation lasts a long time and which have complex genomes. Simple organisms have a better chance of adapting to unfavourable environmental changes (Guttinger, 1994).

Despite numerous initiatives at the international, national, regional and local levels, global and most local environments have continued to proceed toward an apparently inexorable and progressive degradation (di Castri and Hansen, 1992).

The erosion of biological diversity is a relatively recent concern in the international scientific community despite the alarm cries that farsighted naturalists have been uttering for a number of years (Dorst, 1965; IUCN, 1980; Ehrlich and Ehrlich, 1981; Hawkes, 1983; Schonewald-Cox *et al.*, 1983; Wilson, 1985, 1988, 1989; Wilson and Peter, 1988). This concern, which developed mainly in industrialized countries where the multi-faceted pressure exerted by man on natural ecosystems is being felt on an unprecedented scale, is rooted in the fear that a generalized reduction of biodiversity may lead to a loss of operability and stability of ecosystems (Solbrig, 1991b) and, finally, to a loss of resources which might prove vital for man in the future. Growing interest is being shown today in the issues connected with this worrying evolution by the international community, be it political, socioeconomic, legal or scientific. An awareness is emerging of the risks that are liable to be caused by the reduction of the variety of resources that can be used by humanity, whose demographic growth seems difficult to control at present.

The Convention on Biodiversity signed by 156 countries at the Earth Summit in June 1992 in Rio de Janeiro thus shows that conservation of biodiversity is currently regarded as a problem of worldwide scope, even if – mainly on economic and social grounds but also for reasons connected with the difficulty of implementing in practice the general principles it sets out – numerous obstacles must still be overcome before the Convention is generally enforced at the national, regional and local levels.

Some critics denounce the lack of concrete follow-up given by national governments to the Convention on Biodiversity. Although they are right in principle, they forget too quickly the well-known difficulty of translating theory into action, which can easily be explained here by the complexity and diversity of the problems to be solved, affecting as they do the very foundations of our value systems, our ways of life and the organization of our societies. The Rio de Janeiro Earth Summit is an important stage in the history of humanity and it will take time before we see widespread implementation of the recommendations of *Agenda 21*, which summarizes the various points of the strategy to be implemented.

The challenges to humanity

The challenges thrown down to humanity by the loss of biodiversity and the hazards which the reduction in the number of species could pose to future generations have been discussed by numerous authors (Brown, 1981; Ehrlich and Ehrlich, 1981; Ramade, 1981; Ehrlich, 1984; Wilson, 1985, 1988, 1989; Clark and Munn, 1986; Soulé, 1986; Wolf, 1987; Ojeda and Mares, 1989; Reid and Miller, 1989; McNeely *et al.*, 1990; Myers, 1990; Groombridge, 1992; Barbault, 1994). These challenges are at the centre of the line of research currently being pursued in the context of the international collaborative research programme IUBS–SCOPE–UNESCO–MAB 'Diversitas' (Solbrig, 1991b) and are one of the key issues of the UNESCO–MAB programme on the Role of Land–Inland Water Ecotones in Landscape Management and Restoration (Naiman *et al.*, 1989).

The arguments that have been developed in favour of biodiversity conservation by many authors have been summarized by Solbrig (1991a), among others. They depend on a number of different perspectives, all conditioned by a variety of cultural and economic factors (Groombridge, 1992). They form the basis of the Convention on Biodiversity. According to Solbrig (*op. cit.*), it is acknowledged that there are essentially four reasons for conserving species:

- 'there are ethical considerations regarding loss of life and the prerogative of the human species to eliminate other species from this planet;

- there are aesthetic concerns regarding the loss of unique landscapes and species and the corresponding impoverishment of the human experience if it is denied the opportunity to encounter the multi-faceted products of natural selection;

- there are economic speculations about the potential use of species. Organisms whose properties have not yet been investigated may be important as sources of drugs, or as food, or as raw material for the emerging field of biotechnology; and

- there are also scientific arguments for the preservation of species. Many not yet described species may possess novel biological properties that may help us to understand how nature works. They may also play unique roles in the ecosystem'.

An estimation of the socioeconomic benefits accruing from biological diversity conducted by the 'Center for our Common Future' (Keating, 1993) clearly shows the necessity of conserving wild species. It is estimated, for example, that about 4.5% of the GDP of the United States (approximately 87 billion US dollars per year) originates from the collection and catching of wild species. Another example that demonstrates the necessity of conserving genetic diversity: a gene of an Ethiopian variety of barley has made it possible to protect the whole of California's barley production (160 million US dollars per year) against a virus that causes stunting of the plant. In the field of medicine a child suffering from leukaemia in 1960 had only a one-in-five chance of survival. Today that child's chances are four to one, thanks to drugs containing active substances discovered in a variety of periwinkle from Madagascar's tropical rain forest. There are obviously many other examples (Keating, 1993).'...Resources on this planet are finite and they must be managed in a sustainable way if they are to continue to serve as our principal source of sustenance. ...In order to ensure the maximum quantity and quality of renewable natural resources for ourselves and our descendants, we must learn to use resources sustainably' (Solbrig, 1991a). However, as Wilson (1988) and Barbault (1994) rightly stress, we must be wary of an overly utilitarian conception of biodiversity, cultivated to serve special interests, and must emphasize the purely moral reasons for advocating conservationist wisdom in terms of the rights of other people, including plant and animal species other than man.

Scientists estimate that over the next 25 years more than a million species of plants and animals will become extinct (Wilson, 1988; Ehrlich and Wilson, 1991; Soulé, 1991). Even if they are only forecasts and even allowing for the fact that most of the extinctions will be caused in the tropics because of intensive logging, increased attention must be paid worldwide and in all types of ecosystems to identifying, analyzing and quantifying the causes of this disturbing trend so that effective solutions can be proposed for conserving biodiversity.

All human activities affect to varying degrees the environmental conditions at the local, regional and global levels and influence the living components of the geosphere–biosphere system (di Castri and Younès, 1990). The impacts of human activities on biodiversity may be direct or indirect, permanent (cyclical with variable frequency) or occasional, immediate or delayed. They occur through the response of species and communities to changes (local, regional and global scales) in the environmental conditions (depending on their tolerance spectrum) and through the influence of habitat impairment on the interplay of intra- and interspecific biotic relations (Wilcox and Murphy, 1983). In the long term these activities may have an impact on genetic diversity (specific diversity and genetic diversity within the species) through their influence on the evolutive processes of extinction, selection, drift, gene flows and mutation (Ledig, 1988). Habitat fragmentation has the effect of dividing previously continuous populations of species into small sub-populations. If these are sufficiently small, then chance processes lead to raised probabilities of extinction within a relatively short time (Groombridge, 1992).

So species may be exterminated by man through a series of effects and agencies which may be divided into two broad categories: direct (hunting, collection and persecution), and indirect (habitat destruction and modification).

The loss and impairment of natural habitats as well as pollution are universally recognized as the prime causes of loss of biodiversity (Soulé, 1986; Wilson and Peter, 1988; Ulfstrand, 1992). Landscape degradation and reduction of the spatio-temporal heterogeneity of the conditions of life that accompany this phenomenon are suspected to have a direct and far-reaching influence on pools of genes, populations and communities as well as an indirect influence on biotic relations e.g. prey–predators (Andrén and Angelstam, 1988), and pollinators–plants (Jennersten, 1988).

A most disturbing observation concerns the acceleration of the process of species extinction in recent decades (Ehrlich and Ehrlich, 1981; Harris, 1984; Soulé, 1986; Wilson and Peter, 1988; Wilson, 1989).

The ever-increasing demand for resources in terms of land area (agriculture, urbanization, industry, leisure), materials (food, construction materials) and energy from an ever-increasing human population and the attendant array of harmful effects (pollution, degradation, fragmentation and disappearance of habitats) constitute the greatest threats to the integrity of ecosystems and, consequently, to biodiversity.

Even the most cautious demographic forecasts for the human population indicate that the threat of extinction will worsen. According to some authors, conservation of biodiversity has thus become a major challenge that is posed in terms of the survival of mankind (Wilson, 1985; Ehrlich and Ehrlich, 1987; Ehrlich and Wilson, 1991; IUCN/UNEP/WWF, 1991; Ulfstrand, 1992).

A difficult challenge to take up

The key question we face is whether the fundamental changes in our ways of life – on which the conservation of biodiversity ultimately depends to a large extent – will be made quickly enough to effectively counteract and catch up the speed at which the environment and its resources are deteriorating. The room for manoeuvre is very small, and there is no reason to be optimistic. It is an open question whether man's intelligence will gain the upper hand and overcome the brakes and inertia of the world he himself has built.

Seen from this perspective, scientists have a particular responsibility, a central role to play both in order to understand better the 'biodiversity' phenomenon and to be able to draw up clear guidelines for careful resource management; but also and above all in order to communicate their knowledge and to explain to populations the issues concealed behind the faster pace of extinction of animal and plant species. However, it must be realized that an increased public commitment by scientists will be achieved only if the reputation of this type of activity is enhanced. Hitherto it has not only failed to find grace in academic circles but is often even regarded as suspect because it transcends the usual framework that society has set for the scientist.

Understanding and knowledge are the motors, the very preconditions for action. Thus informing populations and raising their consciousness by using language suited to their value systems and ways of life appear to be one of the most effective means of changing mentalities and taking up the challenge to humanity. In this context the training of young people is probably the best chance of seeing the emergence one day of societies that are more respectful of their environment.

In this respect it is striking to observe that, despite broad recognition of the importance of biodiversity in its many aspects (ethical, aesthetic, economic, scientific), the 'biodiversity' phenomenon is far from being understood in all its dimensions. Despite the large body of knowledge which has been accumulated about the biodiversity of living organisms, especially from the eighteenth century onwards, and which has enabled more than 400,000 plant species and 1.5 million animal species to be identified to date (Groombridge, 1992), there is a critical need to understand and in the long term to try to optimize the interacting dynamics of biodiversity and human development. A knowledge of biodiversity patterns and regulating factors thus appears to be one of the major scientific challenges to be taken up if we are to be able to promote a form of management of natural systems and their resources that is geared to sustainable development (Western, 1992).

Furthermore, it remains very difficult to analyse the complexity of the problems connected with species conservation in natural and anthropomorphic ecosystems (Soulé, 1986; di Castri and Younès, 1990). It must take into account, in an integrated multi-scale and multi-criterion approach, the genetic, biochemical, physiological and ecological aspects. In other words, a true ecological synthesis must be carried out at each level of integration of the biotic components on the one hand and at the scale of entire ecological systems on the other.

Recently, several initiatives have been developed to make the work of academic ecologists more relevant to pressing global environmental problems, including loss of biodiversity and more sustainable development (see for example Naiman et al., 1989; di Castri and Younès, 1990; Huntley et al., 1991; Lubchenco et al., 1991; Hansen and di Castri, 1992).

But it must be emphasized that we need to be able to rely on scientists skilled in taxonomy if we are to be able to grasp such a set of problems. Indeed, as Bramley (1994) pointed out: 'It would be difficult to grasp the full extent of species biodiversity without some form of taxonomy and, if biodiversity is to be maintained, a baseline recording of the number and type of individuals already present within a habitat will need to be undertaken. This would allow changes in population to be followed and could be used to compare the functioning of similar ecosystems. However, the usefulness of this information relies heavily on the accurate identification of different species by field biologists. Inaccurate identification could have serious consequences, as species within the same genus can occupy different niches within the same habitat. Indeed, some aquatic and riparian plants show a high level of physiological and morphological plasticity which would tend to increase the likelihood of incorrect identification. The obvious way to reduce error is to employ trained and experienced taxonomists or at least to train biologists in basic taxonomic principles'.

Now one is obliged to note that until recently there was a general trend in undergraduate courses to marginalize the study of taxonomy in favour of other, newer subjects. It is clear that it will not be possible seriously to address biodiversity problems without revitalizing and enhancing the status of taxonomy, which for too long has been considered by some slightly snobbish people to be a kind of 'old-fashioned science' centred too much on description and not enough on the functioning of living organisms. We also need to develop other methods and tools that are able to integrate ecosystem complexity linked to biodiversity at different scales, such as statistical analysis, Geographic Information Systems (GIS) and Remote Sensing (RS) (Haslet, 1990; Stoms, 1992; Stoms and Estes, 1993; Wheeler, 1993).

ABOUT LAND–INLAND WATER ECOTONES

What is a land–inland water ecotone ?

The notion of ecotones was first used by Livingstone (1903) in the sense of an environmentally stochastic stress zone and by Clements (1905) to denote the junction zone between two plant communities, where processes of exchange or competition between neighbouring patches might be readily observed. It is the zone in which the principal species of adjacent plant communities meet their limits. This concept typically refers to the notion of stress zone (tonos = stress) that had already been expressed by de Candolle (1855) as of the first quarter of the nineteenth century in the term 'zone contestée' and is related to the terms 'Kampfzone' and 'Kampfgürtel' used later on by Schröter (1926) to designate the upper limits of forests in the Alps.

Aquatic and terrestrial ecosystems within the same landscape are linked directly by water moving in the hydrologic cycle. Thus the connection between terrestrial and aquatic ecosystems is a functional one and the ecological consequences of this linkage are profound (Likens and Bormann, 1975). There are many kinds of coupling of landscape with the water in relation to geomorphological characteristics of shores and land uses which change along shorelines (Hasler, 1975; Pieczyńska, 1990a; Petts, 1990).

As a zone of contact between terrestrial and aquatic ecosystems, the ecotone may be regarded both as part of the edge of these two ecosystems and as an individualized systemic entity characterized by a special structure and mode of operation that are conditioned by the land–water coupling.

Today, land–water ecotones are recognized as ecological systems of specific and variable abiotic and biotic characteristics different from adjacent ecosystems (di Castri *et al.*, 1988; Holland, 1988; Naiman and Décamps, 1991). They represent places where interactions among patches are especially strong (Vieira da Silva, 1979).

According to Jeffers *et al.* (1989), an ecotone is not simply a boundary or an edge. Whilst an edge or a boundary is a purely structural entity, an ecotone is defined by dynamic processes and in addition to its edge characteristics functions by regulating the flow of materials between patches.

At the landscape level ecotones can be viewed as zones where spatial and temporal rates of change in ecological structure and function are rapid relative to rates across the landscape as a whole (di Castri and Hansen, 1992). Ecotones and landscape patterns are created and maintained by a hierarchy of constraining factors including air mass dynamics, megatopography, local geomorphology, disturbance, competition, plant growth and plant development (Nelson *et al.*, 1990). So ecotones can be viewed as unstable components and sensitive to frequent change. They may exhibit distinct patches in temporal and spatial records of a landscape mosaic (Wissmar and Swanson, 1990). These patches differ in their physicochemical and biological properties from adjacent ecosystems.

As summarized by Petts (Chapter 12), the patches will be of a different type (defined primarily by morphological, sedimentological and vegetational criteria) and different age (reflected by successional criteria for each type). The arrangement of patches within a sector changes over time in response to successional processes and disturbance (by erosion and deposition).

Major properties and characteristics of ecotonal zones have been presented in many papers (di Castri *et al.*, 1988; Naiman *et al.*, 1989; Naiman and Décamps, 1990; Holland *et al.*, 1991; Gosz, 1991, 1992; di Castri and Hansen, 1992; Johnston *et al.*, 1992 etc.).

Ecotones are ecosystems characterized by discontinuities and rates of change in the amounts and directions of the ecological flow. But while the dynamics of ecosystems are essentially stabilized by feedback mechanisms, those of ecotones are inherently unstable but may be self-sustaining in time if not in space (Jeffers *et al.*, 1989). Processes operating in ecotones are little different in principle from those in the upland and aquatic systems they separate. However, many of the processes (e.g. vegetation transpiration, denitrification) and feedbacks are very different in intensity than in adjacent ecosystems (Wetzel *et al.*, 1989). In addition, the frequencies of change are much greater in ecotones than in adjacent ecosystems.

The specific features of land–water ecotones derive from the more or less strong interaction between the three fundamental living environments (water–soil–atmosphere). This multiple relationship, which is variable in time and space, bestows on the land–water ecotone a special structure, mode of operation and evolution that condition the living world and its biodiversity. Shore slope, water-level fluctuation and type of sediment are physical attributes of major importance (Pieczyńska, 1990a; Pieczyńska and Zalewski, Chapter 3). In this complex of factors the hydrology regime within which a land–water ecotone operates is perhaps the most fundamental determinant of its behaviour (Wetzel *et al.*, 1989).

Flooding/drying regimes (degree, periodicity) in river or lake systems are environmental disturbances maintaining the temporal and spatial heterogeneity affecting biodiversity. Water flow rates and patterns determine the transport of nutrients and carbon across the ecotone to adjacent ecosystems. The distribution of water in sediment conditions numerous fundamental processes through the agency of its influence on the other abiotic parameters of the environment (temperature, light, availability of nutrients, oxygen concentration, oxidation–reduction potentials, pH) (Mitsch and Gosselink, 1986; Wetzel et al., 1989). The hydrology regime directly influences the processes of production/decomposition of organic material and indirectly the biotic relations (competition phenomena, predation, etc.).

Aquatic–terrestrial ecotones are characterized by an intense exchange activity (Vieira da Silva, 1979) on account of their instability and variability at different scales of space and time (Naiman and Décamps, 1991). The result of this dynamic is a great heterogeneity of the environmental conditions and a particularly wide spectrum of habitats (Pieczyńska, 1975, 1990a, 1990b; Pieczyńska and Zalewski, Chapter 3; Petts, 1990, Chapter 12). These reflect the mosaic of resource and disturbance patches of different ages and successional stages (Petts *op. cit.*).

The diversity of the habitat as a major contribution to the pattern of species diversity has long been recognized, as fundamentally different habitats promote the colonization by different species (MacArthur, 1972; Cody, 1975). Within certain spatial (minimum area) and functional limits, the more heterogeneous and complex the physical environment, the more complex the plant and animal communities and the higher the species diversity. This factor can be considered at both the macro and micro scale (Krebs, 1985). The level of biodiversity is determined both by external forcing functions (climatic conditions, nutrients load, perturbations, etc.) and weaker internal mechanisms including tolerance capacity, biotic interrelations (di Castri *et al.*, 1988).

The transition zones between land and water ecosystems are characterized by conditions ranging from more or less extensive physico-chemical and biological gradients to relatively spatially limited interfaces that modify flows of water, material (sediments, nutrients), energy and organisms between the ecological systems at variable temporal and spatial scales (Wiens *et al.*, 1985; Naiman *et al.*, 1988a, 1988b; Naiman *et al.*, 1989). Thus land–water ecotone conditions vary in a more or less dramatic step-wise manner that requires of any organism tolerating it a morphological structure or biochemical mechanism which is not found in most related species, and is costly, either in energetic terms or in terms of the compensatory changes in the biology of the organisms that are to accommodate it (Begon *et al.*, 1990). The variety of plants' life forms is a good example of the diversity of the answers and solutions developed to promote an adaptation by individuals to environmental constraints (Juge and Lachavanne, Chapter 5).

Three of the properties of land–water ecotones have been acknowledged as particularly important in the UNESCO–MAB international collaborative research programme on the role of ecotones between continental terrestrial and aquatic ecosystems (Naiman *et al.*, 1989):

- their role in maintaining local, regional and global biological diversity, which is the subject of this book;

- their role as a filter of ecological flows (material, energy, organisms) between terrestrial and aquatic ecosystems. The ecotone may modify the ecology flows moving through these transitional zones, depending both on the external force and their internal characteristics (Correll, 1986; Pinay *et al.*, 1990) and it functions as a buffer zone and consequently; and
- their possible role in the stability of the ecological systems that they delimit.

The last two roles are of particular importance in the context of the problems relating to the protection of water against pollution. Indeed, most industrialized countries have by now developed curative-type strategies aimed mainly at combating pollution, by setting up a variety of (individual or collective) systems for cleaning up domestic or industrial waste water. Now while such measures allow – when enforced effectively – isolated pollution sites to be controlled, they respond very poorly to the need to control diffuse pollution sites of atmospheric and agricultural origin (pesticides, nutrients, solid particles).

Although the best strategy to be implemented to control this pollution is of the preventive type (for example, tackling the problem at source by implementing measures to limit soil erosion phenomena, by limiting the use of nutrients and pesticides), it appears that land–inland water ecotones could, under certain conditions, supplement this strategy by trapping pollutants before they disrupt the surface water.

The issues relating to the biodiversity and filter effect of land–water ecotones are often dealt with separately whereas they are strongly linked. The ability of land–water ecotones to trap certain chemical elements influences biological diversity (for example, through nutrient availability, the toxic effect of certain substances, etc.), through various processes (geomorphology, erosion/sedimentation, production, accumulation/decomposition of organic matter, etc.) which help to determine the structure and diversity of habitats. The heterogeneity of the resultant environmental conditions is an important factor in the growth and stability of plant, invertebrate and vertebrate populations (Wedeles *et al.*, 1992). Conversely, biodiversity, particularly that of plants, will influence the filtering capability of land–water ecotones directly (assimilation by photosynthesis promoted by the utilization of mineral elements in varied root strata) and indirectly (by influencing the mineral and organic content of the substrate).

Ecotone – an ambiguous concept

While the concept of ecotone itself as a stress zone with specific variable abiotic and biotic characteristics between two systems (di Castri and Younès, 1990; Naiman and Décamps, 1991; Holland *et al.*, 1991) is intuitively well perceived and understood by all scientists, its practical application is not unanimously agreed upon.

The explanation of the divergences observed is to be sought in the variety of approaches and scales of ecological speculation. A good example of the various points of view is illustrated in the different chapters of this book.

In the course of time innumerable terms have been coined at various dimensions/scales in ecology and geography, in various languages and various scientific schools (Jenik, 1992; Herben *et al.*, 1992). The concept of ecotone has not escaped this problem, thereby making it difficult to apply and fuelling many confusions.

Since it was introduced by Livingstone (1903) and Clements (1905), the concept of ecotone has been re-evaluated and expanded in the last ten years in the light of landscape ecology introduced by Watt (1947) and developed by Forman (1981), Risser *et al.* (1984), Pickett and White (1985), Forman and Godron (1986), Urban *et al.* (1987), Allen and Hoekstra (1990) and of the hierarchy theory (Allen and Starr, 1982; Allen *et al.*, 1984; O'Neill *et al.*, 1986).

This concept is behind the definition proposed in the context of the collaborative research programme launched by UNESCO–MAB/SCOPE on the Role of Land–Inland Water Ecotones in Landscape Management and Restoration, (Holland, 1988) *'Ecotone is a zone of transition between adjacent ecological systems, having a set of characteristics uniquely defined by space and time scales and by the strength of the interactions between adjacent ecological systems'*. It is proposed that the term 'ecotone' be extended and used to refer to interfaces, edges, transition zones or boundaries between adjacent ecosystems.

Although the fertilization of the ecotone concept by these different theories has helped to make the central meaning of the concept clearer (Hansen and di Castri, 1992), it must be acknowledged that its practical use remains problematic.

In fact, the ambiguity and confusion that surround the ecotone concept have to be sought in the concept itself, which contains two fundamental notions with a partially overlapping meaning:

- the notion of stress, referring to limit conditions that prevail at the extremes of the tolerance spectrum of populations, species and communities (conditions between sub-optimal and lethal values) which corresponds to the etymological meaning of the term ecotone; and

- the notion of transition zone between two related ecological systems at landscape level.

The very general character of the definition proposed in the context of the UNESCO–MAB Ecotones programme (Holland, 1988) is regarded as an advantage by some authors (Holland, 1988; di Castri *et al.*, 1988; Naiman *et al.*, 1989) and as a major disadvantage by others (Van der Maarel, 1990;

Backéus, 1993) who criticize it for leaving room for ambiguities, particularly in connection with the concept of ecocline introduced by Clements (1916).

The proponents of a general concept of the term ecotone, adopting a holistic approach, stress the advantage of applying the term ecotone to any transition zone, whatever the level of biotic and ecosystemic integration under consideration: populations, communities, ecosystems, elements of landscapes and biomes. In this concept explicit consideration of scale is fundamental to studying and understanding ecotones (Décamps and Naiman, 1990; Hansen *et al.*, 1992).

As emphasized by Meentemeyer and Box (1987), Gosz and Sharpe (1989) and Gosz (1991, 1992), landscape ecology – and therefore landscape boundaries – cannot escape dealing with spatial analysis, spatial scale and scale-change effects. A landscape may appear to be heterogeneous at one scale but quite homogeneous at another scale, making spatial scale inherent in the definition of landscape heterogeneity and diversity. This is also true for the other biosystem levels.

So the considerations dealing with ecotonal zones must necessarily be set in a hierarchical context, as conclusions drawn from one hierarchical level cannot be extrapolated to another and therefore the correct level must be investigated to provide the required answers (Farmer and Adams, 1991; Levin, 1992).

But it must be recognized that there are not only advantages in this position and that several arguments can be held up against it, particularly in connection with the difficulties involved in operational utilization of the concept. In fact, in adopting this position, questions as fundamental as 'What is a boundary or ecotone?' 'Where is an ecotone?' have no simple answer other than 'It depends!' (Gosz, 1991).

Several authors, headed by Van der Maarel (1990) suggest that one adhere to the etymological meaning of the word ecotone and that a distinction should be drawn between the concept of ecotone – as an environmentally stochastic stress zone in Livingstone's original sense (1903) to which one can associate the concept of ergocline (Legendre and Demers, 1985) – and the concept of ecocline, a relatively heterogeneous but environmentally more stable gradient zone – which is a concept outlined by Clements (1916) and then developed by Whittaker (1960), Van der Maarel and Westhoff (1964) and Slaviková (1986, quoted by Jenik, 1992), through the concept of coenocline (Jenik, 1992).

Based on the conceptual foundation of the typology of boundaries presented by Van Leeuwen (1966), Van der Maarel (1990) proposed drawing a clear-cut distinction between ecotone and ecocline on the basis of the following criteria:

- in the ecotone situation the fluctuations of one or several abiotic factors are great and create a temporal series of environments that are highly different but individually homogeneous, resulting in a depressing effect on biodiversity since few species are suited to such fluctuations; and

- in the ecocline situation there is a gradual difference in at least one major environmental factor (e.g. water) on an area occupied by an ecosystemic continuum affected by other abiotic factors (e.g. topography) or produced by biotic factors (e.g. interspecific competition). The heterogeneity of the environmental conditions that characterize such a situation allows many species with different ecological requirements to coexist and all the transitional stages can be observed.

The distinction proposed by Van der Maarel is attractive. It is justified if one refers to the very different possible evolution of the two situations. Indeed, in the ecotone situation, the state of pulse stability (Odum, 1971) maintains the ecosystem at some intermediate point in the development sequence resulting in a compromise between youth and maturity imposed by rhythmic, short-term perturbations (hydrological forcing function). The environment is very selective and blocks the development of the progressive ecological successions by cyclical rejuvenation of the life conditions (pulse-stabilized subclimax). In the ecocline situation the environmental conditions, although heterogeneous, are more stable and allow the development of the evolutive series that lead to the climax.

On the other hand, it is not always obvious to draw a distinction between an ecocline characterized mainly by the notion of gradient and an ecotone characterized by both the notions of stress and gradients. Where does an ecotone end? Where does an ecocline start?

In keeping with the broad scope of the definition of ecotone proposed by the UNESCO–MAB/SCOPE (Holland, 1988), di Castri and Hansen (1992) recognized some possible overlap or confusion in relation to other concepts in ecology. They proposed to consider both agents and patterns of change:

- in the situations with abrupt change in spatial patterns leading to non-linear behaviour, the boundary is an ecotone in the narrow sense; and

- in the situations with gradual change in which the functional and structural trends of the systems approach a certain linearity, the boundary is an ecocline or true gradient.

Furthermore, depending on change in time, they proposed to distinguish a sudden change which corresponds to an unpredictable non-incorporated disturbance (di Castri, 1991), characterized by non-linear and chaotic behaviour; a progressive series of changes is what is usually considered to be the property of ecological successions. Evidently all the intermediate situations of agents and patterns of change can be found in nature.

But here again, although these considerations make for a more precise perception of the concepts of ecotone and ecocline, they do not enable us to identify an unquestionable boundary between these two terms.

Thus it appears that, depending on the scale of observation and the force of the interactions considered, it is possible to identify several ecotones (stress zones) within an ecocline and several ecoclines (gradients) within an ecotone.

Did you say ecotone? No, ecocline! Ah, ecotone!? Thus it is essentially a matter of perception!

Although more correct and forming part of an academically more purist approach, we think the relatively restrictive concept of the ecotone defended by Van der Maarel (1990) and others (for example, limitation to the eulittoral zone in stagnant water ecosystems) is of limited operational value, especially from the viewpoint of resource management at the scale of the landscape. Such resource management must obviously consider the shores of the lakes and watercourses as a whole and must take into account all the habitats conditioned by the multiple, direct and indirect interactions between land and water, which operate at different time and space scales.

Toward a hierarchical application of the ecotone concept

It is generally accepted that the littoral of a lake and the bank of a river as an aquatic–terrestrial habitat stretch linearly for kilometres, that is to say, it is extensive in length, however minor in breadth. Depending on shore slope and water-level fluctuation, ecotones vary in size from narrow strips to very broad areas. Land–water ecotones consist of heterogeneous patch mosaics that differ in their physico-chemical and biological properties from adjacent ecosystems.

But, as mentioned by Pieczyńska and Zalewski (Chapter 3), they are studied and described without determination of their spatial limits. Moreover, when they are specified the proposed limits of the ecotone do not always coincide (see the different concepts of the ecotone in the various chapters of this book). As we have seen, this lack of consensus about the ecotone concept reflects both the various perceptions (ecotone considered in the broad or narrow sense) and different approaches (spatial scale, biotic component under consideration). Today there is an urgent need for broad agreement, particularly with a view to being able to provide a theoretical reference framework that is useful or indeed indispensable for comparative studies and management of ecotones. The term ecotone is currently used by the various authors to describe different realities and any attempt to compare the phenomena and mechanisms at work in these land–water transition zones becomes extremely difficult and hazardous.

Whatever the view adopted, there remains the problem of the choice of criteria to be taken into consideration to identify the boundaries of the transitional system between terrestrial and aquatic ecosystems.

Indeed, there is not necessarily a coincidence of boundaries on the edge patterns, depending on the biotic components under consideration (Gosz, 1991). Many species of amphibians, fish, birds or mammals depend on a more extensive territory (Joly and Morand, Chapter 7; Zalewski, Chapter 8; Dobrowolski, Chapter 9; Schröpfer, Chapter 10) than that delimited by the ecotonal zone (even taken in the broad sense) for fulfilling their various vital functions (food, rest, reproduction). The common heron, for example, feeds either in the shallow littoral zone or in the coastal wetlands or again in the meadows of the hinterland, whereas it reproduces and rests in the riparian forest. The pike catches its prey preferably in the littoral zone, in sections of shore rich in water plant communities and reproduces in the low wetlands during the high-water period.

Lachavanne and Juge (in preparation) suggest making a distinction between:

(1) Interfaces which characterize situations separating two adjacent discrete habitats (dominant vertical approach): soil/air, soil/water and water/air characterized by a dramatic step-wise variability, and

(2) Ecotones which characterize situations separating two adjacent patches (dominant horizontal approach): different spatio-temporal levels of narrow-to-broad transition zones (gradients) between aquatic and terrestrial ecosystems, sub-systems or communities inside the ecotone in the broad sense (see below).

They propose to apply the ecotone concept at different scales of space and systemic organization and to adopt a four-level approach for the application of this concept in the transition zone between terrestrial and lake ecosystems (Figure 1.1).

This proposal seeks to take account of the diversity of the concepts (broad sense, narrow sense) and of the approaches (spatial scales, individual biotic components, etc.) proposed in the literature (Augier *et al.*, 1992). Lachavanne and Juge have illustrated their idea by means of the contact zone between terrestrial ecosystems and the lake ecosystem.

In keeping with the UNESCO–MAB definition of ecotone (Holland, 1988), they propose to consider this contact zone as a transitional ecosystem divisible into sub-systems organized in hierarchically nested series. According to the holistic approach (Allen and Starr, 1982; Allen *et al.*, 1984; O'Neill *et al.*, 1986; Urban *et al.*, 1987), it is assumed that systems are more than the sum of their parts and possess new emergent properties which cannot be predicted from analyses of their component parts: systems are hierarchically organized with new properties at each level of the hierarchy. It is acknowledged that the higher levels are generally larger in area, and behave more slowly than the lower levels, which they both constrain and contain. The boundaries are set in such a way that the sub-systems are homogeneous in respect of the criterion selected, whatever

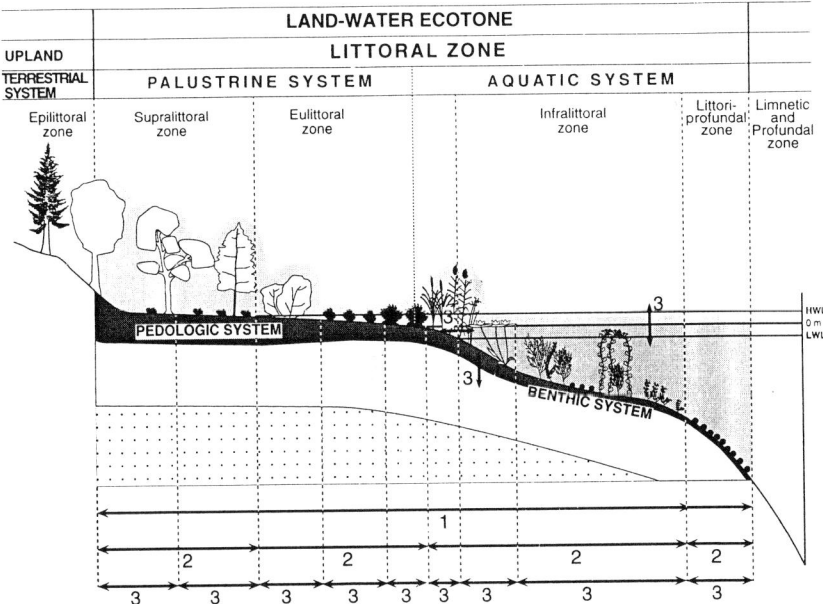

Figure 1.1 Schematic representation of a multi-scale application of the ecotone concept to the circumlacustrine shore; examples of first-, second- and third-degree ecotones (see text). To improve clarity, the fourth-degree ecotones are not represented.

its nature may be. The homogeneity principle has been stated by Zonneveld (1974) and clarified by Van der Maarel (1976), for whom the setting of boundaries is determined by the scale of observation.

The approach is based on the land–water relationship as it emerges through the distribution of species–plant associations as bioindicating elements of systems whose functioning is conditioned by the presence of water (Juge and Lachavanne, Chapter 5).

Joint consideration of the three parameters water–soil–vegetation as multi-scale application criteria of the ecotone concept is justified by the paramount role played by the first two (water–soil) for vegetation and by the multiple ecological role played by the latter (vegetation) for animal species (spatial organization of the ecosystem, food, shelter, rest, reproduction) (Pieczyńska, 1975, 1990a; Pieczyńska and Zalewski, Chapter 3), vegetation being an excellent bioindicator of the land–water relationship and of the past and present perturbations undergone.

In order to be able to make useful comparisons, we suggest considering in a multi-scale approach the ecotone concept as including all the areas of lake shore or river bank driven predominantly by the land–water relationship.

Thus in the case of the lacustrine shore, Lachavanne and Juge (in preparation) propose to consider the following system of ecotones:

(1) The first-degree ecotone is the ecotone in the broad sense whose boundaries are imposed by density-independent factors (principally water). It includes all the littoral systems whose structure and mode of functioning are influenced by the land–water coupling, that is, the supralittoral, eulittoral and infralittoral as per Hutchinson's classification of lake beds (1967). This ecotone is typically a transition zone between terrestrial and aquatic ecosystems at the scale of the landscape. It corresponds to the lake coast as defined by Forel (1892) as 'the region comprising the zone of land that is altered by the lake and the zone of the lake that is altered by the *terra firma*' but at a mesocosm scale, that is, which includes the notion of limnosystem comprising a catchment basin (source of material, nutrients and organisms), watercourses (tributaries) to transport these materials and a lake (recipient). It is the most functional and operational scale of application of the ecotone concept, both as a heterogeneous system, a medium of life and animal and plant diversity and as a zone in which special and suitable management must be considered by the managers.

(2) The second-degree ecotones are included in the first-degree ecotone; they are areas characterized by different hydrological regimes (substrate never, occasionally or permanently flooded). According to Hutchinson's (1967) classification, we can distinguish:

- the supralittoral (substrate never flooded) entirely above water level but subject to spraying by waves;
- the eulittoral (substrate occasionally flooded): between highest and lowest seasonal water levels; often a zone of disturbance by waves breaking on exposed shores; and
- the infralittoral (substrate permanently flooded).

Each of these parts of the shore constitutes a transitional zone between systems of the same hierarchical level. For example, the supralittoral is an ecotonal zone between the epilittoral and the eulittoral, the latter between the supralittoral and the infralittoral, etc.

(3) The third-degree ecotones are included inside both the first- and the second-degree ecotones; they are also areas characterized by a density-independent factor (water) generating different vegetated formations (physiognomic homogeneity). We can distinguish globally wetland forest, wetland herbaceous and hydrophyte formations such as emergent – floating-leaved – and submerged rooted or adnate vegetation. Each of these formations corresponds to different (particularly hydrological) environmental conditions and forms a transitional zone between the other plant formations present within or outside of the

ecotone. For example, wetland herbaceous formations constitute an ecotonal zone between the wetlands, forest and emergent rooted hydrophyte formations.

(4) The fourth-degree ecotones are included within the ecotones of higher levels; they are characterized by density-independent factors and by predominantly density-dependent factors (competitive) between different plant groups/associations.

Such an approach has many advantages, particularly that of bestowing a physical reality on the ecotone (a series of spatially delimited, hierarchically nested patches), which is useful for making comparisons between land–water ecotones.

It provides a reference framework for analyzing the relations of the animal species with the land–water ecotone in connection with its function as a supplier of resources (rest, feeding and reproduction area) and in particular makes it easier to differentiate the ecotone-specific species (those that perform all their vital functions in the ecotone) from those that are non- or partially-specific (those that use the ecotone for one or other of the vital functions).

IMPORTANCE OF LAND–INLAND WATER ECOTONES FOR THE CONSERVATION OF BIODIVERSITY

Biodiversity and land–inland water ecotones

The biogenic capacity of an area is a function of its physicochemical characteristics and of the structure of the mosaic of the various habitats which it exhibits.

The importance of biodiversity in the land–water ecotone depends, as for any ecosystem, on its geographic location (latitude, height), on the geomorphological and hydrological characteristics, on the stage of evolution of the system and on the regime of perturbations (Diamond, 1988). Local community association appears to be under local, as well as regional and global, constraints.

Because of several processes at work within the ecotone itself and because of its proximity and functional ties to the adjacent ecological systems, biodiversity (i.e. the number of species) is relatively high in ecotones (Risser, 1990). Consequently, ecotones are of major importance in maintaining biodiversity and the global gene pool.

Characterized by a great variety of living conditions at varied time and space scales, land–water ecotones develop a particularly broad spectrum of habitats (Hasler, 1975; Likens and Bormann, 1975; Pieczyńska, 1975, 1990a, 1990b; Pieczyńska and Zalewski, Chapter 3; Vieira da Silva, 1979; Clark, 1979; Petts, 1990; Naiman and Décamps, 1991; Hansen and di Castri, 1992). Compared to terrestrial ecosystems, the hydrogeomorphological factors confer to the land–inland water ecotone an additional heterogeneity at the level of soil conditions. This affords the potential of a multitude of constantly renewed ecological niches, which makes possible the coexistence of numerous species and communities of species with distinct ecological requirements.

Numerous authors have recognized the great biological diversity of land–water ecotones (Odum, 1971; Patten *et al.*, 1985; Naiman *et al.*, 1989; Naiman and Décamps, 1990), as well as the role played by these ecotones in maintaining critical habitats for some species (Johnston and Odum, 1956) or as suppliers of sites (refuge, rest) for migratory birds (Holland, 1988). Thus the abundance and survival of certain species are themselves directly linked to the importance of the ecotones in the area under consideration (Ghiselin, 1977). The particular value of river corridors is illustrated by the studies of Knopf (1986) and Knopf *et al.* (1988) who show that although riparian habitat occupies less than 1% of the western North American landscape, it provides habitat for more species of birds than all other habitats combined (see also Dobrowolski, Chapter 9). In a watershed in southeast Alaska, Pollock and Naiman (1994) have shown that collectively, land–water ecotones contain 95% of all plant species found in the watershed, while all uplands contain 21%; riparian ecotones occupy less than five percent of the watershed but contain 69% of all species. Similarly, the riparian zones in the forests of the northwestern United States are recognized as acting as corridors for the movements of several vertebrates (Harris, 1984). These considerations reveal a need for an understanding of ecotone–biodiversity relationships in managing landscapes (Goeden, 1979).

One commonly observed phenomenon in the communities of natural systems is the tendency that they exhibit to be more diversified and denser in transitional zones between ecosystems. It is to Shelford (1913) that we owe the observation of this phenomenon which Leopold (1933) called the 'edge effect'. The causes of this phenomenon are generally attributed, firstly, to the existence of a physicochemical gradient extending over an often restricted space and, secondly, to the co-occurrence of complementary habitats, which is necessary for the accomplishment of various phases of the life cycle of some organisms.

This effect is thus conventionally explained by the presence of:

(1) Eurytopic species, which are characteristic of each of the adjacent ecosystems located in the land–water ecotonal zone, either because they exhibit great tolerance of the environmental factors (especially the hydric factor), or because they find there the resources necessary for fulfilling one or more of their vital functions (shelter, food, reproduction). Thus there are certain animals with a complex life cycle including an aquatic phase, which takes place in land–water ecotonal habitats and that typically require more than one ecological system to survive, such as certain insects, amphibians, fishes, birds, mammals (Diamond,

1973; Fitzpatrick, 1980; Keast and Morton, 1980; Joly and Morand, Chapter 7).

(2) Obligate ecotone species adapted to its fluctuating hydric conditions.

The edge effect is also enhanced during certain periods (seasons, years) by the presence of migratory species and/or typically terrestrial species which use the ecotonal zone only as a corridor for moving along the shores or perpendicular to the shores to fetch drinking water.

Biodiversity is the result of particularly complex interactions emanating from multiple routes linking organisms with each other and with abiotic resources. Taking mainly the studies of Ogden *et al.* (1973), Stein and Magnusson (1976) and Hay (1984) as a basis, Lodge *et al.* (1988) ascertain that the effects of strong interactions may be accentuated or inhibited, depending on the size and spatial arrangement of patches and on direct and indirect interactions among specific food web components. But the amplifying character of the biodiversity of the ecotonal zones (edge effect) does not become operative anywhere or under any conditions (Van der Maarel, 1976). Or rather the edge effect, which always exists, may have an amplifying effect or, on the contrary, a depressing effect on biodiversity. This depends on the nature and force of the environmental constraints and on the regime and amplitude of the disturbances at the different scales of time and space.

Thus Van der Maarel (1976) has shown that ecotones subjected to sudden fluctuations in environmental conditions, as is sometimes the case on the edge of lakes (ecotone = zone corresponding to the eulittoral according to Van der Maarel), are relatively poor in species. This phenomenon can easily be verified by considering the diversity of plant communities in reservoirs, where the water level varies significantly. Van der Maarel proposes to consider three scenarios for developing the edge effect theory: specific diversity in the ecotone may be greater, equivalent to, or lower than that of the adjacent ecological systems, commensurate with the extent of the stress engendered by the alternation of the environmental conditions (aquatic, atmospheric).

At the scale of the landscape, regional biodiversity is maximized at some intermediate level of habitat heterogeneity (Noss, 1983; Joly and Morand, Chapter 7).

Few empirical studies of the edge effect have been carried out to date and the results are often contradictory. The explanation of these contradictions is to be sought in the variety of scales of observation of the phenomenon. For example, the results of the edge effect will be different depending on whether account is taken of the first-degree or second-degree ecotone or one of another degree (Figure 1.1; Juge and Lachavanne, Chapter 5). The effect will also be different depending on whether account is taken of all the taxonomic groups present in the community or only some of them (certain groups of plants or animals). Moreover, the studies that have been conducted are highly sectorial and concern only a limited number of biotic components (vegetation, birds) and usually transitional zones between terrestrial ecosystems (forests/meadows). The living communities of the land–water ecotones are composed of species whose habitats may be very different: intercontinental (example: migratory birds), continental (example: some mammals), regional (example: some fish) or local (example: some invertebrates such as molluscs). Therefore an approach which is both integrated and flexible must necessarily be adopted in order to explain the biodiversity patterns in land–water ecotones.

It must be acknowledged that up to now knowledge has remained inadequate and insufficient to develop a true theory of the edge effect.

Threats to the biodiversity of land–inland water ecotones

Over the past 300 years, man has considerably destroyed and reduced land–inland water ecotones (river, lake), leading to a decrease of local, regional and global (?) biodiversity. An abundant literature exists on threats to the structural and functional integrity of the ecotonal zones especially concerning wetland and alluvial zones on which biodiversity depends (Lee *et al.*, 1975; Pieczyńska, 1976, 1986, 1990a; Burgess and Sharpe, 1981; Lachavanne, 1982, 1985; Vanhecke and Charlier, 1982; Keddy, 1983; Hansen and Walker, 1985; Lachavanne *et al.*, 1986a, 1986b; Mitsch and Gosselink, 1986; Petts *et al.*, 1989; Jeffries and Mills, 1990; Patten, 1990; Petts, 1990; Williams, 1990; Haslam, 1991; Boon *et al.*, 1992; Greene and Tunstall, 1992; Amoros and Petts, 1993; Bhownik, 1993; Goldberg, 1994). It is usual practice to distinguish threats at the (i) global (in connection with climatic changes and air pollution), (ii) regional (alteration of the natural characteristics of the catchment basin, pollution caused by human activities) and (iii) local scales (land use, isolated cases of pollution or degradation).

In several regions of the world, lake or fluvial wetlands are considered low value land since in their normal condition they cannot be used for most agricultural activities or urban development. There has been increasing pressure to drain marshes to provide higher value land for suburban development (Lee *et al.*, 1975).

In temperate regions the factors involved in the reduction of the size, transformation, degradation and loss of specificity of ecotones have been identified as prime causes of the progressive disappearance of the fauna and flora they shelter. Lachavanne *et al.* (1985) have identified the following impairing factors in Swiss lakes: control of the lakes' water level for hydroelectric requirements and navigation, the spread of built-up areas (dwellings, roads, jetties), shore filling and consolidation facilities to prevent erosion or flooding (walls, rock filling), the construction of ports and utilization of riparian land for agriculture (drainage of ecotonal wetlands), tourism and leisure (camping sites, public beaches, lake access) and tree plantations for timber production (e.g. *Populus* sp.) constitute the main threats to the biodiversity of land–water ecotones.

More locally, other factors such as dredging for the exploitation of gravel, intensive utilization of various resources (fibres, fishing), direct human pressure (maintenance, trampling, flower picking, etc.), fire or inland water transport can also play a decisive harmful role. To these must be added general degradation of the quality of the habitat due to domestic, industrial and agricultural pollution and to accelerated eutrophication of the water (Lachavanne *et al.*, 1986b; Pieczyńska, 1986, 1990a).

The arrival of the industrial age and the control of water resources in western countries during the last century, the development of intensive agriculture in the 1950s and the emergence of the mass leisure society in the 1970s were thus the important stages in the process of endangering species, leading to the erosion of biodiversity in land–water ecotones. All processes of landscape fragmentation and ecotone restructuring are likely to be exposed to an unprecedented acceleration in time (di Castri and Hansen, 1992). All these changes of human origin exert a strong influence on the structure and dynamics of ecotones in rivers and lakes (Holland and Burk, 1982; Sedell and Frogatt, 1984; Fortuné, 1988; Petts, 1990), and therefore on biodiversity. The degradation and/or restriction and/or fragmentation of suitable sites appear to be the most important phenomena implicated in endangering ecotonal biodiversity.

The extent of the damage caused to date to the structural and functional integrity of land–water ecotones and the loss of specificity of the resultant biotopes are particularly extensive in industrialized countries and in some regions of developing countries with a high population density. For example, in Switzerland about 90% of the wetlands have disappeared since the start of the century (Böhlen, 1990) and already twenty years ago only 37% of the shorelines of the 32 main lakes was still in a natural or nearly natural state (Département Fédéral Suisse de Justice et Police, 1975). In the case of Lake Geneva, where the mean population density of the riparian zone is close to 1000 inhabitants/km^2, this proportion falls to as low as less than 10% (Lachavanne and Juge, 1993).

Ecotones are strategically placed between two ecological systems so as to pick up any disturbance (natural or resulting from human activities) that may occur in either of the adjacent ecological systems (Dyer *et al.*, 1989). Because the influences of degradation factors acting in the two adjacent ecosystems are accumulated in ecotones, it is possible to speak of a 'nuisance edge effect'.

One of the central problems is that man has greatly attenuated or even eliminated the impact of natural perturbations or disasters that generate varied habitats in land–water ecotones so as to utilize the resources and to ensure his safety. Many lakes and watercourses in industrialized countries today have been enclosed in a concrete corset (Holland and Burk, 1984; Sedell and Frogatt, 1984; Johnston *et al.*, 1992). These interventions lead to the partial or complete elimination of the dynamic processes of rejuvenation of the ecosystem and, consequently, to a reduction of the variety of habitats at the local and regional scales.

Currently, conservation of biodiversity can thus frequently only be guaranteed by taking direct action in the ecosystem (restoration and replenishment of nature) (Petts, Chapter 12). Thus paradoxically the diversity of the habitats and species persists only thanks to human intervention. A good example is provided in Switzerland by the conservation of the diversity of habitats on the southern shore of Lake Neuchâtel, which regularly requires various interventions (digging ponds, mowing the wetlands, undergrowth clearing, etc.) to avoid forestation corresponding to the region's climactic stage.

Relations between man and natural systems are often referred to in terms of a threat to species. However, it should be noted that man not only constitutes a factor for erosion of biodiversity but that he may, involuntarily, promote it locally or in a region by diversifying the habitats. The creation of ecotones in homogeneous ecosystems, by allowing colonization by species that are shore-line specialists, is an example of this (Holland, 1988). Thus the clearing of forests in temperate countries, which has been accentuated since Roman times to provide arable land, allowed the introduction of herbaceous plants which would not have been able to develop in forest ecosystems (Landolt, 1991).

Another example is provided by the positive consequences for biodiversity connected with the initial phases of enriching water with nutrients, when the system changes from an oligotrophic to a meso-eutrophic state under the influence of inputs of an anthropogenic origin (Lachavanne, 1985; Juge and Lachavanne, Chapter 5).

Finally, the introduction of new species into an ecosystem may just as easily lead to an increase of diversity in the case of a vacant ecological niche as to a reduction of diversity when the species introduced is more competitive and eliminates one or more of the species present by means of the biotic relations.

Di Castri and Hansen (1992) identified some predictions of the most obvious changes in ecotones in general (and consequently in land–water ecotones) in terms of biodiversity – firstly, because of landscape fragmentation due to unprecedented attacks on natural resources and later on because of climate change on a planetary scale. They include: (1) a dramatic decrease in biodiversity, from both ecological and genetic viewpoints, (2) an unprecedented rate of local and subsequently global extinction of species and (3) the disruption of the existing species assemblies with the possibility of spreading and colonization of new invading species more suited to the new environmental conditions (di Castri, 1991).

The factors affecting biodiversity act through the agency of processes that are genetic, physiological (variation in growth, metabolism, fertility, etc.), demographic (variations in mortality and age-specific fertility rates) and ethological (variations in space occupancy, utilization of trophic resources, rhythms of activity and social organization of populations).

THE LAND–INLAND WATER ECOTONE: A LIFE-SIZED LABORATORY FOR STUDYING THE PATTERNS AND REGULATING FACTORS OF BIODIVERSITY

Land–inland water ecotones have to receive more attention in the future

Despite their importance and obvious benefit for the conservation of biodiversity, land–inland water ecotones have received only limited attention in the past both from terrestrial ecologists and limnologists. This is attributable mainly to the fact that researchers have focused their studies on homogeneous terrestrial and aquatic systems and to the increased difficulty of studies of interface environments characterized by great instability of the physico-chemical conditions at varied scales of space and time.

More and more ecologists are convinced that an expanded ecotone theory will help us to understand and manage landscapes in relation to the ability of the ecotone to control ecological flows (energy, sediment, nutrients, organisms) and interactions among patches (di Castri *et al.*, 1988; Naiman *et al.*, 1989; Risser, 1990; Naiman and Décamps, 1990; di Castri and Hansen, 1992). 'Studying landscapes without consideration of ecotones may be as fruitless as trying to understand cellular behaviour without a knowledge of all membranes' (Hansen and di Castri, 1992).

Issues relating to the influence of interactions and exchanges across heterogeneous landscapes on biotic and abiotic processes (Hasler, 1975; Wiens *et al.*, 1985), to the management of this heterogeneity (Risser, 1985) and to the quality of continental waters (Lauga *et al.*, 1986; Johnston and Naiman, 1987; Décamps *et al.*, 1988; Pringle *et al.*, 1988; Wetzel, 1990) have been addressed. Several studies have attempted to quantify the flows of material and energy between terrestrial and aquatic ecosystems (Hasler, 1975; Holland *et al.*, 1991; Hillbricht-Ilkowska and Pieczyńska, 1993). But the study of the role played by these processes on the biodiversity of land–water ecotones or conversely, the role played by the presence of a more or less varied flora and fauna on these processes, is scarcely addressed. Several theories are questioned, in some cases for theoretical reasons, in others because of frequent exceptions or simply because of a lack of knowledge of the underlying process (Harper, 1977, quoted in Schulze and Zwölfer, 1987).

The diversity of a number of taxonomic groups such as higher plants or large vertebrates, for example, can be deduced from the inventories drawn up since the eighteenth century. But the lack of precise information about the environmental conditions of the ecosystem does not allow one to gain an understanding of the determinants of the evolution of the biodiversity reconstituted in this way. Some ecosystems have been studied in greater detail than others, mainly for economic reasons connected with the resources that they supply or on public health grounds or due to the danger threatening them. In the circumlacustrine or fluviatile land–inland water ecotonal zones, more sustained attention has been paid to the wetlands, for example, than to the submerged littoral zones (Pieczyńska, 1975, 1990a; Gopal and Masing, 1990 etc.). This research has made it possible to specify the intra- and inter-ecosystemic distribution of species and their distribution zone. The overall result of the studies conducted in the past is an accumulation of sectorial observations (species lists, distribution maps, etc.) which may be interesting but there has been no attempt to integrate the data into a theory of biological diversity of ecotones.

Gopal and Masing (1990) have reviewed all worldwide relevant data about land–water ecotones for wetlands and have highlighted the very incomplete nature of the knowledge that has been gathered to date about land–water ecotones. Some tentative approaches at the scale of individual communities have been carried out but there are still not enough of them and many results have yet to be substantiated or confirmed (Holland, 1988; Naiman and Décamps, 1990; Patten, 1990; Holland *et al.*, 1991). Indeed, numerous ecological arguments that have been put forward, even if they are valid, are still based on an extremely restricted body of data (di Castri and Younès, 1990) and do not allow predictive models to be developed. In particular, little is known about factors regulating the functioning of ecotones and diversity.

Gaps in function of organisms or communities of the ecosystem have to be identified and filled until there is a good understanding of all species in the major groups which contribute to species interactions, ecosystems processes and biodiversity (Carney, Chapter 11). A multi-author research programme that spans a range of scales, techniques and disciplines and that subdivides the community into recognizable categories (zones, guilds, elements of food-web, forms) is thought to be the best way to tackle the questions of biological diversity. But the ecosystemic approach to biodiversity remains very difficult. Combining the problem of species diversity with the problem of ecosystem functioning remains a very complicated task. Both are time-dependent processes which operate on a short-term (ecological) and a long-term (evolutionary) scale (Sukopp and Trepl, 1987).

In any case the fact that there are two systems (terrestrial and aquatic) with different structural and functional characteristics makes it difficult to study and develop general principles concerning the dynamics of land–water systems.

The complex topic of biological diversity can be addressed in many ways at the population, community and system levels (Simpson, 1988; Solbrig, 1991a). So any single explanation for species diversity will certainly be inadequate since species diversity is generally recognized as a product of interacting forces that vary in importance in both space and time. The knowledge to be acquired must be sufficient to provide a theoretical framework capable of correctly interpreting the data relating to biodiversity in land–inland water ecotones, to shed light on the choices of management and restoration measures which will guarantee its long-term conservation.

Key biodiversity-related questions

A re-evaluation and conceptualization of theory concerning biodiversity and ecotones appear to be necessary to explain better the patterns observed in nature (di Castri and Hansen, 1992).

A set of priority hypotheses and recommendations dealing with biodiversity at the genetic, species-to-community, and ecosystem levels was presented by Solbrig (1991a) in the Report of the IUBS–SCOPE–UNESCO Harvard Forest Workshop on Biodiversity (Petersham, MA, USA, June 27–July 1, 1991).

Several items of theoretical and empirical data lead one to think that ecotones may be particularly appropriate for testing new ecological ideas because of the intensity of interacting processes (Holland, 1988; Naiman et al., 1989; Risser, 1990; di Castri and Younès, 1990; Holland and Risser, 1991; di Castri and Hansen, 1992). In particular, because ecotones usually represent relatively steep gradients in abiotic and biotic variables, these ecological systems permit the testing of mechanisms contributing to biodiversity and controls on the flow of materials across the landscape. Risser (1990) suggests that ecological questions such as interdependence of spatio-temporal scales for explaining biodiversity, invasion by plants and animals, questions relating to the continuum concept and ecotones as sensitive indicators of global change could be studied successfully in ecotones. Understanding biodiversity and what it means fundamentally and the factors that generate and govern it is the best way to ensure its conservation and the functions which it performs within the systems.

The borderline conditions that prevail at the boundaries between related aquatic and terrestrial systems and the spatially tight series of very different habitats and a disturbance-driven dynamic make land–water ecotones a subject of choice, a life-sized laboratory for observing numerous ecological phenomena relevant to an understanding of the patterns of biodiversity (Wiens et al., 1985; di Castri et al., 1988; Naiman et al., 1988a). It may be expected that studying the influence of borderline conditions on individuals, populations or communities will make it possible to detect and reveal more easily than in more stable conditions certain fundamental processes and mechanisms of the determinism of biodiversity.

Several questions of both theoretical interest (understanding of biodiversity regulating factors: relative importance of abiotic and biotic factors, role of vital attributes of species that are characteristic of ecotones) and applied interest (ecotones management) can be identified (Naiman et al., 1989). In particular, several controversial ecological questions are capable of being analysed in land–water ecotones with a reasonable chance of success.

Edge effect. We know now that the edge effect – the tendency for communities to be more dense and often more diverse in transition zones between communities (Leopold, 1933; Odum, 1971) – does not apply to all taxa or to all ecotones (e.g. Van der Maarel, 1976; Harris, 1984). Thus there is an urgent need for understanding ecotone–biodiversity relationships in managing landscapes (Goeden, 1979). In what type of ecotone and under what conditions does the edge effect appear? In cases in which the edge effect does occur, how can biodiversity be affected by the extent (length, breadth, thickness) and quality of ecotones (Ghiselin, 1977; Lovejoy *et al.*, 1979)? At the landscape level, how can frequency of ecotones affect biodiversity? What is the evolution of the edge effect in time? How do man's activities affect the edge effect?

Ecological successions – Vegetation continuum. Defined as the non-seasonal, directional and continuous pattern of colonization and extinction on a site by species populations, occurring at a highly variable time scale (Begon *et al.*, 1990) and as the result of mechanisms such as facilitation, tolerance and inhibition (Connell and Slatyer, 1977), ecological successions present several characters, some of which are not unanimously agreed upon by ecologists. Evolution in time of the pattern of communities is caused by the dynamic processes of deaths, replacements and microsuccessions driven by allochthonous and autochthonous forces (Horn, 1974). However, the relative importance of these two forces on vegetation changes and the succession of wetlands towards a regional climax, still remains a controversial subject (Van der Valk, 1981, 1987; Mitsch and Gosselink, 1986).

In connection with biodiversity it is generally accepted that as a result of the increase in the environment's heterogeneity, the specific diversity of biocenoses, biochemical diversity and evenness increase throughout the successions.

Whereas certain characters of ecological successions, such as the decrease in the production/biomass ratio, the increase in the complexity of trophic networks, the contraction of niche amplitude and the change over from r selection to K selection are unanimously recognized, others, in contrast, like the increase in specific diversity and the diversity–stability relationship in ecosystems are still the subject of controversy.

Clearly these controversies are often generated by mixtures of the scale of observation. For example, it is often assumed that areas with so-called climax ecosystems will be more diverse than areas at earlier successional stages. However, an area with a mosaic of systems at different successional stages (which is typically the state of land–water ecotones), will probably be more diverse than the same area at climax provided that each system occupies a sufficiently large area of its own (Groombridge, 1992). Moreover, even if the concept of succession is well documented, the evident fact that the sequence of successions leading, for example, to forest is autogenous is not yet very clear (Mitsch and Gosselink, 1986).

Under certain conditions (gently sloping banks, substrate suitable for plants to take root) lake ecotones can thus provide sites of choice for the study of succession phenomena. They offer an opportunity to analyse the characteristics of the various stages of succession and the processes and mechanisms responsible for progressive or regressive evolution of communities in common climatic conditions and to highlight the relative importance of the autogenic forces (centre of ecotonal zones) and allogenic forces (edge of ecotonal zones) at work (Juge *et al.* in preparation).

In connection with fluctuations of the water level (seasonal, inter- and multi-annual variations) a part of the transitional ecotonal system (e.g. infralittoral in a lake system) is continuously at the beginning of the aggradation phase (Ulrich, 1987; see also pulse-stabilized subclimax after Odum, 1971). This characteristic allows one to track and analyse what Whittaker (1967) called the cyclo-climaxes in which progressive and regressive successions alternate.

By bringing together in an often restricted area a large spectrum of successional stages ranging from pioneer aquatic groups to sub-climactic or climactic forest groups (palustrine and terrestrial), lacustrine ecotones provide an excellent example of space–time representation of ecological successions. In the classical view of successions, the zoning of plant groups according to the hygric gradient (groups of wetlands) and the hydric gradient (aquatic groups) can represent at the same time the different stages of succession in time, as has been suggested by Clements (1916, 1936) and illustrated for example by Moor (1969). Several lines of evidence reviewed by Mitsch and Gosselink (1986) and Van der Valk (1987) seem to show (1) that zonation, which indicates an environmental gradient to which individual species are responding, does not necessarily indicate succession and (2) that the idea of a regional terrestrial climax for wetlands is inappropriate. Thus ecotones also provide a subject of choice for discussing the controversial question relating to the vegetation continuum concepts (individualistic hypothesis of Gleason, 1917 developed into the continuum concept by Whittaker, 1967; McIntosh, 1980).

Finally, land–water ecotones are favourable for the study of a scarcely addressed issue, *viz.* the relationship between the diversity of plants and the diversity of animals. Whereas plant successions are well documented, ecological successions, including animal populations, are hardly documented at all. There are probably some elements of fauna or part of its life cycle specific to the various stages of development of phytocoenoses, but others can adapt to, or even be dependent on, various stages of development of phytocenoses for all or some of their vital functions. A better understanding of the relationship between the diversity of plants (species, structural diversity) and the diversity of animals is indispensable to ensuring adequate management of land–water ecotones and of ecosystems in general.

Effects of perturbation. More and more authors agree in thinking that a particularly promising way of obtaining an insight into ecosystems is to analyse the effects of perturbations at different scales of time and space (Mooney and Godron, 1983; Sousa, 1984; Pickett and White, 1985; Schulze and Zwölfer, 1987; Turner, 1987; Pickett *et al.*, 1989; Rykiel, 1989a, 1989b; Townsend, 1989; di Castri and Younès, 1990; Wissmar and Swanson, 1990; Gosz, 1991; Swanson *et al.*, 1992). As Schulze and Zwölfer (1987) emphasized, 'it is difficult to detect the operation of a machine that is working perfectly'. To study resilience, hysteresis responses and the temporal circumstances of change in ecosystems, we need perturbations. Natural and anthropogenic perturbations are, in fact, major sources of biodiversity gains or losses in most ecosystems (di Castri and Younès, 1990).

The changes of state of land–water ecotones due (i) to flow rate controls in lotic systems (Minshall, 1988; Roux *et al.*, 1989; Amoros, 1991; Amoros and Petts, 1993) or to water-level fluctuations control in lentic systems (Keddy and Reznicek, 1986; Gasith and Gaafny, 1990) and (ii) to the impairment and fragmentation of habitats under the multivarious influence of man (destruction of biotopes linked to development of the banks) and the various forms of man-made pollution (eutrophication, toxic pollution) constitute on-going 'experiments' which, although fortuitous and uncontrolled, provide conditions favourable for the study of the processes that condition the structure and functioning of ecological systems, and consequently their biodiversity.

Ecotones enable us to verify in particular the model proposed by Connell (1978), according to which the lack of perturbation and, at the other extreme, disaster-type perturbations seriously prejudice diversity; the latter being promoted by conditions that are intermediate in terms of frequency and importance (see also Sousa, 1979, 1984; Paine and Levin, 1981; Ward and Stanford, 1983). They also provide a particularly favourable framework for studying the effects of anthropogenic fragmentation of landscapes and the degree of habitat connectivity on biodiversity (effects on organisms' transversal migrations between aquatic ecosystems and the uplands, corridor function, etc.).

OBJECTIVES OF THIS BOOK

What is the role and the importance of terrestrial–aquatic ecotones in maintaining local, regional and global biodiversity? This was one of the four key questions proposed at the Sopron Symposium (Hungary, May, 1988) in order to improve our understanding of the role of ecotones in aquatic landscape management (Naiman *et al.*, 1989). This highlighted point represents one of the principal challenges facing aquatic ecologists and managers of lake and river banks in the future.

This book presents the papers given at a special session devoted to 'Biodiversity in Land–Inland Water Ecotones' at the 25th Congress of the International Association of Theoretical and Applied Limnology (SIL), held in Barcelona from August 21–27, 1992. In addition, two invited papers were joined in order to supplement different aspects of biodiversity in land–water ecotones. It is a contribution by UNESCO–MAB to the 'Diversitas' programme launched jointly by the IUBS–SCOPE and UNESCO–MAB and aims to document various important issues, particularly issues and hypotheses concerning biodiversity under the MAB Ecotones programme connected with the role of ecotones in aquatic landscape management (Naiman et al., 1989). This book sets out to document several issues relating to biodiversity in the land–water ecotones and in particular to:

- give an overview of current knowledge of biodiversity in (natural and anthropized) land–water ecotones;

- improve our understanding of the importance of land–water ecotones for maintaining local, regional and global biodiversity and in particular to verify the hypothesis of the MAB Ecotones programme. 'Land–inland water ecotones are often characterized by higher biological diversity than adjacent patches, consequently the frequency of ecotones across a landscape directly affects biodiversity in a predictable manner' (Naiman et al., 1989);

- improve our understanding of the factors controlling biodiversity (abiotic and biotic, including diseases) and of their incidence on the different biotic components (in particular to document the scale at which organisms perceive and utilize natural resources) and on the different communities they form;

- better identify the role of biodiversity in the functioning of transition systems between ecosystems;

- identify the main relevant scientific issues to be addressed in research on biodiversity in ecotones, the ultimate aim being to contribute to the development of a general theory of biological diversity;

- determine the nature and the incidence of human impacts (loss and fragmentation of habitat, pollution, climate change) on biodiversity and to investigate the processes and mechanisms involved;

- provide data and relevant information which is useful for making recommendations to improve the management of ecotonal zones and to promote biodiversity; and

- draw useful conclusions in order to guide education and teachers and to alert the public to biodiversity conservation problems.

This book obviously also shares the main objectives of the 'Diversitas' programme as summarized by di Castri and Younès, 1990, *viz*: (1) to identify scientific issues that require international cooperation on the role of biodiversity in ecosystems' function, (2) to address general questions about how knowledge of species and ecosystems' diversity can contribute to global ecology and (3) to investigate how species diversity contributes to the functioning of the system. The ultimate aim is to build the scientific bases that are indispensable for well-informed management of ecosystems from the viewpoint of sustainable conservation of biodiversity.

This book comprises several papers describing various aspects of the biodiversity in land–water ecotones, principally in temperate countries. All the authors have been involved in investigating and giving the state of the art for each major biotic component. The individual papers exemplify their approach to this question: do different biotic groups have specific patterns of biodiversity in relation to their biology and ecology characteristics or do they obey common principles and laws?

This book is organized into 14 chapters. In Chapter 1 we attempt to provide a general framework for the book by explaining why it is of great importance to study the biodiversity of land–inland water ecotones with the ultimate objective of promoting better conservation of biodiversity. Dynamics of diversity and its expression over gradients and boundaries are introduced in Chapter 2 by Margalef, who examines especially the play between feedback loops and time–space dependence of diversity. The habitat complexity as a factor of biodiversity in land–water ecotones is evaluated by Pieczyńska and Zalewski (Chapter 3). In Chapter 4 Aragno and Ulehlova review and discuss microbial diversity and potentially important bacterial function in a macro- and microecotone context in relation to oxic and anoxic environments that characterize ecotonal zones. Chapters 5 to 10 are devoted to diversity of different biotic components in land–water ecotones: plants by Juge and Lachavanne, invertebrates by Giudicelli and Bournaud, amphibians by Joly and Morand, fish by Zalewski, birds by Dobrowolski and mammals by Schröpfer. In Chapter 11 Carney discusses land–water biodiversity from a landscape (catchment basin) perspective for tropical countries and in Chapter 12 Petts sets out the scientific basis for conserving diversity along river banks. By discussing different aspects (political, scientific) of the biodiversity crisis, Stearns attempts to place science in perspective (Chapter 13). Finally, Lachavanne and Juge summarize the different major points of the book relative to the patterns and regulating factors of biodiversity in land–water ecotones from the conservation viewpoint.

Thus this book is organized so as to make it possible to give a view of the patterns and regulating factors of biodiversity in land–water ecotones that is both differentiated (different taxonomic groups) and synthetic (communities).

ACKNOWLEDGEMENTS

I thank R. Juge, E. Castella and A. Lehmann for their stimulating remarks and K. Whiteley for translating the text into English.

REFERENCES

Allen, T.F.H. and Hoekstra, T.W. (1990). The confusion between scale-defined levels and conventional levels of organization in ecology. *Journal of Vegetation Science*, **1**, 5–12

Allen, T.F.H., O'Neill, R.V. and Hoekstra, T.W. (1984). Interlevel relations in ecological research and management: some working principles from hierarchy theory. US Department of Agriculture, Forest Service General Technical Report RM-1100. Rocky Mountains Forest and Range Experiment Station, Fort Collins, CO

Allen, T.F.H. and Starr, T.B. (1982). *Hierarchy: Perspectives for Ecological Complexity*. University of Chicago Press, Chicago, IL

Amoros, C. (1991). Changes in side-arm connectivity and implications for river system management. *Rivers*, **2**, 105–12

Amoros, C. and Petts, G.E. (eds.) (1993). *Hydrosystèmes Fluviaux. Collection d'Ecologie 24*. Masson, Paris

Andrén, H. and Angelstam, P. (1988). Elevated predation rates as an edge effect in habitat islands: experimental evidence. *Ecology*, **69**, 544–7

Aragno, M. and Ulehlova, B. (1996). Microbial diversity and functions in land–inland water ecotones. In Lachavanne, J.-B. and Juge, R. (eds.) *Biodiversity in Land–Inland Water Ecotones*, pp. 81–108. UNESCO/Parthenon, Paris/Carnforth

Augier, P., Baudry, J. and Fournier, F. (1992). *Hiérarchies et Echelles en Ecologie*. Agence de Coopération Culturelle et Technique, Paris.

Backéus, I. (1993). Ecotone versus ecocline: vegetation zonation and dynamics around a small reservoir in Tanzania. *Journal of Biogeography*, **20**, 209–18

Barbault, R. (1994). *Des Baleines, des Bactéries et des Hommes*. Editions Odile Jacob

Begon, M., Harper, J.L. and Townsend, C.R. (eds.) (1990). *Ecology, Individuals, Populations and Communities*. 2nd edn. Blackwell Scientific Publications, Cambridge, MA

Bhowmik, N. (1993). Effects of natural and man-made events on the land–water interfaces of large river basins. In Gopal, B., Hillbricht-Ilkowska, A.L. and Wetzel, R.G. (eds.) *Wetlands and Ecotones. Studies on Land–Water Interaction*, pp. 101–22. National Institute of Ecology, New Delhi

Böhlen, B. (1990). Préface. In *Zones Humides de Suisse. Sauvegarde et Entretien*, Office Fédéral de l'Environnement, des Forêts et du Paysage (OFEFP), Berne

Boon, P.J., Calow, P. and Petts, G.E. (eds.) (1992). *River Conservation and Management*. John Wiley, Chichester

Bramley, J. (1994). Biodiversity in freshwater ecosystems: the need for an aquatic plant taxonomy course? *Freshwater Biological Association*, **4**(3), 216–8

Brown, J.H. (1981). Two decades of homage to Santa Rosalia: toward a general theory of diversity. *American Zoologist*, **21**, 577–88

Burgess, R.C. and Sharpe, D.M. (eds.) (1981). *Forest Island Dynamics in Man-Dominated Landscapes*. Springer Verlag, New York

Carney, H.J. (1996). Ecotonal biodiversity and sustainability in unique tropical landscapes. In Lachavanne, J.-B. and Juge, R. (eds.) *Biodiversity in Land–Inland Water Ecotones*, pp. 233–248. UNESCO/Parthenon, Paris/Carnforth

Clark, J. (1979). Freshwater wetlands: habitats for aquatic invertebrates, amphibians, reptiles and fish. In Greeson, P.E., Clark, J.R. and Clark, J.E. (eds.) *Wetland Values and Functions: the State of our Understanding*, pp. 330–43. Proceedings of the national symposium on wetlands (1979). American Water Resources Association, Minneapolis, MN

Clark, W.C. and Munn, R.E. (eds.) (1986). *Sustainable Development of the Biosphere*. IIASA, Laxenburg, Austria and Cambridge University Press, UK

Clements, F.E. (1905). *Research Methods in Ecology*. University Publishing Company, Lincoln, NE

Clements, F.E. (1916). *Plant Succession: an Analysis of the Development of Vegetation*. Publication 242, Carnegie Institution of Washington, Washington, DC

Clements, F.E. (1936). Nature and structure of the climax. *Journal of Ecology*, **25**, 253–84

Cody, M.L. (1975). Towards a theory of continental species diversities: bird distributions over Mediterranean habitat gradients. In Cody, M.L. and Diamond, J.M. (eds.) *Ecology and Evolution of Communities*, pp. 214–57. Belknap Press of Harvard University Press, Cambridge, MA

Connell, J.H. (1978). Diversity in tropical rain forests and coral reefs. *Science*, **199**, 1302–10

Connell, J.H. and Slatyer, R.B. (1977). Mechanisms of successions in natural communities and their role in community stability and organization. *American Naturalist*, **111**, 1119–44

Correll, D.L. (ed.) (1986). *Watershed Research Perspectives*. Smithsonian Institution Press, Washington, DC

Décamps, H., Fortuné, M., Gazelle, F. and Pautou, G. (1988). Historical influence of man on the riparian dynamics of a fluvial landscape. *Landscape Ecology*, **1**, 163–73

Décamps, H. and Naiman, R.J. (1990). Towards an ecotone perspective. In Naiman, R.J. and Décamps, H. (eds.) *The Ecology and Management of Aquatic–Terrestrial Ecotones*, pp. 1–5. MAB Book Series 4. UNESCO/Parthenon, Paris/Carnforth

De Candolle, A. (1855). *Géographie botanique raisonnée, ou exposition des faits principaux et des lois concernant la distribution géographique des plantes de l'époque actuelle*. V. Masson, Paris

Département Fédéral Suisse de Justice et Police. (1975). *Rives des lacs, partie A: Objectifs et contenu de l'aménagement des rives des lacs. Le délégué à l'aménagement du territoire*. Office Fédéral des Imprimés, Berne

Diamond, J.M. (1973). Distribution ecology of New Guinea birds. *Science*, Vol. 179

Diamond, J.M. (1988). Factors controlling species diversity: overview and synthesis. *Annals of the Missouri Botanic Garden*, **75**, 117–29

Di Castri, F. (1991). Ecosystem evolution and global change. In Solbrig, O.T. and Nichols, G. (eds.) *Perspectives on Biological Complexity* (Monograph Series No 6), pp. 189–217. IUBS Press, Paris

Di Castri, F. and Hansen, A.J. (1992). The environment and development crises as determinants of landscape dynamics. In Hansen, A.J. and di Castri, F. (eds.) *Landscape Boundaries: Consequences for Biotic Diversity and Ecological Flows*, pp. 3–18. Ecological Studies 92, Springer Verlag, New York

Di Castri, F., Hansen, A.J. and Holland, M.M. (eds.) (1988). A new look at ecotones: emerging international projects on landscape boundaries. *Biology International*, Special Issue, **17**, 1–163

Di Castri, F. and Younès, T. (1990). Ecosystem function of biological diversity. *Biology International*, Special Issue, **22**, 1–20

Dobrowolski, K.A. (1996). Bird diversity in ecotonal habitats. In Lachavanne, J.-B. and Juge, R. (eds.) *Biodiversity in Land–Inland Water Ecotones*, pp. 205–221. UNESCO/Parthenon, Paris/Carnforth

Dorst, J. (1965). *Avant que Nature Meure*. Delachaux & Niestlé, Neuchâtel

Dyer, M.J., Eybergen, F., Jongman, R., Odum, W., Pinto, L., Pract, R. and Rusek, J. (1989). The role of land/inland water ecotones in studying global change. In Naiman, R.J., Décamps, H. and Fournier, F. (eds.) Role of Land–Inland Water Ecotones in Landscape Management and Restoration: a Proposal for Collaborative Research, pp. 69–72. *MAB Digest No 4*, UNESCO, Paris

Ehrlich, P.R. (1984). The structure and dynamics of butterfly populations. In Vane-Wright, R.I. and Ackery, P.R. (eds.) *The Biology of Butterflies*, pp. 25–40. Academic Press, London

Ehrlich, P.R. and Ehrlich, A.H. (1981). *Extinction: the Causes and Consequences of the Disappearance of Species*. Random House, New York

Ehrlich, A.H. and Ehrlich, P.R. (1987). *Earth*. Methuen, London

Ehrlich, P.R. and Wilson, E.O. (1991). Biodiversity studies: science and policy. *Science*, **235**, 758–61

Farmer, A.M. and Adams, M.S. (1991). The nature of scale and the use of hierarchy theory in understanding the ecology of aquatic macrophytes. *Aquatic Botany*, **41**, 253–61

Fitzpatrick, J.W. (1980). The wintering of North American tyrant flycatchers in the neotropics. In Keast, J.A. and Morton, E.S. (eds.) *Migrant Birds in the Neotropics: Ecology, Behaviour, Distribution and Conservation*. Smithsonian Institution Press, Washington, DC

Forel, F.A. (ed.). (1892). *Le Léman. Monographie Limnologique*, Tome 1. Slatekine Reprints, Genève

Forman, R.T.T. (1981). Interaction among landscape elements: a core of landscape ecology. *Proceedings of the International Congress of the Netherlands Society Landscape Ecology. Veldhoven 1981*, pp. 35–48. PUDOC, Wageningen

Forman, R.T.T. and Godron, M. (1986). *Landscape Ecology*. John Wiley, New York

Fortuné, M. (1988). Historical changes of a large river in an urban area: the Garonne River in Toulouse. *Regulated Rivers: Research and Management*, **2**, 179–86

Gasith, A. and Gaafny, S. (1990). Effects of water level fluctuation on the structure and function of the littoral zone. In Tilzer, M. and Serruya, C. (eds.) *Large Lakes. Ecological Structure and Function,* pp. 156–71. Springer Verlag, New York

Ghiselin, J. (1977). Analysing ecotones to predict biotic productivity. *Environmental Management,* **1**, 235–8

Giudicelli, J. and Bournaud, M. (1996). Invertebrate biodiversity in land–inland water ecotonal habitats. In Lachavanne, J.-B. and Juge, R. (eds.) *Biodiversity in Land–Inland Water Ecotones,* pp. 143–160. UNESCO/Parthenon, Paris/Carnforth

Gleason, H.A. (1917). The structure and development of the plant association. *Torrey Botanical Club Bulletin,* **44**, 463–81

Goeden, G.B. (1979). Biogeographic theory as a management tool. *Environmental Conservation,* **6**, 27–32

Goldberg, E.D. (1994). *Coastal Zone Space. Prelude to Conflict?* UNESCO Publishing Environment and Development, Paris

Gopal, B. and Masing, V. (1990). Biology and ecology. In Patten, B. (ed.) *Wetlands and Shallow Continental Water Bodies.* Vol. 1. *Natural and Human Relationships,* pp. 91–239. SPB Academic Publishing, The Hague

Gosz, J.R. (1991). Fundamental ecological characteristics of landscape boundaries. In Holland, M.M., Risser, P.G. and Naiman, R.J. (eds.) *Ecotones: the Role of Landscape Boundaries in the Management and Restoration of Changing Environments,* pp. 8–31. Chapman and Hall, New York and London

Gosz, J.R. (1992). Ecological functions in a biome transition zone: translating local responses to broad-scale dynamics. In Hansen, A.J. and di Castri, F. (eds.) *Landscape Boundaries: Consequences for Biotic Diversity and Ecological Flows,* pp. 55–75. Ecological Studies 92, Springer Verlag, New York

Gosz, J.R. and Sharpe, P.J.H. (1989). Broad-scale concepts for interaction of climate, topography and biota at biome transitions. *Landscape Ecology,* **3**, 229–43

Greene, C.H. and Tunstall, S.M. (1992). The amenity and environmental value of river corridors in Britain. In Boon, P.J., Calow, P. and Petts, G.E. (eds.) *River Conservation and Management,* pp. 423–41. John Wiley, Chichester

Groombridge, B. (ed.) (1992). *Global Biodiversity. Status of the Earth's Living Resources.* Chapman and Hall, New York and London

Güttinger, H. (1994). Jusqu'à quel point les organismes vivants et les écosystèmes supportent-ils le stress? Analyse écotoxicologique. *EAWAG News,* **36 F**, 9–12

Hansen, A.J. and di Castri, F. (eds.) (1992). *Landscape Boundaries: Consequences for the Biotic Diversity and Ecological Flows.* Ecological Studies 92, Springer Verlag, New York

Hansen, A.J., Risser, P.G. and di Castri, F. (1992). Epilogue: biodiversity and ecological flows across ecotones. In Hansen, A.J. and di Castri, F. (eds.) *Landscape Boundaries: Consequences for Biotic Diversity and Ecological Flows,* pp. 423–38. Ecological Studies 92, Springer Verlag, New York

Hansen, A.J. and Walker, B.H. (1985). The dynamic landscape: perturbations, biotic response, biotic patterns. *South African Institute of Ecology Bulletin,* **4**, 5–14

Harper, J.L. (1977). *Population Biology of Plants.* Academic Press, New York

Harris, L.D. (1984). *The Fragmented Forest: Island Biogeography Theory and the Preservation of Biotic Diversity.* University of Chicago Press, Chicago

Haslam, S.M. (1991). *River Plants of Western Europe. The Macrophyte Vegetation of Water-courses of the European Economic Community*. Cambridge University Press, Cambridge

Hasler, A.D. (ed.) (1975). *Coupling of Land and Water Systems*. Springer Verlag, Berlin

Haslet, J.R. (1990). Geographic information systems: a new approach to habitat definition and the study of distributions. *Tree*, **5**(7), 214–8

Hawkes, J.G. (1983). *The Diversity of Crop Plants*. Harvard University Press, Cambridge, MA

Hay, M.E. (1984). Patterns of fish and urchin grazing on Caribbean coral reefs: are previous results typical? *Ecology*, **65**, 446–54

Herben, T., Prach, K. and Rusek, J. (1992). Actual topics in ecotone research: data and concepts. *Ekologia (CSFR)*, **11**(3), 325–7

Hillbricht-Ilkowska, A.L. and Pieczyńska, E. (eds.) (1993). *Nutrient Dynamics and Retention in Land–Water Ecotones of Lowland, Temperate Lakes and Rivers. Development in Hydrobiology*. Kluwer Academic Publishers

Holland, M.M. (compiler) (1988). SCOPE/MAB technical consultations on landscape boundaries. Report of a SCOPE/MAB workshop on ecotones. In di Castri, F., Hansen, A.J. and Holland, M.M. (eds.) A New Look at Ecotones: Emerging International Projects on Landscape Boundaries. *Biology International*, Special Issue, **17**, 47–106

Holland, M.M. and Burk, C.J. (1982). Relative ages of western Massachusetts oxbow lakes. *Northeastern Geology*, **4**(1), 23–32

Holland, M.M. and Burk, C.J. (1984). The herb strata of three Connecticut River oxbow swamp forests. *Rhodora*, **86**(848), 397–415

Holland, M.M. and Risser, P.G. (1991). The role of landscape boundaries in the management and restoration of changing environments: introduction. In Holland, M.M., Risser, P.G. and Naiman, R.J. (eds.) *Ecotones: the Role of Landscape Boundaries in the Management and Restoration of Changing Environments*, pp. 1–7. Chapman and Hall, New York and London

Holland, M.M., Risser, P.G. and Naiman, R.J. (eds.) (1991). *Ecotones: the Role of Landscape Boundaries in the Management and Restoration of Changing Environments*. Chapman and Hall, New York and London

Horn, H.S. (1974). The ecology of secondary succession. *Annual Review of Ecology and Systematics*, **5**, 25–37

Huntley, B.J., Ezcurra, E., Fuentes, E.R., Funjii, K., Grubb, P.J., Habek, W., Harger, J.R.E., Holland, M.M., Levin, S.A., Luchenko, J., Mouney, H.A., Neronov, V., Noble, I., Pulliam, H.R., Ramakrishnan, P.S., Risser, P.G., Sala, O., Sarukhan, J. and Sombroek, W.G. (1991). A sustainable biosphere: the global imperative (The International Sustainable Biosphere Initiative). *Ecology International*, **20**, 1–14

Hutchinson, G.E. (1967). A Treatise on Limnology. Vol. II. *An Introduction to Lake Biology and Limnoplankton*. John Wiley, New York

IUCN (1980). *World Conservation Strategy*. International Union for the Conservation of Nature and Natural Resources, Gland

IUCN/UNEP/WWF (1991). *Caring for the Earth. A Strategy for Sustainable Living*. IUCN, Gland

Jeffers, J., Bretschko, G., El Daoushi, F., Fernald, E., Hadley, M., Hansen, A., Jolankai, G., Kharchenko, T., Lachavanne, J.B., Lauga, J., Risser, P., Shugart, H.H., Stanford, J., Suzuki, M., Vervier, P., Webber, P. and Wissmar, R.C. (1989). Theoretical considerations of the ecotone concept. In Naiman, R.J., Décamps, H. and Fournier, F. (eds.) Role of Land–Inland Water Ecotones in Landscape Management and Restoration: a Proposal for Collaborative Research. pp. 55–9. *MAB Digest No 4*. UNESCO, Paris

Jeffries, M. and Mills, D. (1990). *Freshwater Ecology*. Bellhaven Press, London

Jenik, J. (1992). Ecotone and ecocline: two questionable concepts in ecology. *Ekologia (CSFR)*, **11**(3), 243–50

Jennersten, O. (1988). Pollination in *Dianthus deltoides* (Caryophyllaceae): effects of habitat fragmentation on visitation and seed set. *Conservation Biology*, **2**, 359–66

Johnston, C.A. and Naiman, R.J. (1987). Boundary dynamics at the aquatic–terrestrial interface: the influence of beaver and geomorphology. *Landscape Ecology*, **1**, 47–57

Johnston, D.W. and Odum, E. (1956). Breeding bird populations in relation to plant succession on the Piedmont of Georgia. *Ecology*, **37**, 50–62

Johnston, C.A., Pastor, J. and Pinay, G. (1992). Quantitative methods for studying landscape boundaries. In Hansen, A.J. and di Castri, F. (eds.) *Landscape Boundaries: Consequences for the Biotic Diversity and Ecological Flows*, pp. 107–25. Ecological Studies 92, Springer Verlag, New York

Joly, P. and Morand, A. (1996). Amphibian diversity and land–water ecotones. In Lachavanne, J.-B. and Juge, R. (eds.) *Biodiversity in Land–Inland Water Ecotones*, pp. 161–182. UNESCO/Parthenon, Paris/Carnforth

Juge, R., Buttler, A., Clerc, C., Buche, M., Cornali, P. and Lachavanne, J.B. (1997). Pattern of plant diversity along the land–water ecotone in a temperate deep lake, in preparation

Juge, R. and Lachavanne, J.B. (1996). Patterns and regulation of plant diversity in lacustrine ecotones. In Lachavanne, J.-B. and Juge, R. (eds.) *Biodiversity in Land–Inland Water Ecotones*, pp. 109–141. UNESCO/Parthenon, Paris/Carnforth

Keast, A. and Morton, E.S. (eds.) (1980). *Migrant Birds in the Neotropics: Ecology, Behavior, Distribution and Conservation*. Smithsonian Institution Press, Washington, DC

Keating, M. (1993). Sommet de la terre (1992). Un programme d'action. Centre pour notre avenir à tous, Genève

Keddy, P.A. (1983). Shoreline vegetation in Axe Lake Ontario: effects of exposure on zonation patterns. *Ecology*, **64**, 331–44

Keddy, P.A. and Reznicek, A.A. (1986). Great Lakes vegetation dynamics: the role of fluctuating water levels and seeds. *Journal of Great Lakes Research*, **12**, 25–36

Knopf, F.L. (1986). Changing landscapes and the cosmopolitism of the eastern Colorado avifauna. *Wildlife Society Bulletin*, **14**, 132–42

Knopf, F.L., Johnson, R.R., Rich, T., Samson, F.B. and Szaro, R.C. (1988). Conservation of riparian ecosystems in the United States. *The Wilson Bulletin*, **100**, 272–84

Krebs, C.J. (1985). *Ecology: the Experimental Analysis of Distribution and Abundance*, 3rd edn. Harper & Row, New York

Lachavanne, J.B. (1982). Influence de l'eutrophisation des eaux sur les macrophytes des lacs suisses: résultats préliminaires. In Symoens, J.J., Hooper, S.S. and Compère, P. (eds.) *Studies on Aquatic Vascular Plants*, pp. 333–9. Royal Botanical Society of Belgium, Brussels

Lachavanne, J.B. (1985). The influence of accelerated eutrophication on the macrophytes of Swiss lakes: abundance and distribution. *Verhandlungen Internationale Vereinigung für Theoretische und Angewandte Limnologie*, **22**, 2950–5

Lachavanne, J.B. and Juge, R. (1993). Integrating environmental considerations into the management of Lake Geneva, Switzerland. In *OECD Coastal Zone Management: Selected Case Studies*, pp. 137–64. Paris

Lachavanne, J.-B. and Juge, R. (1996). Land–inland water ecotones as transitional systems of particularly high biodiversity: towards a synthesis. In Lachavanne, J.-B. and Juge, R. (eds.) *Biodiversity in Land–Inland Water Ecotones*, pp. 277–296. UNESCO/Parthenon, Paris/Carnforth

Lachavanne, J.B. and Juge, R. Application of ecotone concept to lake shores, in preparation

Lachavanne, J.B., Juge, R. and Noetzlin, A. (1986a). Evolution des macrophytes du Léman (rives genevoises 1972–1984). *Sciences de l'Eau*, **5**, 419–33

Lachavanne, J.B., Juge, R., Noetzlin, A. and Perfetta, J. (eds.) (1985). *Zustand, Erhaltung und Schutz der Ufer des Vierwaldstättersees*. Ed. Bundesamt für Forstwesen und Landschaftschutz, Aufsicht Kommission Vierwaldstättersee und Universität Genf

Lachavanne, J.B., Juge, R., Perfetta, J., Noetzlin, A. and Lods-Crozet, B. (1986b). Etude chorologique et écologique des macrophytes des lacs suisses en fonction de leur altitude et de leur niveau trophique, 1976–1985. Rapport FNRS, Berne

Landolt, E. (1991). Plantes vasculaires menacées en Suisse. Listes rouges nationales et régionales. Office fédéral de l'environnement, des forêts et du paysage (OFEFP), Berne

Lauga, J., Décamps, H. and Holland, M.M. (1986). *Land Use Impacts on Aquatic Ecosystems. Proceedings of the Toulouse Workshop*. MAB–UNESCO and PIREN–CNRS, April (1986)

Ledig, F.T. (1988). The conservation of diversity in forest trees. *BioScience*, **38**(7), 471–9

Lee, G.F., Bentley, E. and Amundson, R. (1975). Effects of marshes on water quality. In Hasler, A.D. (ed.) *Coupling of Land and Water Systems*, pp. 105–27. Springer Verlag, Berlin

Legendre, L. and Demers, S. (1985). Auxiliary energy, ergoclines and aquatic biological production. *Naturaliste Canadien*, **112**, 5–14

Leopold, A. (1933). *Game Management*. Charles Scribner's Sons, New York

Levin, S.A. (1992). The problem of pattern and scale in ecology. *Ecology*, **73**, 1943–67

Likens, G.E. and Bormann, F.H. (1975). An experimental approach to New England landscapes. In Hasler, A.D. (ed.) *Coupling of Land and Water Systems*, pp. 7–29. Springer Verlag, Berlin

Livingstone, B.E. (1903). The distribution of the upland societies of Kent County, Michigan. *Botanical Gazette*, **35**, 36–55

Lodge, D.M., Barko, J.W., Strayer, D., Melack, J.M., Mittelbach, G.G., Howarth, R.W., Menge, B. and Titus, J.E. (1988). Spatial heterogeneity and habitat interactions in lake communities. In Carpenter, S.R. (ed.) *Complex Interactions in Lake Communities,* pp. 181–208. Springer Verlag, New York

Lovejoy, T.E., Bierregaard, R.O. Jr., Rylands, A.B., Malcolm, J.R., Quintela, C.E., Harper, L.H., Brown, K.S. Jr., Powell, A.H., Powell, G.V.N., Schubart, H.O.R. and Hays, M.B. (1986). Edge and other effects of isolation on Amazon forest fragments. In Soulé, M.E. (ed.) *Conservation Biology: the Science of Scarcity and Diversity,* pp. 257–85. Sinauer Associates, Sunderland, MA

Lubchenco, J., Olson, A.M., Brubaker, L.B., Carpenter, S.R., Holland, M.M., Hubbel, S., Levin, S.A., MacMahon, J.A., Matson, P.A., Melillo, J.M., Mooney, H.A., Peterson, C.H., Pulliam, H.R., Real, L.A., Regal, P.J. and Risser, P.G. (1991). The Sustainable biosphere initiative: an ecological research agenda. *Ecology,* **72**, 371–412

MacArthur, R.H. (1972). *Geographical Ecology: Patterns in the Distribution of Species.* Harper & Row, New York

Margalef, R. (1996). Dynamics of diversity and its expression over gradients and boundaries. In Lachavanne, J.-B. and Juge, R. (eds.) *Biodiversity in Land–Inland Water Ecotones,* pp. 47–59. UNESCO/Parthenon, Paris/Carnforth

McIntosh, R. (1980). The background and some current problems of theoretical ecology. *Synthese,* **43**, 195–255

McNeely, J.A., Miller, K.R., Reid, W.V., Mittermeier, R.A. and Werner, T.B. (1990). *Concerning the World's Biological Diversity.* IUCN, WRI, CI, WWF-US, the World Bank, Gland and Washington, DC

Meentemeyer, V. and Box, E.O. (1987). Scale effects in landscape studies. In Turner, M.G. (ed.) *Landscape Heterogeneity and Disturbance,* pp. 15–34. Springer Verlag, New York

Minshall, G.W. (1988). Stream ecosystem theory: a global perspective. *Journal of the North American Benthological Society,* **7**, 263–88

Mitsch, W.J. and Gosselink, J.G. (1986). *Wetlands.* Van Nostrand Reinhold, New York

Mooney, H.A. and Godron, M. (eds.) (1983). *Disturbance and Ecosystems.* Springer Verlag, New York

Moor, M. (1969). Zonation und Sukzession am Ufer stehender und fliessender Gewässer. *Vegetatio Acta Geobotanica,* **17**

Myers, N. (1990). The biodiversity challenge: expanded hot-spots analysis. *The Environmentalist,* **10**(4), 243–55

Naiman, R.J. and Décamps, H. (eds.) (1990). *The Ecology and Management of Aquatic–Terrestrial Ecotones.* MAB Book Series 4. UNESCO/Parthenon, Paris/Carnforth

Naiman, R.J. and Décamps, H. (1991). Landscape boundaries in the management and restoration of changing environments: a summary. In Holland, M.M., Risser, P.G. and Naiman, R.J. (eds.) *Ecotones: the Role of Landscape Boundaries in the Management and Restoration of Changing Environments,* pp. 130–7. Chapman and Hall, New York and London

Naiman, R.J., Décamps, H. and Fournier, F. (eds.) (1989). Role of Land–Inland Water Ecotones in Landscape Management and Restoration: a Proposal for Collaborative Research. *MAB Digest 4,* UNESCO, Paris

Naiman, R.J., Décamps, H., Pastor, J. and Johnston, C.A. (1988a). The potential importance of boundaries to fluvial ecosystems. *Journal of the North American Benthological Society*, **7**, 289–306

Naiman, R.J., Holland, M.M., Décamps, H. and Risser, P.G. (1988b). A new UNESCO programme: research and management of land–inland water ecotones. In di Castri, F., Hansen, A.J. and Holland, M.M. (eds.) A New Look at Ecotones: Emerging International Projects on Landscape Boundaries. *Biology International*, Special Issue, **17**, 107–36

Nelson, R.W., Ambasht, R.S., Shardendu, R.S., Amoros, C., Begg, G.W., Bonetto, A.A., Wais, I.R., Dister, E., Wenger, E., Finlayson, C.M., Handoo, J.K., Pandit, A.K., Mavuti, K.M., Parish, D. and Savey, P. (1990). In Kusler, J. and Day, S. (eds.) River floodplain and delta wetlands management team: a project of the world wetlands partnership. *Proceedings of the International Symposium on Wetlands and River Corridor Management* (Charleston, South Carolina, USA, July 5–9, 1989), pp. 75–82. Association of the Wetland Managers, Berne, New York

Noss, R.F. (1983). A regional landscape approach to maintain diversity. *BioScience*, **33**, 399–409

Odum, E. (1971). *Fundamentals of Ecology*. 3rd edn. W.B. Saunders, Philadelphia

Ogden, J.C., Brown, R.A. and Salesky, N. (1973). Grazing by the echinoid *Diadema antillarum* Phillippi: formation of halos around West Indian patch reef. *Science*, **182**, 715–7

Ojeda, R. and Mares, M.A. (1989). The biodiversity issue and Latin America. *Revista Chilena de Historia Natural*, **62**, 185–91

O'Neill, R.V., DeAngelis, D.L., Waide, J.B. and Allen, T.F.H. (1986). A hierarchical concept of ecosystems. *Monographical Population Biology* 23, Princeton University Press, Princeton, NJ

Paine, R.T. and Levin, S.A. (1981). Intertidal landscapes: disturbance and the dynamics of pattern. *Ecological Monographs*, **51**, 145–78

Patten, B.C. (ed.) (1990). *Wetlands and Shallow Continental Water Bodies*. Vol. 1. *Natural and Human Relationships*. SPB Academic Publishing, The Hague

Patten, B.C., Jørgensen, S.E., Gopal, B., Kvet, J., Loeffler, H., Svirezhav, Y. and Tundisi, J. (1985). Ecotones: an edge approach to gene pool preservation and management in the biosphere. *Prospectus for a new SCOPE programme from the Scientific Advisory Committee for wetlands and shallow continental water bodies*. Athens, GA

Petts, G.E. (1990). The role of ecotones in aquatic landscape management. In Naiman, R.J. and Décamps, H. (eds.) *The Ecology and Management of Aquatic–Terrestrial Ecotones*, pp. 227–62. MAB Book Series 4. UNESCO/Parthenon, Paris/Carnforth

Petts G.E. (1996). Scientific basis for conserving diversity along river margins. In Lachavanne, J.-B. and Juge, R. (eds.) *Biodiversity in Land–Inland Water Ecotones*, pp. 249–268. UNESCO/Parthenon, Paris/Carnforth

Petts, G.E., Roux, A.L. and Moller, H. (eds.) (1989). *Historical Changes of Large Alluvial Rivers in Western Europe*. John Wiley, Chichester

Pickett, S.T.A., Kolasa, J., Armesto, J.J. and Collins, S.L. (1989). The ecological concept of disturbance and its repression at various hierarchical levels. *Oikos*, **54**, 129–36

Pickett, S.T.A. and White, P.S. (eds.) (1985). *The Ecology of Natural Disturbance and Patch Dynamics.* Academic Press, New York

Pieczyńska, E. (1975). Ecological interactions between land and the littoral zones of lakes (Masurian Lakeland, Poland). In Hasler, A. (ed.) *Ecological Studies. Analogy and Synthesis,* pp. 263–76. Springer Verlag, Berlin

Pieczyńska, E. (ed.) (1976). *Selected Problems of Lake Littoral Ecology.* Warsaw University, Warsaw

Pieczyńska, E. (1986). Littoral communities and Lake Eutrophication. In Lauga, J., Décamps, H. and Holland, M.M. (eds.) *Land Use Impacts on Aquatic Ecosystems. Proceedings of the Toulouse Workshop,* pp. 105–17. MAB–UNESCO and PIREN–CNRS, Paris, April, 1986

Pieczyńska, E. (1990a). Lentic aquatic–terrestrial ecotones: their structure, functions and importance. In Naiman, R.J. and Décamps, H. (eds.) *The Ecology and Management of Aquatic-Terrestrial Ecotones,* pp. 103–40. MAB Book Series 4. UNESCO/Parthenon, Paris/Carnforth

Pieczyńska, E. (1990b). Littoral habitats and communities. In Jørgensen, S.E. and Loeffler, H. (eds.) *Guidelines of Lake Management.* Vol. 3. *Lake Shore Management,* pp. 39–71. ILEC/UNEP, Japan

Pieczyńska, E. and Zalewski, M. (1996). Habitat complexity in land–inland water ecotones. In Lachavanne, J.-B. and Juge, R. (eds.) *Biodiversity in Land–Inland Water Ecotones,* pp. 61–79. UNESCO/Parthenon, Paris/Carnforth

Pinay, G., Décamps, H., Chauvet, E. and Fustec, E. (1990). Functions of ecotones in fluvial systems. In Naiman, R.J. and Décamps, H. (eds.) *The Ecology and Management of Aquatic–Terrestrial Ecotones,* pp. 141–70. MAB Book Series 4. UNESCO/Parthenon, Paris/Carnforth

Pollock, M.M. and Naiman, R.J. (1994). Hydrologic control of plant diversity in land–water ecotones. *Proceedings of the International Workshop on the Ecology and Management of Aquatic–Terrestrial Ecotones.* University of Washington, Seattle. Feb. 14–19, 1987. Abstract

Pringle, C.M., Naiman, R.J., Bretschko, G., Karr, J.R., Oswood, M.W., Webster, J.R., Welcomme, R.L. and Winterbourn, M.J. (1988). Patch dynamics in lotic systems: the stream as a mosaic. *Journal of the North American Benthological Society,* 7, 503–24

Ramade, F. (1981). *Ecologie dans Resources Naturelles.* Masson, Paris

Reid, W.V. and Miller, K.R. (1989). *Keeping Options Alive: the Scientific Basis for Conserving Biodiversity.* World Resources Institute, Washington, DC

Risser, P.G. (compiler) (1985). Spatial and temporal variability of biospheric and geospheric processes: research needed to determine interactions with global environmental change. The International Council of Scientific Union Press, Paris

Risser, P.G. (1990). The ecological importance of land–water ecotones. In Naiman, R.J. and Décamps, H. (eds.) *The Ecology and Management of Aquatic–Terrestrial Ecotones,* pp. 7–21. MAB Book Series 4. UNESCO/Parthenon, Paris/Carnforth

Risser, P.G., Karr, J.R. and Forman, R.T.T. (1984). *Landscape Ecology: Directions and Approaches. Illinois Natural History Survey.* Special Publication No 2

Roux, A.L., Bravard, J., Amoros, C. and Pautou, G. (1989). Ecological changes of the French Upper Rhône River since 1750. In Petts, G.E., Roux, A.L. and Moller, H.

(eds.) *Historical Changes of Large Alluvial Rivers in Western Europe*, pp. 323–50. John Wiley, Chichester

Rykiel, E.J. Jr. (1989a). Ecological disturbance theory. In Singh, M.G. (ed.) *Systems and Control Encyclopedia – Theory, Technology, Applications*, pp. 1311–18. Pergamon Press, Oxford

Rykiel, E.J. Jr. (1989b). Ecological disturbances. In Singh, M.G. (ed.) *Systems and Control Encyclopedia – Theory, Technology, Applications*, pp. 1318–21. Pergamon Press, Oxford

Schonewald-Cox, C.M., Chambers, S.M., MacBryde, F. and Thomas, L. (eds.) (1983). *Genetics and Conservation: a Reference for Managing Wild Animal and Plant Populations*. Benjamin-Cummings, London

Schröpfer, R. (1996). Mammal diversity in inland–water ecotone habitats. In Lachavanne, J.-B. and Juge, R. (eds.) *Biodiversity in Land–Inland Water Ecotones*, pp. 223–31. UNESCO/Parthenon, Paris/Carnforth

Schröter, C. (1926). *Das Planzenleben der Alpen*. Zürich

Schulze, E.-O. and Zwölfer, H. (eds.) (1987). *Potential Limitations of Ecosystem Analysis*. Ecological Studies 61, Springer Verlag, Berlin

Sedell, J.R. and Froggatt, J.L. (1984). Importance of streamside forests to large rivers: the isolation of the Willamette River, Oregon, USA, from its floodplain by snagging and streamside forest removal. *Verhandlungen Internationale Vereinigung für Theoretische und Angewandte Limnologie*, **22**, 1828–34

Shelford, V.E. (1913). The reaction of certain animals to gradients of eve porting power and air. A study in experimental ecology. *Biological Bulletin*, **25**, 79–120

Simpson, B.B. (1988). Biological diversity in the context of ecosystem structure and function. *Biology International*, **17**, 15–17

Slaviková, J. (1986). *Plant Ecology* (in Czech). Prague

Solbrig, O.T. (ed.) (1991a). *From Genes to Ecosystems: a Research Agenda on Biodiversity*. IUBS–SCOPE–UNESCO, Cambridge, MA

Solbrig, O.T. (1991b). Biodiversity: Scientific Issues and Collaborative Research Proposals. *MAB Digest 9*, UNESCO, Paris

Soulé, M.E. (ed.) (1986). *Conservation Biology. The Science of Scarcity and Diversity*. Sinauer Associates, Inc., Sunderland, MA

Soulé, M.E. (1991). Conservation: tactics for a constant crisis. *Science*, **253**, 744–50

Sousa, W. (1979). Disturbance in the marine intertidal boulder fields: the nonequilibrium maintenance of species diversity. *Ecology*, **60**, 1225–35

Sousa, W. (1984). The role of disturbance in natural communities. *Annual Review of Ecology and Systematics*, **15**, 535–91

Stearns, S.C. (1996). Biodiversity: a review of the scientific issues. In Lachavanne, J.-B. and Juge, R. (eds.) *Biodiversity in Land–Inland Water Ecotones*, pp. 269–75. UNESCO/Parthenon, Paris/Carnforth

Stein, R.A. and Magnusson, J.J. (1976). Behavioral response by crayfish to a fish predator. *Ecology*, **57**, 751–61

Stoms, D.M. (1992). Effects of habitat map generalization in biodiversity assessment. *Photogrammetric Engineering and Remote Sensing*, **58**(11), 1587–91

Stoms, D.M and Estes, J.E. (1993). A remote sensing research agenda for mapping and monitoring biodiversity. *International Journal of Remote Sensing*, **14**(10), 1839–60

Sukopp, V. and Trepl, L. (1987). Extinction and naturalization of plant species as related to ecosystem structure and function. In Schulze, E.-O. and Zwölfer, H. (eds.) *Potential Limitations of Ecosystem Analysis*, pp. 245–76. Ecological Studies 61, Springer Verlag, Berlin

Swanson, F.J., Wondzell, S.M. and Grant, G.E. (1992). Landforms, disturbance and ecotones. In Hansen, A.J. and di Castri, F. (eds.) *Landscape Boundaries: Consequences for Biotic Diversity and Ecological Flows*, pp. 304–23. Ecological Studies 92, Springer Verlag, Berlin

Townsend, C.R. (1989). The patch dynamics concept of stream community ecology. *Journal of the North American Benthological Society*, 8, 36–50

Turner, M.G. (ed.) (1987). *Landscape Heterogeneity and Disturbance*. Springer Verlag, New York

Ulfstrand, S. (1992). Biodiversity – how to reduce its decline. *Oikos*, **63**, 3–5

Ulrich, B. (1987). Stability, elasticity and resilience of terrestrial ecosystems with respect to matter balance. In Schulze, E.D. and Zwölfer, H. (eds.) *Potentials and Limitations of Ecosystem Analysis*, pp. 11–49. Ecological Studies 61, Springer Verlag, Berlin

Urban, D.L., O'Neill, R.V. and Shugart, H.H. Jr. (1987). Landscape ecology. *BioScience*, **37**, 119–27

Van der Maarel, E. (1976). On the establishment of plant community boundaries. *Berichte der Deutschen Botanischen Gesellschaft*, **89**, 415–43

Van der Maarel, E. (1990). Ecotones and ecoclines are different. *Journal of Vegetation Science*, **1**, 135–8

Van der Maarel, E. and Westhoff, V. (1964). The vegetation of the dunes near Oostvoorne, Netherlands. *Wentia*, **12**, 1–61

Van der Valk, A.G. (1981). Succession in wetlands: a Gleasonian approach. *Ecology*, **62**, 688–96

Van der Valk, A.G. (1987). Vegetation dynamics of freshwater wetlands: a selection review of the literature. *Archiv für Hydrobiologie, Beiheft Ergebnisse der Limnologie*, **27**, 27–39

Vanhecke, L. and Charlier, G. (1982). The regression of aquatic and marsh vegetation and habitats in the north of Belgium between 1904 and 1980: Some photographic evidence. In Symoens, J.J., Hooper, S.S. and Compère, P. (eds.) *Studies on Aquatic Vascular Plants,* pp. 410–11. Royal Botanical Society of Belgium, Brussels

Van Leeuwen, C.G. (1966). A relational theoretical approach to pattern and process in vegetation. *Wentia*, **15**, 25–46

Vieira da Silva, J. (1979). Introduction à la théorie écologique. *Collection d'Ecologie* 14. Masson, Paris

Ward, J.V. and Stanford, J.A. (1983). The intermediate-disturbance hypothesis: an explanation for biotic diversity patterns in lotic ecosystems. In Fontaine, T.D. and Bartell, S.M. (eds.) *Dynamics of Lotic Ecosystems*, pp. 347–56. Ann Arbor Science, Ann Arbor, MN

Watt, A.S. (1947). Pattern and process in the plant community. *Journal of Ecology*, **35**, 1–22

Wedeles, C.H.R., Meisner, J.D. and Rose, M.J. (1992). Wetland science research needs in Canada. North American Wetlands Conservation Council (Canada), Ontario

Western, D. (1992). The biodiversity crisis: a challenge for biology. *Oikos*, **63**, 29–39

Wetzel, R.G. (1990). Land–water interface: metabolic and limnological regulators. *Verhandlungen Internationale Vereinigung für Theoretische und Angewandte Limnologie*, **24**, 6–24

Wetzel, R., Adams, M., Almassy, A., Dinka, M., Eiseltova, M., Fustec, E., Gibert, J., Gopal, B., Leichtfried, M., Livingstone, R.J., Mathieu, J., Parma, S., Pastor, J., Pieczyńska, E., Roberts, G., Sagova, M., Thompson, K. and Uerhoog, F. (1989). Nutrient, energy and water flows through land–water ecotones. In Naiman, R.J., Décamps, H. and Fournier, F. (eds.) Role of Land–Inland Water Ecotones in Landscape Management and Restoration: a Proposal for Collaborative Research. *MAB Digest No 4*, pp. 61–4. UNESCO, Paris

Wheeler, D.J. (1993). Commentary: linking environmental models with geographic information systems for global change research. *Photogrammetric Engineering and Remote Sensing*, **59**(10), 1497–501

Whittaker, R.H. (1960). Vegetation of the Siskiyou Mountains, Oregon and California. *Ecological Monographs*, **30**, 279–338

Whittaker, R.H. (1967). Gradient analysis of vegetation. *Biological Revue*, **42**, 207–64

Wiens, J.A., Crawford, C.S. and Gosz, J.R. (1985). Boundary dynamics: a conceptual framework for studying landscape ecosystems. *Oikos*, **45**, 421–7

Wilcox, B.A. and Murphy, D.D. (1983). Conservation strategy: the effects of fragmentation on extinction. *American Naturalist*, **125**, 879–87

Williams, M. (ed.) (1990). *Wetlands: a Threatened Landscape*. Basil Blackwell, UK

Wilson, E.O. (1985). The biological diversity crisis. *BioScience*, **35**, 700–6

Wilson, E.O. (ed.) (1988). *Biodiversity*. National Academy Press, Washington, DC

Wilson, E.O. (1989). Threats to biodiversity. *Scientific American*, **261**(3), 108–16

Wilson, E.O. and Peter, F.M. (eds.) (1988). *Biodiversity*. National Academy Press, Washington, DC

Wissmar, R.C. and Swanson, F.J. (1990). Landscape disturbances and lotic ecotones. In Naiman, R.J. and Décamps, H. (eds.) *The Ecology and Management of Aquatic–Terrestrial Ecotones*, pp. 65–89. MAB Book Series 4. UNESCO/Parthenon, Paris/Carnforth

Wolf, E.C. (1987). *On the Brink of Extinction: Conserving the Diversity of Life*. Worldwatch Paper 78, Washington, DC

Zalewski, M. (1996). Fish diversity and ecotonal habitat. In Lachavanne, J.-B. and Juge, R. (eds.) *Biodiversity in Land–Inland Water Ecotones*, pp. 183–203. UNESCO/Parthenon, Paris/Carnforth

Zonneveld, I.S. (1974). On abstract and concrete boundaries, arrangement and classification. In Sommer, W.H. and Tüxen, R. (eds.) *Tatsachen und Probleme der Grenzen in der Vegetation*. 431 pp. Bericht Symposium International Verein Vegetationskunde. 1968, 17–43. J. Cramer, Lehre

CHAPTER 2

DYNAMICS OF DIVERSITY AND ITS EXPRESSION OVER GRADIENTS AND BOUNDARIES

Ramon Margalef

INTRODUCTION

Diversity is an intuitive concept which became explicit at least 60 years ago. It refers to regularities manifested in the way individuals are distributed into species, or total biomass falls into species biomass. Situations are ranged between assemblages that are very diverse, very rich in species (Amazonia-like) and those in which a very limited number of species make most of the active biomass, like monocultures. Diversity shifts steadily, between the advancing and receding edges of multipopulational change. It may be impossible to define proper replicates and to stabilize variances in any quantitative measure of diversity. Papers by Motomura (1932), Fisher *et al.* (1943), Margalef (1958a, 1958b, 1969, 1974, 1989, 1991a, 1991b, 1994), Pielou (1975), Frontier (1985), Lurie and Wagensberg (1984), Magurran (1988), and many others, deal with diversity indices and some of the difficulties associated with quantitative approaches.

Specific diversity of ecosystems is computed and expressed with reference to the representatives of selected taxonomic groups. Nature is so rich that it is impossible or impractical to work numerically with the whole range of organisms present in any ecosystem. Species diversity may be considered as a measure of *the information* accumulated, generated or readjusted through population dynamics. Populations change and their demographical parameters shift and actual values and dynamics of diversity change accordingly.

The use of the word biodiversity (Solbrig, 1991) has expanded rapidly over the last few years and applies to the richness of genetic systems in any conveniently defined segment of the biosphere. It is also a measure of information and it changes more slowly than the local diversities that it supports and allows. Relationships between diversity and biodiversity have been assumed, but not explored. Both refer to irreversible properties of complex systems. These pages attempt also to contribute and build the necessary links between biodiversity and diversity, coming, in this paper, from the side of diversity. The dynamics of diversity contribute to the generation, regulation or demise of global biodiversity. Contemporary ecologists are very emphatic about the sure risk of losing biodiversity, but less keen to explore the mechanism of control represented by dynamics of diversity.

The study of diversity has to include the isolating effects of space, that creates or helps to create gradients and boundaries. Recognition of boundaries answers further to practical needs arising in the task of making maps, also in relation to remote sensing.

Diversity is directly associated with the flow of energy through the (eco)system (Odum, 1975) and relates to concepts (like ascendancy, in Ulanowicz, 1986, for instance) that concern the way the paths for energy flow through the system become structured. Simulation provides much help in the analysis of these relations.

Frequently, diversities computed for the same ecosystems and with reference to diverse taxonomic groups are positively correlated, but this is not necessary. Nowadays it is increasingly common to also refer to and compute the diversity of taxonomic categories, defined usually by the size of their respective components, a procedure that is well suited to plankton research, when particle counting and measuring is automatically done.

A DYNAMIC MODEL

In an heuristic approach, two extreme situations may be:

(1) A chemostat or flow culture, in which populations of bacteria, yeast or unicellular algae are grown; a fast flowing stream exemplifies the natural situation most closely related to the ideal. In all these populations, initially seeded by a multispecies or polygenic mixture, number of individuals present (N) may tend to infinity, but under relatively strong and sustained flow, the number of species or of genotypes (S) goes down and eventually may end as just one, before washing out, more as a result of the dynamics of mixed populations than because of some precise limitation of genetic material. In our artificial river, the presence of walls would allow the development of fixed or sessile populations of lower turnover (from bacteria up, see Marshall, 1976) and increasing diversity, and this is a representative example of a sharp gradient or of a boundary, with a well defined physical background.

(2) An ideal artificial system can be represented by a museum display case, behind which is a mildly maniacal curator, always ready to exchange a duplicate piece of the collection for any specimen of a species not yet represented in the cabinet. A similar example could be 'Noah's ark' or a zoo. In all these situations the trend is $N \to S$.

If $S = N^k$, then k ($0 < k < 1$) becomes an index of diversity as good as many previously proposed (Magurran, 1988; such indices have the purpose of quantifying what started as the subjective impression or estimate of the naturalist). It is not wise to speculate about the possibility of computing an average k for the whole Earth. But if such a global index were accepted just for speculative purposes, k would be highly dependent on timing, some periods leading to accelerated flow of biomass and decrease of diversity, and others favouring the accumulation of relatively 'lazy' biomass and its genetic differentiation. It is plainly obvious that diversity will increase slowly, step by step, but at any time it can decrease precipitously. The general irreversibility or asymmetry of change in relation to time is very well expressed in the sudden catastrophes so notorious in ecological succession.

THE PLAY BETWEEN TWO FEEDBACK LOOPS

Two ideal feedback loops can explain the mutual regulation of the number of species S (in an approach that is substantially coincident with the one proposed in 1967 by MacArthur and Wilson) and the number of individuals N, or the total biomass, that follows any appropriate form of the asymptotic or logistic curve of Verhulst–Pearl, applied here to the whole biomass.

The two expressions may be used with suitable and convenient degrees of complication. However, I propose to adopt the most simple formulae for the species, $dS/dt = cS^m - dS^{m'}$, and for the individuals or 'biomass', $dN/dt = aN^h - bN^{h'}$. Assuming a stationary state, $dS/dt = 0$, $dN/dt = 0$, and transforming to log $(bc/ad) = B$, we obtain $k = B\,(h' - h)/(m' - m)$, and k is another index of diversity, that in this particular expression compares the effectiveness of two feedback loops.

If biomass is allowed to increase through the addition of another individual of an existing species, then diversity drops. If the available resources do not allow an increase in biomass, but the environment does not oppose genetic differentiation, then in most situations, 'biodiversity' would increase. The same conclusion can be reached working, step-by-step, through the expression of Brillouin that uses the information theory approach (Margalef, 1974).

This approach simplifies and summarizes much discussion on the meaning of the dynamics of diversity, as expressed by any convenient index. But the simple image that emerges has to be completed for the particular ecosystems that are dominated by organisms which are apt to reorganize the physical environment. These master exosomatic energy through the production of exosomatic structures. Such capacity appeared independently in several evolutionary peaks. Examples are provided by stromatolites, wood in trees, the masses of calcium carbonate in coral reefs, termite nests, houses, towns, and highways in mankind. Such dominant species may push down diversity in systems around them (Andersen, 1992).

TIME-DEPENDENCE OF DIVERSITY, AS EXPRESSED IN SUCCESSION

Change in nature can be either self-organizing (accretion of information, succession) or more or less catastrophic in response to disturbance. Self-organizing mechanisms increase information and diversity; disturbance is expected to lower diversity. This explains why successional change and increase in diversity are generally associated. Examples are everywhere and I present in Figure 2.1 some data on phytoplankton from the Ria de Vigo, a bay in NW Spain, well-known for the red tides (of diatoms) that often culminate periods of succession. Red tides occur when an inflow of nutrients stimulates

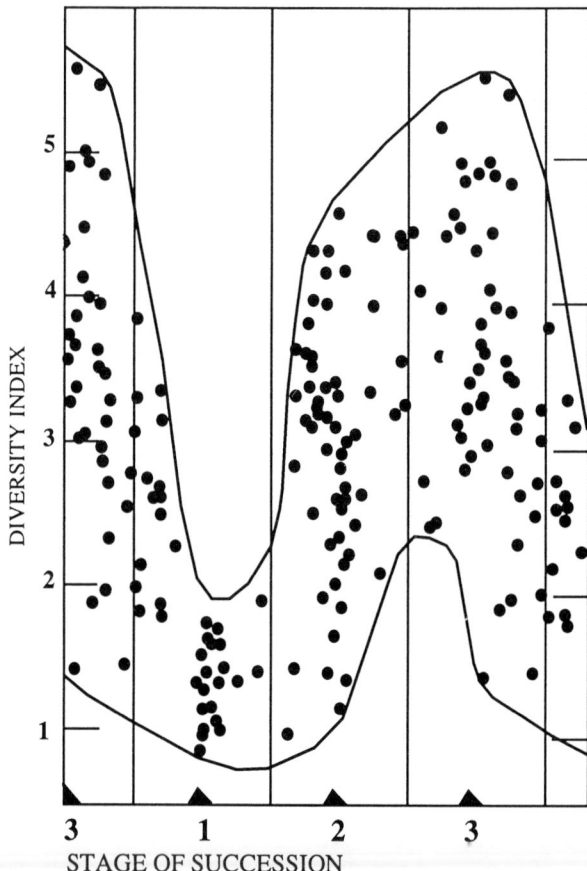

Figure 2.1 Relation between stage of ecological succession and an index of diversity. Stages 1–2–3 refer to superficial waters in the Ria de Vigo, NW Spain, according to combined oceanographic and populational criteria, for the whole summer of 1955. The diversity index that has been used is equal to $(S-1)/\log_e N$, S being number of species and N total number of cells (Margalef, 1958b)

phytoplankton growth and energy flow through the system, and may end by driving down diversity in some areas.

Further explanations of departures from a pattern in diversity along a successional gradient related to diatom development are associated with strong upwelling or vertical mixing, at the start of ordinary planktonic successions (Margalef, 1967). This is paradigmatic of comparable situations when a whole family of plants or of insects has evolved and diversified in adaptation to transitory initial stages of regular successions. In general the negative correlation between P/B and diversity is sustained.

DIVERSITY AND SPACE

Diversity is a quantitative property that refers to a sample, and usually is computed on the individuals belonging to a given taxonomic group. Diversity is expected to change as a function of space, as when the explored space is enlarged. Any index of diversity is sensitive to the extension of the sample, giving a spectrum of diversity that starts with a diversity of zero – one single individual of any species – and increases in an indefinite way. Diversity is really a function of space. Considerable discussion has been going on into the eventual tendency to flatten or plateau at some value of the extension of the continuous space or of the corresponding biomass, over which the community extends. Obviously, there are different styles of diversity. The spectrum is rather flat in a corn field, but increases indefinitely in ecosystems like the tropical rain forest.

It is hard to recognize inflexion points (Frontier, 1985) along the diversity spectrum, a task often associated with the wish to define 'minimal areas'. The situation has to be accepted with all its complexity and not generalized depending on local or personal experience, like selecting a defined slope of S against N (or against area A) as a criterion to select a minimal area (or minimal volume) appropriate for the study of vegetation or plankton.

Such attempts may answer the useful purpose of recognizing some more or less sharp distinction between the active part of the ecosystem and a more resting fraction of it, in the form of a segregated seed or germ bank that may be of high diversity. This more passive segment may occupy a peripheral position with respect to the most active part, like the banks of a river in relation to potamoplankton, or hedges in relation to crops in the field, or the twilight layers in the sea and lakes in relation to the top and more illuminated volumes of water.

In such situations, diversity often increases sharply when passing into the germ bank. This is best exemplified by the great number and variety of cysts or resting stages present in the sediment of temporary ponds, lakes, marine bays and in the banks of rivers. These contain information on past or seasonal biological activity and are stores of biodiversity.

There may be many ways to lay down transects and to combine and interpret the measures of diversity obtained in associated or overlapping spaces. Spectra of diversity may help to detect sharp boundaries, especially when comparing spectra centred in successive points along a transect. Many possibilities can be imagined, but most of them involve so much work that rarely is it found to be justified. The preparation and analysis of diversity spectra over three dimensions, as would be necessary in the case of plankton, is rarely attempted and never carried very far, but in theory it should not be different, except that the vertical dimension z has to retain its quite distinct meaning in comparison to the dimensions x, y, routinely associated with the horizontal plane.

BOUNDARIES

Ideal boundaries with reference to any property follow surfaces that are the site of the most strong gradients. The direction of the gradients should be normal to the surface of the boundary at each point of the surface. Here I will consider only boundaries projected over a plane, which are not easily defined in relation to plankton. Boundaries are not equal; the degree of their visibility is related, in part, to their geometry. The style of each boundary may be approached and described in fractal terms and is related to the style of diversity and of the spectra of diversity on either side, and is also manifested as breaks or singularities in spectra of diversity drawn across the boundary.

A simple expression of the degree of complexity of a boundary is the relation between its actual length, measured along the boundary as closely as the selected unit steps will allow (F), and the rectilinear distance measured between the two end points (L) of the measured stretch, all in the same unit steps, $L = F^\mu$. Analogous approaches could be proposed for application to three dimensional boundaries.

Boundaries may be classed as: (1) rather straight, as in a vegetation patch sharply bounded by a path, a field, or a water body, or as the limit that stabilizes the area between the rain forest and any area exploited or controlled by humans; (2) sinuous boundaries that tend to dissolve into the background, like those traceable over extensions covered by mature vegetation and that may be recognized only on the basis of floristic composition. Dutch plant ecologists (van Leeuwen, 1966) have recognized both extreme sorts of boundaries as representing opposite trends, and have named the boundaries that tend to be more straight *limes convergens*, with μ close to 1, and the more sinuous as *limes divergens* and, if ideally stretched, very long boundaries, in which $0 < \mu < 1$. $\mu = 0.5$ might characterize the intermediate situation in which the boundary is convoluted. Perhaps such a situation may be taken as a dividing point for any convenient classification of boundaries, if needed. However, succession might drive ecological boundaries to infinite lengths ($\mu \to 0$).

Dynamics of diversity, gradients and boundaries

Diversity and boundaries are human abstractions projected on nature and a considerable degree of correspondence between quantification and distribution of diversity and the quality of the boundaries is expected. Diversity is frequently different on both sides of an ideal boundary, but the boundary itself is common to both sides, and the properties of the boundary are unique.

Looking for a physical analogy that could provide some insight, boundaries in ecosystems can be compared with sections of the shapes created by differences in surface tensions between fluids in contact. They depend on surface energy of the fluids: if differences are great, boundaries tend to shorten, and μ is 1 or close to 1; in boundaries between fluids with similar surface tension, μ is expected to be lower, and the boundaries may become less tightly drawn. This is a fruitful analogy, since relations exist between the degree of sinuosity of the boundaries and dynamic properties of the ecosystems in contact, that eventually find expression in their respective diversities. It is to be expected that low diversity systems – crop fields, although a 'semi-artificial' case – tend to have short boundaries, whereas highly sinuous or intricate boundaries are more commonly found between very complex systems of high diversity.

The example of Figure 2.2 and Tables 2.1 and 2.2 refers to a situation observed between Barcelona and Mallorca in February 1990. Phytoplankton were growing actively – at rather low diversity – and a sharp front (an ergocline) existed in the area. Diversity was at its lowest in the neighbourhood of the front, which is also the most fertile place, but specific composition of the phytoplankton was quite different at either side. The spatial differences in diversity persisted with little deformation after the position of the hydrographic structures shifted during a few days of rough weather.

Between areas of different diversity, boundaries tend also to straighten (μ close to 1). Important differences in diversity are most commonly associated with exchanges dependent on gradients of organization and productivity: when one system exploits the neighbour, the boundary in between often tends to straighten.

If such relations hold, the most significant differences between ecosystems may consist in their ratios between production and biomass (P/B). A higher P/B ratio is ordinarily associated with low diversity, and may signify a surplus of product, which can be exported across a high tension boundary. At the other side of the same boundary we might have an ecosystem with a lower P/B ratio or lower turnover and a functional or dominant position. In the example of Figure 2.2, both systems were very productive close to the common boundary, although with quite different dominants. This example is also of interest for the different function of coccolithophorids and diatoms in relation to the carbon cycle, because of their different utilization of CO_2, which would suggest that one relatively larger fraction of carbon is immobilized as calcium carbonate only at the SE side of the boundary.

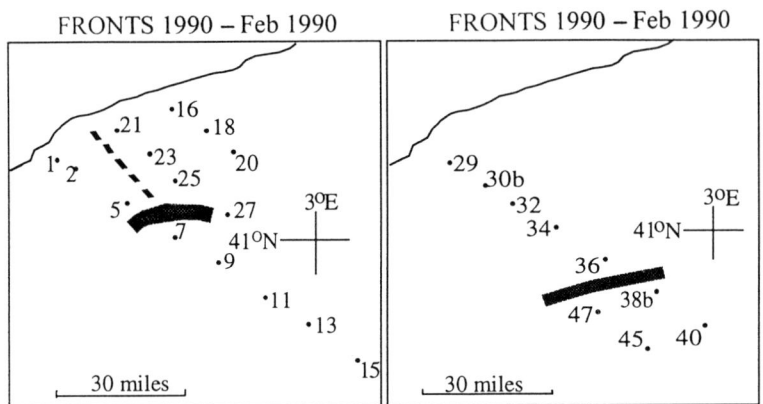

Figure 2.2 Oceanographic stations in CRUISE FRONTS 90, before (left) and after (right) the spell of rough weather. February 1990

QUESTION OF NAMES

It can be assumed that, as succession proceeds, diversity increases, eventually boundaries become sinuous and are finally lost. As the actual length of the boundary increases, perhaps the total 'energy' (or any analogue of it) involved in the boundary remains constant. When neighbouring spaces are subject to parallel or common successional change, boundaries cease to be sites of net exchange, and become sinuous or are lost, they become *limes divergens* in the ecological sense, although the adjacent areas are converging.

Ecotone is a zone of potential which exists, for instance, between grasslands and other more complex ecosystems. In the same sense the gradient between a stream and its bank is an ecotone. Such frontiers may be typical ecotones, as are the *ergoclines* (Legendre and Demers, 1985). Shelford (1963) states explicitly that an ecotone is a community that is transitional between two biomes or other large units and which is not a serial transition community. This might imply that ecotones are not meant to exist between successional stages, and the term *ecocline* has been applied for such intermediate and transitional positions, a decision that may apply in the example of a river, but perhaps much less in other situations in which one part of the system is repressed, usually through exploitation. The term ecocline has been used in the equivalent sense of *limes divergens*.

DIVERSITY AND BIODIVERSITY, THE EXAMPLES OF STREAMS AND PLANKTON

Biodiversity is a relatively new word that has proved more useful than diversity in our cultural world. It refers to the whole genetic treasure of nature. Diversity attempts only to explore quantitatively, and in a somewhat

Table 2.1 Diversity of 1388 pooled samples of phytoplankton (nano + microplankton, 5–500 μm) that represent the actual population in 5.1 l of water, sampled in the photic zone (0–100 m depth) of NW Mediterranean between 1960 and 1991. Numbers refer to cells which are classed as individual units. This avoids the problems posed by cormophytes where it is difficult to define the unit. Total number of identified cells was 195,983, with an average of 38.3 cells ml^{-1}. Species are ranked by decreasing total cell number, and only the names of the first 21 are given. 131 species in the tail of the original list are not included here (Margalef, 1994). This tail is continued for a further 10 columns without naming species

Station no.	Species										
21192	*Chaetoceros curvisetus*						135	56	34	18	9
16039	*Emiliania huxleyi*						135	56	34	18	8
15077	*Solenicola setigera*						129	55	34	18	8
13595	*Nitzschia delicatissima*						124	51	33	18	8
8553	*Chaetoceros compressus*						120	49	33	18	8
6630	*Phaeocystis poucheti*						120	49	32	17	8
5334	*Cryptomonas pseudobaltica*						114	48	31	17	8
5254	*Rhodomonas* cf. *baltica*						114	48	29	17	8
4779	*Chaetoceros vixvisibilis?*						113	48	29	16	8
4547	*Leptocylindrus danicus*						111	45	29	16	8
4505	*Nitzschia pseudofraudulenta*						109	45	28	15	7
4303	*Haptophyceae*						106	45	28	15	7
4293	*Plagioselmis* sp.						102	44	27	15	6
4191	*Platymonas* sp.						99	44	27	15	6
4092	*Paraphysomonas* sp.						97	44	27	14	6
3795	*Asterionellopsis glacialis*						94	43	27	14	6
3516	*Lauderia (Schroederella) delicatula*						93	43	25	13	6
2723	*Chaetoceros laciniosus*						88	43	25	12	6
2956	*Cyclotella* cf. *caspia*						80	42	24	12	6
3733	*Chaetoceros socialis*						80	43	34	13	6
2613	*Thalassionema nitzschioides*						79	42	24	11	5
2059	942	644	380	282	187	75	42	23	11	5
2046	893	633	356	282	187	74	41	22	11	5
1988	833	601	354	266	187	70	41	22	11	5
1897	831	587	354	259	176	67	38	21	11	5
1857	817	562	352	248	171	67	37	21	10	5
1624	771	546	338	247	170	66	37	20	10	5
1604	751	527	338	241	166	66	37	20	10	5
1386	749	430	332	212	163	63	36	20	10	5
1324	740	422	326	211	160	61	36	20	10	5
1145	736	413	319	205	150	60	35	19	10	4
1063	698	412	314	202	145	59	35	19	10	4
1047	674	393	298	195	144	58	35	19	9	4
1009	658	384	283	191	140	57	35	18	9	4

clumsy way, some synoptic properties associated with the waxing and waning of populations drawn from the store of biodiversity.

Imagine that the biosphere, according to time and place, uses components drawn from the treasure of biodiversity and plays with them in the game that we call dynamics of populations. This is always an experiment in evolution

Table 2.2 Distribution of phytoplankton in the photic zone, NW Mediterranean, February 1990, before and after a rapid shift towards SE in the distribution of a well-defined hydrographic discontinuity. See Table 2.1 for details

Stations 1–5 22 samples, 44 ml water					Stations 6–15 43 samples, 86 ml water				
2240	*Chaetoceros curvisetus*				492	*Emiliania huxleyi*			
439	*Schroederella delicatula*				380	*Chaetoceros curvisetus*			
213	Chrysophyceae				157	Chrysophyceae			
160	*Emiliania huxleyi*				154	*Paraphysomonas?*			
115	*Chaetoceros laciniosus*				110	*Gymnodinium* sp.			
84	*Thalassiosira* sp.				93	*Chaetoceros socialis*			
76	54	42	25		84	55	43	29	
75	50	41	24		79	54	42	28	
62	45	35	23		72	54	40	25	
62	45	35	22		63	53	39	25	
60	43	34	18		58	46	37	25	

Stations 16–27 62 samples, 124 ml					Stations 29–36 50 samples, 100 ml				
5175	*Chaetoceros curvisetus*				3555	*Chaetoceros curvisetus*			
1596	*Schroederella delicatula*				1293	*Chaetoceros laciniosus*			
1356	*Chaetoceros laciniosus*				1092	*Schroederella delicatula*			
732	*Chaetoceros socialis*				455	*Chaetoceros socialis*			
486	*Chaetoceros compressus*				357	*Emiliania huxleyi*			
418	Chrysophyceae				352	Chrysophyceae			
398	169	106	98		298	146	81	58	
272	166	104	87		183	119	76	57	
251	154	102	84		155	109	71	56	
193	139	100	76		152	100	68	53	
169	117	100	65		147	98	60	35	

Stations 38–47 51 samples, 102 ml				
1402	*Emiliania huxleyi*			
485	*Chaetoceros curvisetus*			
228	Chrysophyceae			
215	*Paraphysomonas?*			
172	*Chaetoceros laciniosus*			
111	*Gymnodinium* sp.			
108	78	59	47	
107	76	56	43	
90	70	49	38	
87	62	48	37	
79	61	47	34	

and eventually gives back results, including some novelties, to the general store of biodiversity.

An instructive example is provided along the gradient that links the centre of a river with the soil of its banks, or with the slower flow between the water and the gravel bed. Life in free-flowing water is like the flow culture used as a

Table 2.3 Examples of low diversity blooms (sharp spikes in restricted spaces) during the situation depicted in Table 2.2 and in the precise points given in Figure 2.1. February 1990, NW Mediterranean

Station 5; 30 m depth, in 2 ml		Station 47; 50 m depth, in 2 ml	
389	*Chaetoceros curvisetus*	300	*Emiliania huxleyi*
47	*Lauderia (Schroederella) delicatula*	33	*Chaetoceros curvisetus*
10	*Gymnodinium* sp.	13	*Umbilicosphaera sibogae*
9	*Nitzschia delicatissima*	13	*Chaetoceros laciniosus*
9	*Gymnodinium* sp.	8	*Gymnodinium* sp.
8	*Emiliania huxleyi*	8	Chrysophyceae
7	*Nitzschia pseudofraudulenta*	7	*Lauderia (Schroederella) delicatula*
6	*Paraphysomonas?*	4	*Chrysochromulina* sp.
5	*Chaetoceros socialis*	4	*Gymnodinium* sp.
5	*Chaetoceros compressus*	3	*Syracosphaera* sp.
4	*Thalassiosira* sp.	3	*Chaetoceros lorenzianus*
3	2 2 1	3	2 2 2 2 2 1

paradigm of the generative process that leads to very low diversity. The banks of the river replicate the example of the museum display case of our initial imaginary model. The soil contains both active and resting forms of thousands of genotypes that are potential colonizers of the rivers, both of the pebbles and sand of the bottom (potamobenthos), and of the free-flowing water (potamoplankton) at different times of the year, and different water levels. This example illustrates the counter-current and complementary exchange between the source of biodiversity – going from the bank to the centre of the river – and the supply of excess production and eventually of selected genetic novelty from the flowing water to the banks. Within the same species, the eventual differentiation between adhering forms and forms that develop faster in open water starts with the bacteria.

A further example is provided by the developing populations of phytoplankton of seas and lakes in water fertilized by vertical mixing or by upwelling. Strong mixing introduces a relatively high diversity (expressed in a flat spectrum), recruited from the twilight zone, that usually does not last, or, if more persistent, it is only so because of the contribution of many species whose development is simultaneously stimulated in the nutrient-rich environment. But this stage does not last and most plankton blooms are of low diversity, represented by 'spikes' that contrast with the more 'flat' distributions general in less dynamic situations (Table 2.3). The fertilized water masses often have lateral contacts with other water masses that have been less recently disturbed, are more stratified and have planktonic populations that express high and rather diagonal diversity spectra. The topological style of developing patchiness combines diversity and boundaries in the expected way.

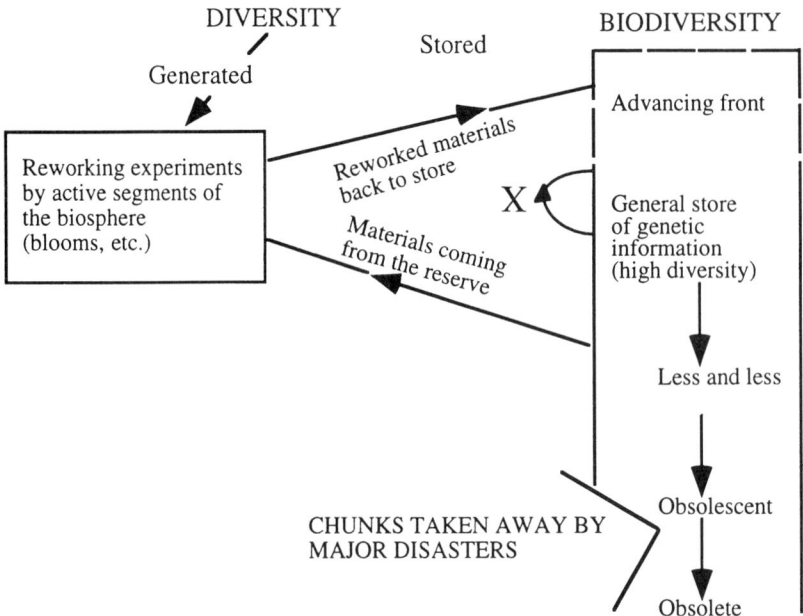

Figure 2.3 The distinction between diversity and biodiversity supported by fashion and the media may be, after all, acceptable. Diversity is represented by the real structure of the assemblages of genotypes that are active and always processing and reprocessing biodiversity. X is the place of a very important loop

Other illustrations could be added easily on life in the soil, or caves, always in relation to important ecological boundaries. Figure 2.3 attempts to visualize the basic relations involved.

REFERENCES

Andersen, A.N. (1992). Regulation of 'momentary' diversity by dominant species in exceptionally rich ant communities in the Australian seasonal tropics. *American Naturalist*, **140**, 401–20

Fisher, R.A., Corbet, A.S. and Williams, C.B. (1943). The relation between the number of individuals and the number of species on a random sample of animal population. *Journal of Animal Ecology*, **12**, 42–58

Frontier, S. (1985). Diversity and structure in aquatic ecosystems. *Oceanogr. Mar. Biol., Annual Review*, **23**, 253–312

Legendre, L. and Demers, S. (1985). Auxiliary energy, ergoclines and aquatic biological production. *Naturaliste Canadien*, **112**, 5–14

Lurie, D. and Wagensberg, J. (1984). An extremal principle for biomass diversity in Ecology. In Lamprecht, I. and Zotin, A.I. (eds.) *Thermodynamics and Regulation of Biological Process*, pp. 259–73. Walter de Gruyter and Co., Berlin

MacArthur, R.H. and Wilson, E.O. (1967). *The Theory of Island Biogeography*. Princeton University Press, Princeton, NJ

Magurran, A.E. (1988). *Ecological Diversity and its Measurement*. Croom Helm, London

Margalef, R. (1958a). Information theory in ecology. *General Systems*, **3**, 36–70

Margalef, R. (1958b). Temporal succession and spatial heterogeneity in phytoplankton. In Buzzati-Traverso, A.A. (ed.) *Perspectives in Marine Biology*, pp. 323–47. University of California Press, CA

Margalef, R. (1967). Some concepts relative to the organization of plankton. *Marine Biology Annual Review*, **5**, 257–89. Barnes edition, Allen and Unwin, London

Margalef, R. (1969). Diversidad del fitoplancton de red en dos areas del Atlantico. *Investigación Pesquera*, **33**, 275–86

Margalef, R. (1974). *Ecologia*. Omega, Barcelona

Margalef, R. (1989). On diversity and connectivity, as historical expressions of ecosystems. *Coenoses*, **4**, 121–6

Margalef, R. (1991a). Reflexiones sobre la diversidad y significado de su expresión cuantitativa. *Diversidad Biológica/Biological Diversity*, pp. 105–12. Fundación Ramón Areces, Madrid

Margalef, R. (1991b). *Teoria de los Sistemas Ecológicos*. Publicacions University of Barcelona

Margalef, R. (1994). Through the looking glass: How marine phytoplankton appears through the microscope when graded by size and taxonomically sorted. *Scientia Marina*, **58**, 87–101

Marshall, K.C. (1976). *Interfaces in Microbial Ecology*. Harvard University Press, Cambridge, MA and London

Motomura, I. (1932). *Doobutugatu Zassi*, **44**, 379–83

Odum, E.P. (1975). Diversity as a function of energy flow. In van Dobben, W.H. and Lowe-McConnell, R.H. (eds.) *Unifying Concepts in Ecology*, pp. 11–14. W. Junk, Publishers, The Hague and PUDOC, Wageningen

Pielou, E.C. (1975). *Ecological Diversity*. John Wiley, New York

Shelford, V.E. (1963). *The Ecology of North America*. University of Illinois Press, Urbana, IL

Solbrig, O.T (ed.) (1991). *From Genes to Ecosystems: a Research Agenda for Biodiversity*. IUBS–SCOPE–UNESCO, Cambridge, MA

Ulanowicz, R.E. (1986). *Growth and Development. Ecosystems Phenomenology*. Springer Verlag, New York

van Leeuwen, C.G. (1966). A relational theoretical approach to pattern and process in vegetation. *Wentia*, **15**, 25–46

CHAPTER 3

HABITAT COMPLEXITY IN LAND–INLAND WATER ECOTONES

Ewa Pieczyńska and Maciej Zalewski

INTRODUCTION

Land–inland water interface zones – ecotones – are recognized as ecological systems of specific and variable abiotic and biotic characteristics (Naiman and Décamps, 1990; Holland *et al.*, 1991). Ecotones vary in size from narrow strips to very broad areas. Their spatial limits are often difficult to determine, for example, when small headwater streams and large rivers with wide floodplains are compared (Thorpe, 1991). In most cases, land–inland water ecotones are studied and described without determination of their spatial limits. As Naiman and Décamps (1991) pointed out, the identification of a particular ecotone depends on the question being asked or the problem to be solved. When literature data are compared, it should also be taken into account that various terminologies are used in studies of similar types of ecotones. For example, lake shallows overgrown by helophytes are described in various papers as: littoral, eulittoral, wetland, wetter beach, shore zone etc.

This paper attempts to determine the origin of ecotone habitat heterogeneity and its importance in supporting biotic diversity and in controlling the flows of materials between water bodies and adjacent terrestrial patches.

SIZE AND PHYSICAL CHARACTERISTICS OF STREAMSIDE AND LAKESIDE ECOTONES

Ecotone size primarily depends on flooding and drying regime. Water-level fluctuations in lakes and rivers move the shoreline over a distance which depends on the slope of the shore terrace and the adjacent terrestrial areas.

In most natural lakes (especially small ones), the periodically flooded zone is relatively narrow. It is wider in large post-glacial lakes with extensive shallow areas. In the largest Polish lake – Lake Sniardwy – the yearly differences in water level of 41 cm create an average movement of the shoreline of 53 m, with a maximum of 296 m. In lakes with especially unstable water levels these values can be much higher.

The shoreline movement varies not only between lakes but also within one water body. Depending on the bottom configuration, plant cover and detritus

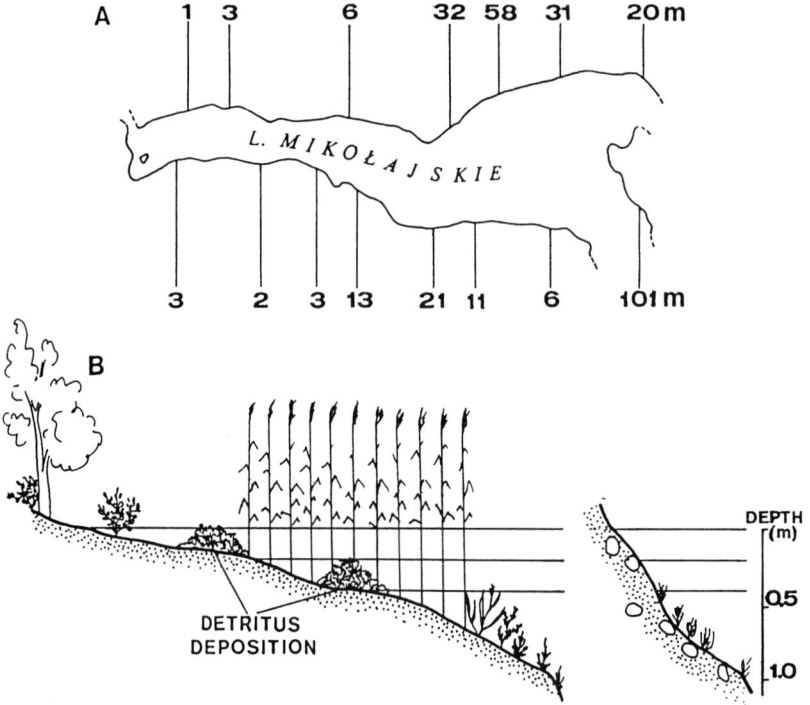

Figure 3.1 (A) Distance (m) between shorelines during the highest and lowest seasonal water levels (41 cm of yearly differences) at 15 sites in Lake Mikolajskie. (B) Influence of the shore slope, water-level fluctuations and detritus deposition on the ecotone size and complexity

accumulation, water-level fluctuations may have a different effect on the creation or the loss of various microhabitats within the shore zone. Examples for Lake Mikolajskie, Poland are shown in Figure 3.1.

The shore zone which periodically floods and dries up is especially wide in the cases of some large rivers and wetlands. Salo (1990) distinguishes several processes of various time-scales which result in ecotone changes and patch formations. They range from megaform processes ($>10^4$ y – evolution of catchment basins) to microform processes (<1 y – annual flood regime, water flow). The author points out that fluvial dynamics create patches varying from <1 m² to tens of thousands of square kilometers. River fluctuations also result in the formation of short-lasting lagoons or permanent cutoff lakes (former channels)

Wetlands[1] occur in the transitory zones between water bodies and uplands but they also occupy separate areas (sometimes very large). They have various

---

[1] Depending on the classification system, similar types of ecosystems are defined as wetlands, marshes or swamps.

Table 3.1 Range of variation of composition of littoral bottom in five lakes (% of sediment volume) (from Rybak, 1980 changed)

Sediment fraction	Lake				
	Majcz W.	Inulec	Glebokie	Zelwazek	Jorzec
Detritus (<2.5 mm)	7–92	9–79	1–88	19–74	7–49
Plant fragments (>2.5 mm)	2–24	1–38	>1–35	15–80	>1–26
Sand	>1–87	2–78	9–53	0–40	0–88
Gravel and stones	>1–25	0–20	0–50	0	0–3
Mollusc shells	0–17	0–69	0–17	0–19	0–76

internal patches and well differentiated external boundaries (Holland *et al.*, 1990). Water-level fluctuations change wetland size and their internal structures, resulting in the appearance and disappearance of aquatic patches. Van der Valk and Davis (1979) show visible cyclical changes in the proportions of open water and emergent vegetation cover in a typical prairie marsh (Eagle Lake, Iowa, USA). The area is overgrown with emergent vegetation changes due to both the hydrological regime and biotic factors (muskrat and insect damage). Day *et al.* (1988) on the basis of studies of five marshes along the Ottawa River, Canada show that the main factors controlling vegetation composition are water depth, spring floods and the fertility gradient produced by waves and flowing water.

A large variety of soils and bottom sediments are noticed in ecotones – rocky, stony, sandy and muddy shores are observed along rivers and lakes.

Table 3.1 presents a varying range of bottom sediment compositions in the shore–littoral region of five Masurian lakes, Poland (30 sites in each lake were analysed). The proportions of all compared components vary greatly at particular sites, from even less than 1 to more than 80% of sediment volume.

The presence of shore-side lagoons especially increases habitat complexity in land–inland water ecotones and results in patch discontinuity. Such patches can be isolated or connected with a main lake or river. Patches arise and disappear or change size and shape seasonally.

Figure 3.2 presents visible seasonal changes in distribution of small lakeside pools on one site at Lake Mikolajskie. They result from water-level fluctuations (and shoreline movement) as well as changes in detritus accumulation. These astatic patches are characterized by a set of physicochemical properties which visibly differ from that in adjacent lake littoral water (Pieczyńska, 1972).

Erixon (1979) shows the ecological importance of riverside lagoons of the river Vindelälven (Sweden). The author demonstrates the complexity of factors which are responsible for lagoon specificity. Water-level fluctuations

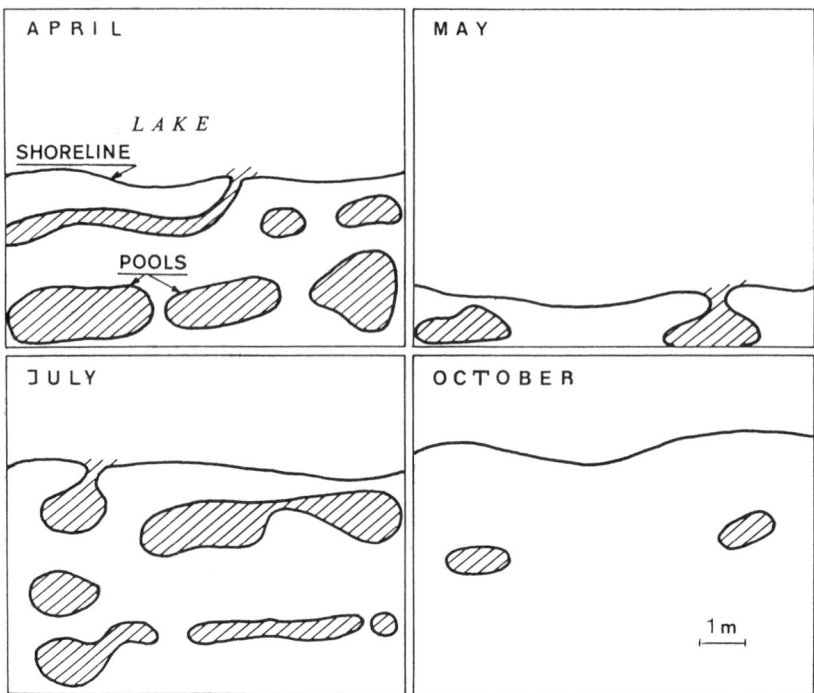

Figure 3.2 Seasonal changes in shoreline position and distribution of near shore pools on one site of Lake Mikolajskie

in the river result in seasonality of lagoon isolation which changes lotic to lentic conditions and *vice versa*.

External physical factors (air temperature, wind) are of greater importance in habitat property regulation in ecotones than in deeper parts of water bodies, where the ratio of water volume to surface area is high.

Under the influence of wind-induced water movement, the whole water volume is mixed in near-shore shallows and consequently bottom deposits are disturbed. Wind is usually very changeable in direction and intensity which results in visible short time changes in ecotone habitats. Straskraba and Pieczyńska (1970) show a much higher range of variation of light conditions in littoral sites overgrown by reeds or shaded by trees during windy days than during calm ones.

Light access to the ecotone is an important characteristic directly affecting physical conditions, macrophyte and algal community structure and productivity, thus influencing food web dynamics. For example, in rivers it was demonstrated that the light access modifies stream temperature (Holtby, 1988), and alters algal community structure (Zalewski *et al.*, 1991).

Both emergent parts of shores and near shore shallows freeze in winter in most of the sites in the temperate zone. The period of freezing changes

Habitat complexity

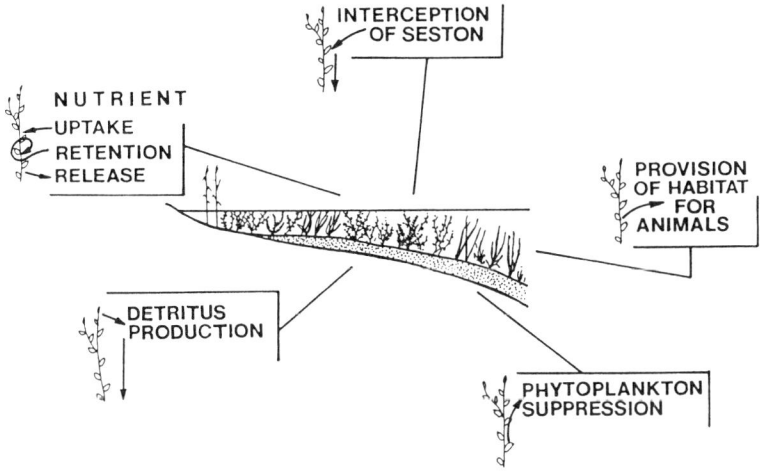

Figure 3.3 Major effects of macrophytes on nutrient cycling and ecological processes in a water body

depending on climate but, for example, reaches six months in the case of rivers in north Sweden (Erixon, 1979). It is well documented that freezing influences the physical structure of the habitat, the chemistry of water and bottom deposits (it changes the movement of nutrients between sediment and water) and the occurrence of macrophytes (Edwards *et al.*, 1986; Renman, 1989).

BIOLOGICAL ASPECTS OF ECOTONE HETEROGENEITY

Bank and near-shore aquatic plants influence most physical features of ecotone habitats. They regulate and differentiate light conditions (shading by near-shore terrestrial and aquatic vegetation), water flow (macrophytes 'thaw' wave action and water current), and the structure of soils and bottom sediments (effect of root growth and plant dying, accumulation of plant detritus). In sites dominated by macrophytes, most of the ecological processes are determined by their structure and biological functions. Multiple macrophyte functions (Figure 3.3) also have direct and indirect effects on the processes in the scale of the whole water body. This effect is limited in advanced stages of eutrophication when visible decreases in macrophyte biomass and cover are noticed (Lachavanne, 1985; de Nie, 1987).

Macrophytes hold a central role in nutrient cycling. Uptake of nutrients by roots and/or shoots, their accumulation in plant tissue, as well as release by actively growing plants and from decaying plants determine existing nutrient levels in the environment. Depending on the site of nutrient uptake (water or bottom sediments), macrophytes may decrease the concentration of nutrients in the water or they may act as a pump of nutrients from the bottom sediments to the water (Carignan, 1982; Wetzel, 1990).

The dynamics of nutrient accumulation in plant tissues and their release during plant senescence and decomposition show visible seasonal variability which, in the scale of a whole water body, is generally determined by the proportion of evergreen plant species and species of short-lasting green biomass. Since a large variety of life forms of bank and littoral plants is noticed, the effects of macrophytes on nutrient dynamics vary in different shore types. The increase of carbonate precipitation under the influence of macrophyte photosynthesis is a well known phenomenon. The marl encrustations on plants are frequently massive and may exceed the weight of the macrophytes (Wetzel, 1983).

Horizontal transport of water is responsible for the exchange of nutrients between shore and limnetic zones. Aside from well known wind-driven circulation, convective circulation induced by night-time cooling of littoral waters plays an important role in this exchange. James and Barko (1991), on the basis of measurements of movement of fluorescent dye, demonstrated intensive convective circulation in Eau Galle Reservoir, USA, and they showed its importance for phosphorus exchange between littoral and pelagic zones.

Interception of seston by dense macrophyte stands is also an important process. It was demonstrated in field experiments (Pieczyńska, 1993) that within packets of submerged macrophytes, artificial plants, or nylon nets exposed in the littoral zone, seston concentration was 2 to 10 times higher (after 24 h of exposure) as compared with adjacent sites. Piwocki (unpublished) observed that slowing down the water movement within the macrophyte beds resulted in an increase of sedimentation rate of seston which was about three times higher than in a neighbouring littoral site devoid of macrophytes.

Presence of vegetation visibly increases biodiversity. There are various types of plant and animal communities directly and indirectly connected with vegetation. Macrophytes provide essential conditions for breeding, spawning, rearing and feeding of animals – permanent ecotone inhabitants as well as visitors from adjacent terrestrial patches (amphibians, reptiles, mammals, birds) or aquatic ones (see other chapters in this volume). Submerged macrophytes and submerged parts of emergent plants provide a substrate for colonization by many taxa of algae and sessile aquatic animals. The amount of this substrate can visibly vary (Table 3.2) depending on plant density and the differentiation of leaf structure of particular species.

The direct and indirect impact of macrophytes on phytoplankton (whose high biomass is one of the most unfavourable symptoms of eutrophication) is of great importance. Interactions between these two biota are very complex and bi-directional, and involve competition for nutrients, shading and allelopathy effects (Phillips *et al.*, 1978; Sand-Jensen and Borum, 1991).

Heterogeneously distributed macrophytes form a mosaic of ecotone patches along the shores of lakes and rivers. The key factor which regulates

Table 3.2 Macrophyte surface area suitable for colonization by periphyton (Lake Mikolajskie, July–September). After Pieczyńska (1968 and further data) and Kowalczewski (1975)

Macrophytes	Plant surface area (m^2 m^{-2} of littoral surface)	
	Average	Maximum
Phragmites australis*	0.7	2.8
Potamogeton lucens	2.1	4.1
Potamogeton perfoliatus	1.1	2.4
Elodea canadensis	8.1	38.2
Myriophyllum spicatum	9.6	49.8

*submerged parts of plants

plant diversity in ecotones is water-level fluctuation. Keddy and Reznicek (1986) showed that existing shore vegetation in the Great Lakes, North America depends strongly upon fluctuations in water level which increase the area of this vegetation, and the diversity of vegetation types and plant species. Pieczyńska (1996) shows that creation of experimental pools with fluctuating water level visibly increases macrophyte species richness in the shore zone of Lake Ros, Poland.

Not only living vegetation but also litter of plant origin plays an important role in the spatio-temporal organization of ecotone habitats. Plant detritus (standing dead plant biomass, detrital sediments, litter accumulated on the shoreline) occurs in various quantities and proportions.

Pieczyńska (1993) shows three extreme types of lake shore zones that differ in detritus retention depending on input level and decomposition rate. Ecotones characterized by high levels of detritus accumulation include both shores devoid of (or with poor) aquatic vegetation but with great input of terrestrial litter, and shore zones with a rich aquatic vegetation, especially dominated by emergent plants. In both of these types, plant material resistant to decomposition dominates and retention rate is high. An especially high level of detritus accumulation is observed on the water body margins and in near-shore shallows where autochthonous and allochthonous detritus may cover the bottom with a thick layer. In contrast, a very low detritus accumulation is observed on sites devoid of macrophytes and terrestrial litter input, on which algae are primary sources of organic matter and dead organisms are quickly decomposed.

Aside from macrophytes, other ecotone organisms which also occur in great quantities can influence ecosystem functioning. These are for example, mats of filamentous algae which provide a very specific and changeable habitat for many littoral organisms. There is also *Dreissena polymorpha* (Pall.) which, when occurring in masses, differentiate the littoral bottom. Its filtration, fecal, and pseudofecal production influence to a great extent nutrient cycling and seston abundance (Stanczykowska, 1984; Reeders and Bij de Vaate, 1990).

ECOTONES AND FEEDBACK PROCESSES REGULATING ECOSYSTEM DYNAMICS

It seems that sound ecological understanding of ecotone function should be based on two components – structural and functional – which play highly overlapping roles. Structural characteristics are mostly related to the regulatory properties of the external physical factors, which regulate nutrient dynamics in the system. This is, for example, light access to the photic zones or wind reduction by trees along lake banks, as a factor which may stabilize the epilimnion water and may in turn have an effect on the internal load intensity. The internal structural factors, such as complexity of littoral habitat, which may modify biotic interactions within the system are also of great importance.

Due to regulations driven by functional factors, many ecological processes (e.g. 'top down effects') are initiated in ecotones. The cascade effects in the aquatic ecosystems initiated by an ecotone zone can be determined by three major patterns. First (b→a), when biotic factors in an ecotone generate a chain of abiotic reactions within the ecosystem. Second, when biotic factors originated from an ecotone initiate biotic responses in the system (b→b), and third, when abiotic ecotone factors initiate a cascade of biotic reactions in the ecosystem (a→b).

An example in which biotic factors initiate abiotic processes (b→a) is the effect of macrophytes on nutrient cycling in the ecosystem (see previous section). The biotic link (b→b) has been shown, for example, by Kraska *et al.* (1990) who demonstrated a significant reduction of phytoplankton under the influence of submerged macrophytes in Lake Budzynskie (Poland). The lake has two visibly distinct parts. The first is overgrown by dense stands of *Myriophyllum verticillatum* L. and the second is devoid of macrophytes. Three years of studies showed that summer phytoplankton biomass as well as chlorophyll concentrations were three to five times lower in the macrophyte dominated part of the lake. This effect was also demonstrated by a smaller scale field experiment, where in the chambers (3×2 m) filled with water of high nutrient content (3 chambers with multi-species macrophyte assemblage and 3 with woody debris) heavy algal blooms were recorded only in the chambers without macrophytes (Zalewski, unpublished).

These relations (b→b) may also be exemplified by the case of biomanipulation. According to the biomanipulation concept (Shapiro, 1977; Gulati *et al.*, 1990) the water quality might be significantly improved – by reduction of algal blooming – using predatory fish communities as a tool for reducing the pressure of planktivorous fish on zooplankton which controls phytoplankton biomass. A high complexity of littoral ecotones is usually necessary to maintain a high density of efficient predators (e.g. pike, *Esox lucius* L.), which in a simple habitat is characterized by high cannibalism (Grimm and Backx, 1990).

As shown by Moss (1990), macrophytes provide essential refuges for cladocerans, efficiently grazing on phytoplankton. They are safe against fish predation when they migrate to the macrophyte beds during the daytime. They move out of the vegetation at night-time to graze in the open water. Under some circumstances, detritus of macrophyte origin may serve as important food for cladocerans. Thus, macrophytes help maintain a large population of grazers which reduce algal biomass (Shapiro, 1977).

The shoreline ecotones may be the starting point of the abiotic–biotic feedback mechanisms (a→b) especially in highly fluctuating environments of the early ecological succession where the biotic community is unbalanced. Examples of such systems are reservoirs where, due to water-level variability, typical littoral vegetation does not occur and its role, to a certain extent, is played by the limited area of periodically flooded terrestrial vegetation (Ploskey, 1985).

During the long-term investigation of Sulejów Reservoir, Poland (Zalewski et al., 1990a, 1990b), it was found that the most important factor determining water quality by reducing density of large efficient filtrators (Daphnia) was perch (*Perca fluviatilis* L.) fry (YOY – young of the year) whose density depended mostly on water level during reproductive and post-reproductive periods (Figure 3.4).

During years of high and stable water level, the elimination of Daphnia by numerous perch (YOY) was very intense. The drastic reduction of filtrators (e.g. from approximately $16\,mg\,l^{-1}$ to 1–$2\,mg\,l^{-1}$ of their biomass) influenced the amount of suspended matter which increased two fold, and up to ten times in the case of algae. The elimination of large zooplankton by perch also resulted in reducing the growth and winter survival of pike-perch (*Stizostedion lucioperca* L.) which depend on zooplankton as their first food. The loss of even one generation of this short-lived (easily catchable) predator species, reduces the biotic control on limnetic planktivorous fish density. The significant correlation between water level and perch reproductive success indicates the possibilities of regulating water quality and fishery yield by manipulating the use of ecotone resources by fish fry.

Another important aspect of abiotic–biotic feedback processes (a→b) is connected with the terrestrial components of the ecotone – tall trees surrounding lakes – which, especially in the case of small ecosystems situated in valleys, may have negative effects on biomanipulation efficiency (Zalewski and McQueen, unpublished). This is because the presence of trees stabilizes epilimnetic water and, by reducing internal nitrogen supply, directs phytoplankton succession toward the occurrence of non-grazable forms (e.g. Cyanobacteria which can use nitrogen from the atmosphere) (Reynolds, 1987).

Finally, the question arises: in which types of ecosystems will either abiotic or biotic regulatory mechanisms initiated at ecotones appear? Zalewski and Naiman (1985) suggested that in different types of ecosystems both mechanisms will appear in different proportions. The prevalence of abiotic or biotic

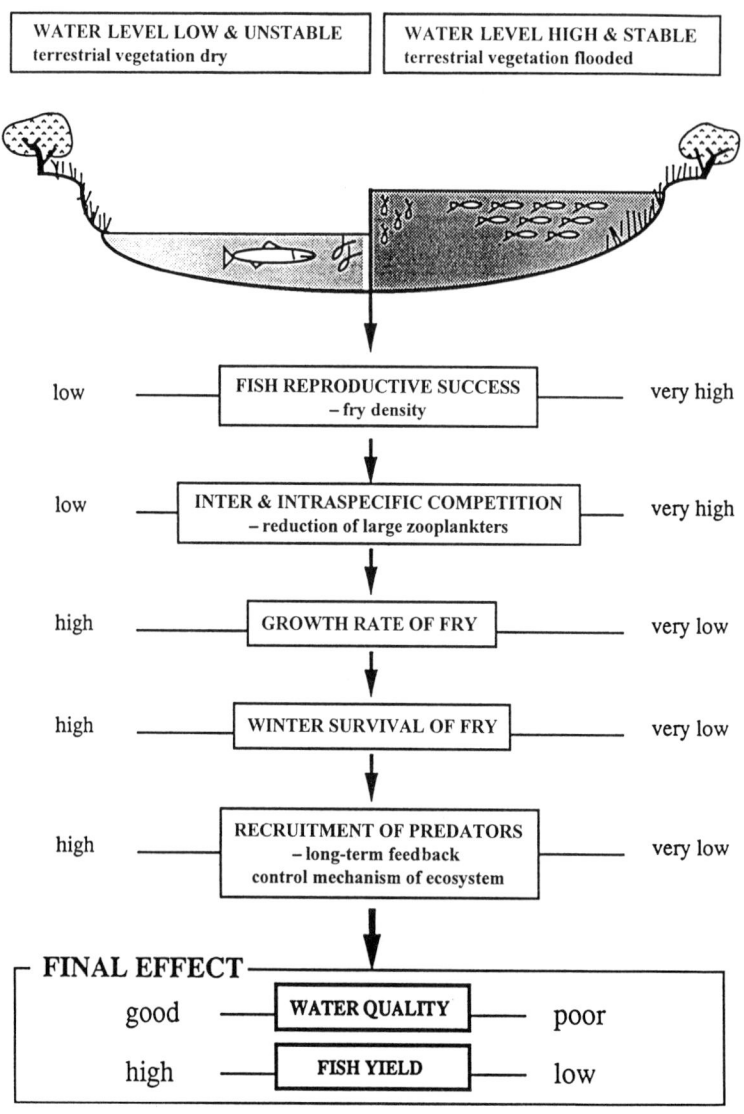

Figure 3.4 Fish community dynamics and water quality as a result of changes in shoreline ecotone complexity due to variations of water level in a lowland reservoir (after Zalewski, 1991)

regulation depends primarily on climate and morphoedaphic conditions (Ryder *et al.*, 1974). However, the nutrient load and the ratio of shore length to the water area, in some situations, may play a prevailing role (Biro and Voros, 1988).

HUMAN INFLUENCES, MANAGEMENT, RESTORATION

Anthropogenic changes to ecotones

Ecotones are important components of landscape which regulate matter and energy flow across ecosystems. The loss of their heterogeneity and biological richness usually has detrimental effects on adjacent ecosystem functioning. Although sometimes human activity on the landscape, watershed and ecosystem level, leads to creation of new ecotones, in most cases management practices result in loss of ecotone heterogeneity. These are: changes in hydrological regime, commercial harvest of plants, introduction of exotic fishes, embankment constructions, stream channel regulations, sewage input, urbanization and others.

The current literature provides many examples of destruction of ecotones on various scales – from large geographical regions (e.g. loss of wetlands in the United States – Olson, 1992) to the case studies of individual water bodies (Pieczyńska *et al.*, 1988; Kira and Uda, 1990). Ecotone destruction usually has detrimental effects on the majority of ecological processes, not only in the ecotone itself, but also in adjacent systems. For example, in streams and rivers, when banks are unprotected by riparian vegetation, soil erosion increases turbidity through rapid runoff from agricultural areas (Schlosser and Karr, 1981). The runoff increases siltation which modifies fish community structure by reducing typical riffle species – benthic insectivores and herbivores (Berkman and Rabeni, 1987).

Channelization virtually eliminates riparian ecotone complexity, thus it reduces the retention of nutrients and organic matter in streams. Munn and Prepas (1986) demonstrated that in some types of streams, 68% of annual phosphorus load is transported downstream during 12 days of spates. Consequently, sedimentation at the floodplain ecotones occurs mainly in the downstream section. It can be expected that organic matter which, due to ecotone losses, is not retained in a channelized stream or incorporated into nutrient spiraling or sedimented in floodplain areas will be transported to the reservoir or river estuary (Schiemer and Zalewski, 1992). This accelerates eutrophication of reservoirs and coastal waters.

Channelization, where woody debris and tree roots are removed, may not only reduce stressful behavioural interaction between fishes but also seriously reduce the amount of available food. Benke *et al.* (1984) demonstrated that in a subtropical river, 80% of drift invertebrates, which are a basic food for many fish species, originate from submerged log, branch and root habitats.

In lakes, pollution and extensive eutrophication in most cases result in the decrease of biomass of macrophytes and their species diversity, especially of submerged ones (Lachavanne, 1985; de Nie, 1987; Jørgensen and Löffler, 1990). As macrophytes play various important functions, their loss has a detrimental effect on ecological processes in the whole lake system.

Filtering function of ecotones – management

It is now widely accepted that ecotones, because of their buffer function, should be taken into account in ecosystem protection and restoration programmes. But the buffer capacity of ecotones varies among various sites and is strongly related to the system heterogeneity. A variety of natural, reconstructed or created ecotone types – lake littoral systems, riparian vegetation zones, wetlands or marshes – can contribute significantly to water quality (Naiman and Décamps, 1990; Holland et al., 1991; Olson, 1992).

Existing literature provides information on a variety of vegetation systems designed to remove pollutants. The effectiveness of pleustonic plants (Culley et al., 1981) and artificial stands of emergents (Gersberg et al., 1984, 1986) are widely discussed. Vegetation itself can be effective, but for enhancement of nutrients or pollutant removal, usually various additional operations are necessary. Gersberg et al. (1984) reported that nitrogen can be effectively removed from water which passes through artificial macrophyte stands, but the addition of an external carbon source is necessary (in the presented case, simple addition of mulched plant biomass was sufficient).

Restored ecotones or artificially constructed ones are designed for the protection of various natural water bodies against point and non-point pollution. Examples include the reconstruction of wetlands in the vicinity of the Zala River – the largest tributary to Lake Balaton, Hungary (Szilagyi et al., 1990), construction of artificial marshes for protection of Lake Jackson, USA (Fernald and Cason, 1986), as well as watershed scale restoration project of Des Plaines River in USA which includes experimental wetland construction (Mitsch, 1992). The wide range of effectiveness of the variety of constructed wetlands (multiple or single wetlands located in upstream or downstream reaches, terrace wetlands in steep landscape and others) was shown by Mitsch (1992).

Although the filtering capacity of ecotones usually refers to the function of vegetation, it is presently obvious that a large variety of physical structures and ecological processes in ecotones are responsible for the control of pollution. Aside from macrophytes, the role of other biota (plankton and periphyton), sedimentation and processes within sediments can be of great importance for ecotone buffering capacity.

Restored and constructed ecotones, aside from their function of pollution or erosion control, play very important roles in the creation of habitats valuable for various plants and animals, including rare and demanding species. Knight (1992) points out that the important benefit of the construction of wetlands for non-point source pollution control is the enhancement of animal populations. The increase in the diversity of plant species and their growth forms provides new niches for animals. High detritus accumulation increases the importance of the detrital food chain in which many species are involved. Additional niches can also be created by fluctuating water levels.

From this and other data it can be concluded that heterogeneous shore habitats play an important role in the proper functioning of lakes and rivers by influencing their biostability. For management purposes, it is necessary to determine the minimal and optimal size of protected or constructed ecotones, their shape and internal structure. Ahola (1990) presents a project on the creation of vegetated buffer zones for ditches, brooks, rivers and lakes within the protection programme of the Vantaa river basin, Finland. On the basis of a detailed analysis of existing ecotones, the project determines the minimal necessary size of proposed buffer zones (5–20 m) or buffer strips (2–5 m) as well as their internal structures. Some authors recommend 10–30 m as an optimal width for the riparian zones efficient in the process of restoration of lowland streams (Petersen et al., 1992). Other literature data (Knauer and Mander, 1989) indicate that the most intensive filtration of nutrients from the catchment appears within the 7 m belt from the stream bank. Thus such ecotone width seems to be a necessary minimum. In general, in rivers, the width of the ecotone buffer zone should increase with the river size, catchment slope, variability of hydraulic regime and the intensity of farming in the catchment areas.

Looking from the landscape processes perspective, the land–water interface plays an important role in the maintenance of biodiversity, not only of aquatic organisms but also of birds, mammals, reptiles and others. Thus the maximum width of buffer zone will be important in creating feeding places, refugia and reproductive areas for both aquatic and terrestrial organisms.

It is obvious that ecotones are, and will be, the subject of unfavourable changes within various management practices in watersheds. Thus the minimization of these detrimental effects is of primary importance. Right now, due to limited data, it is still difficult to provide precise guidelines for the creation of an optimal, intermediate complexity of land–water ecotones (as defined by Zalewski et al., 1994). But on the basis of existing information it might be suggested that the northern banks of streams and rivers should be covered by riparian trees which improve aquatic habitats' diversity, organic matter supply and retention, and also accumulate nutrients provided from the catchment area and from upstream. On the southern banks, 30–60% of the bank length of the riparian canopy should be removed to improve light access which stimulates primary production, thus reducing nutrient spiraling length and consequently accelerating stream self-purification. However, the gaps should be filled with small bushes and herbaceous plants to maintain the filtering effect (Figure 3.5).

Various engineering strategies which include special operations for retaining the beneficial effects of shore zones are proposed in the literature – for lake littoral (Jørgensen and Löffler, 1990), riparian zones (Petts, 1990) and various kinds of wetlands (Olson, 1992). They include guidelines for the locations of artificial channels or embankments (including material used for their construction), creation of additional shore structures, replacement of

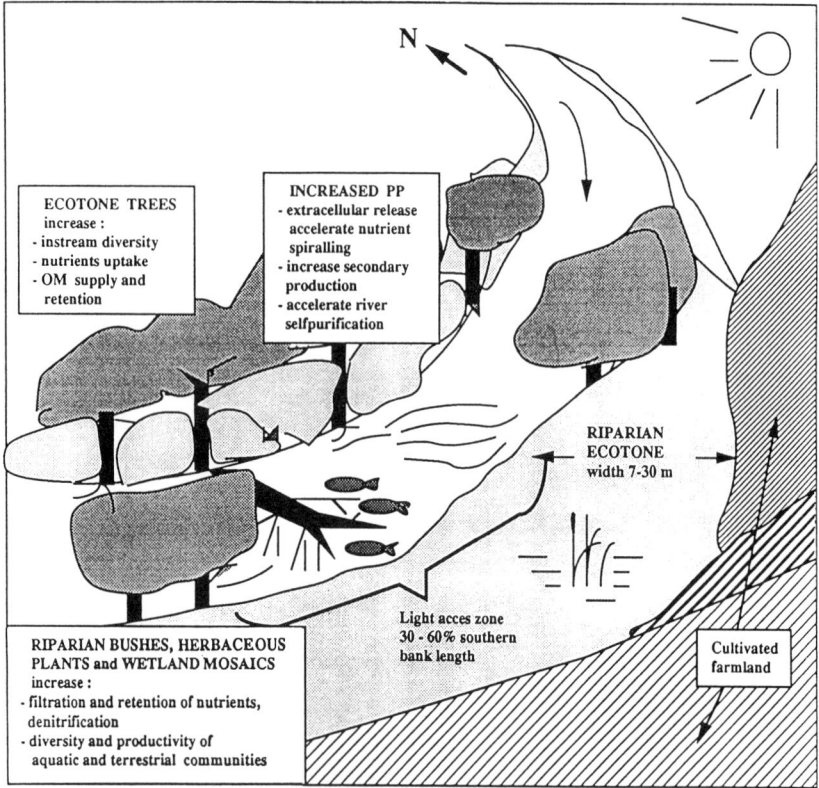

Figure 3.5 Example of suggested intermediate complexity of riparian ecotone for riverine ecosystem conservation and maintenance of biodiversity

artificial structures for erosion control by living vegetation and others. In many cases, even simple procedures may be very effective to maintain the beneficial functions of shore systems.

CONCLUSIONS

(1) Spatio-temporal variability of the land–water ecotones depends primarily on the shore slope, flooding and drying regime as well as depositional and erosional processes. The presence of permanent or periodical shoreside lagoons visibly increases habitat complexity and results in the discontinuity of ecotone habitats.

(2) Bank and near-shore macrophytes hold a central position in the spatial organization of the habitat. They also influence nutrient cycling and sedimentation rate, and they serve as essential sites for rich assemblages of plants and animals. Since a large variety of macrophyte

life forms as well as their mosaic distribution is observed, the effects of macrophytes on ecological processes vary not only among water bodies but also within the shore zones of an individual lake or river.

(3) There is a visible relationship between habitat complexity and biodiversity in ecotones. High biodiversity is noticed in shore zones with rich and differentiated vegetation and high levels of detritus accumulation. A decrease in animal species richness (both of permanent inhabitants and temporary visitors) is usually observed with a decrease in plant cover. Shores which are devoid of vegetation, sandy or stony, and exposed to waves have especially poor biotas.

(4) Ecotones regulate the flow of water and materials across the landscape. Vegetation plays an important role in the ecotone filtering function but several other biotic and abiotic processes are also involved (biotic function of plankton and periphyton, sedimentation). Ecotones may also enhance the effectiveness of restoration practices in main river channels or offshore parts of lakes as they initiate various feedback mechanisms in a water body. Ecotone importance increases with increasing habitat diversity. Thus ecotone management must consider the protection of their habitat complexity and their biodiversity.

REFERENCES

Ahola, H. (1990). Vegetated buffer zone examinations on the Vantaa River basin. *Aqua Fennica*, **20**, 65–9

Benke, A.C., Van Arsdall, T.C. Jr., Gillespie, D.M. and Parrish, F.K. (1984). Invertebrate productivity in a subtropical blackwater river. *Ecological Monographs*, **54**, 25–63

Berkman, H.E. and Rabeni, C.F. (1987). Effect of siltation on stream fish communities. *Environmental Biology of Fishes*, **18**, 285–94

Biro, P. and Voros, L. (1988). Relationship between the yield of bream, *Abramis brama* L., chlorophyll-a concentration and shore length: water area ratio in Lake Balaton, Hungary. *Aquaculture and Fisheries Management*, **19**, 53–61

Carignan, R. (1982). An empirical model to estimate the relative importance of roots in phosphorus uptake by aquatic macrophytes. *Canadian Journal of Fisheries and Aquatic Sciences*, **39**, 243–7

Culley, D.D. Jr., Rejmankova, E., Kvet, J. and Frye, J.B. (1981). Production, chemical quality and use of duckweed (Lemnaceae) in aquaculture, waste management and animal feeds. *Journal of the World Mariculture Society*, **12**, 27–49

Day, R.T., Keddy, P.A., McNeill, J. and Carleton, T. (1988). Fertility and disturbance gradients: a summary model for riverine marsh vegetation. *Ecology*, **69**, 1044–54

Edwards, A.C., Creasey, J. and Cresser, M.S. (1986). Soil freezing effects on upland stream solute chemistry. *Water Research*, **20**, 831–4

Erixon, G. (1979). Environment and aquatic vegetation of riverside lagoon in northern Sweden. *Aquatic Botany*, **6**, 95–109

Fernald, E.A. and Cason, J.H. (1986). Development of an artificial marsh in Tallahassee, Florida: Lake Jackson, a case study. In Lauga, J., Décamps, H. and Holland, M.M. (eds.) *Land use impacts on aquatic ecosystems: the use of scientific information,* pp. 229–41. MAB–UNESCO and PIREN–CNRS, Toulouse

Gersberg, R.M., Elkins, B.V. and Goldman, C.R. (1984). Use of artificial wetlands to remove nitrogen from wastewater. *Journal Water Pollution Control Federation,* **56,** 152–6

Gersberg, R.M., Elkins, B.V., Lyon, S.R. and Goldman, C.R. (1986). Role of aquatic plants in waste water treatment by artificial wetlands. *Water Research,* **20,** 363–88

Grimm, M.P. and Backx, J.J.G.M. (1990). The restoration of shallow eutrophic lakes and the role of northern pike, aquatic vegetation and nutrient concentration. *Hydrobiologia,* **200/201,** 557–66

Gulati, R.D., Lammens, E.H.R.R., Meijer, M.L. and van Donk, E. (eds.) (1990). *Biomanipulation – Tool for Water Management.* Kluwer Academic Publishers, Belgium

Holland, M.M., Risser, P.G. and Naiman, R.J. (eds.) (1991). *Ecotones. The Role of Landscape Boundaries in the Management and Restoration of Changing Environments.* Chapman and Hall, New York, London

Holland, M.M., Whigham, D.F. and Gopal, B. (1990). The characteristics of wetland ecotones. In Naiman, R.J. and Décamps, H. (eds.) *The Ecology and Management of Aquatic–Terrestrial Ecotones,* pp. 171–98. UNESCO/Parthenon, Paris/Carnforth

Holtby, L.B. (1988). Effects of logging on stream temperatures in Carnation Creek, British Columbia, and associated impacts on the coho salmon (*Oncorhynchus kisutch*). *Canadian Journal of Fisheries and Aquatic Sciences,* **45,** 192–5

James, W.F. and Barko, J.W. (1991). Estimation of phosphorus exchange between littoral and pelagic zones during nighttime convective circulation. *Limnology and Oceanography,* **36,** 179–87

Jørgensen, S.E. and Löffler, H. (eds.) (1990). Guidelines of Lake Management. Volume 3. *Lake Shore Management.* International Lake Environment Committee Foundation and United Nations Environment Programme, Otsu, Japan

Keddy, P.A. and Reznicek, A.A. (1986). Great Lakes vegetation dynamics: the role of fluctuating water levels and buried seeds. *Journal of Great Lakes Research,* **12,** 25–36

Kira, T. and Uda, T. (1990). Shore management at Lake Biwa. In Jørgensen, S.E. and Löffler, H. (eds.) Guidelines of Lake Management. Volume 3. *Lake Shore Management,* pp. 153–69. International Lake Environment Committee Foundation and United Nations Environment Programme, Otsu, Japan

Knauer, N. and Mander, U. (1989). Untersuchungen über die Filterwirkung verschidener Saumbiotope an Gewässern in Schleswig-Holstein. 1. Mitteilung: Filterung von Stickstoff und Phosphor. *Zeitschrift für Kulturtechnik und Landentwicklung,* **30,** 365–76

Knight, R.L. (1992). Ancillary benefits and potential problems with the use of wetlands for nonpoint source pollution control. *Ecological Engineering,* **1,** 97–113

Kowalczewski, A. (1975). Periphyton primary production in the zone of submerged vegetation of Mikolajskie Lake. *Ekologia Polska,* **23,** 509–43

Kraska, M., Szyszka, T. and Szczepanowski, P. (1990). Ksztaltowanie struktur planktonu przez makrofity w Jeziorze Budzynskim i Jelonek. In Kajak, Z. (ed.)

Funkcjonowanie Ekosystemów Wodnych, ich Ochrona i Rekultywacja. II. *Ekologia Jezior, ich Ochrona i Rekultywacja. Eksperymenty na Ekosystemach,* pp. 36–43. Wydawnictwo SGGW-AR, Warszawa, Poland. (in Polish)

Lachavanne, J.B. (1985). The influence of accelerated eutrophication on the macrophytes of Swiss lakes: abundance and distribution. *Verhandlungen Internationale Vereinigung für Theoretische und Angewandte Limnologie*, **22**, 2950–5

Mitsch, W.J. (1992). Landscape design and the role of created, restored and natural riparian wetlands in controlling nonpoint source pollution. *Ecological Engineering*, **1**, 27–47

Moss, B. (1990). Engineering and biological approaches to the restoration from eutrophication of shallow lakes in which aquatic plant communities are important components. *Hydrobiologia*, **200/201**, 367–77

Munn, N. and Prepas, E. (1986). Seasonal dynamics of phosphorus partitioning and export in two streams in Alberta. *Canadian Journal of Fisheries and Aquatic Sciences*, **43**, 2464–71

Naiman, R.J. and Décamps, H. (eds.) (1990). *The Ecology and Management of Aquatic–Terrestrial Ecotones*. UNESCO/Parthenon, Paris/Carnforth

Naiman, R.J. and Décamps, H. (1991). Landscape boundaries in the management and restoration of changing environments: a summary. In Holland, M.M., Risser, P.G. and Naiman, R.J. (eds.) *Ecotones. The Role of Landscape Boundaries in the Management and Restoration of Changing Environments,* pp. 130–7. Chapman and Hall, New York, London

Nie de, H.D. (1987). The decrease in aquatic vegetation in Europe and its consequences for fish populations. *European Inland Fisheries Advisory Commission*, Occasional Paper 19, FAO, Rome

Olson, R.K. (1992). Evaluating the role of created and natural wetlands in controlling nonpoint source pollution. *Ecological Engineering*, **1**, xi–xv

Petersen, R.C.L.B., Petersen, M. and Lacoursiere, J. (1992). A building-block model for stream restoration. In Boon, P.J., Calow, P. and Petts, G.E. (eds.) *River Conservation and Management,* pp. 293–309. John Wiley, Chichester

Petts, G.E. (1990). The role of ecotones in aquatic landscape management. In Naiman, R.J. and Décamps, H. (eds.) *The Ecology and Management of Aquatic–Terrestrial Ecotones,* pp. 227–61. UNESCO/Parthenon, Paris/Carnforth

Phillips, G.L., Eminson, D. and Moss, B. (1978). A mechanism to account for macrophyte decline in progressively eutrophicated fresh-waters. *Aquatic Botany*, **4**, 103–26

Pieczyńska, E. (1968). Dependence of the primary production of periphyton upon the substrate area suitable for colonization. *Bulletin de l'Academie Polonaise des Sciences*, Cl. II, **16**, 165–9

Pieczyńska, E. (1972). Ecology of the eulittoral zone of lakes. *Ekologia Polska*, **20**, 637–732

Pieczyńska, E. (1993). Detritus and nutrient dynamics in the shore zone of lakes: a review. *Hydrobiologia*, **251**, 49–58

Pieczyńska, E. (1996). Habitat heterogeneity and biodiversity in the shore zone of water bodies. *Acta Hydrobiologica*, **37**, 29–35

Pieczyńska, E., Ozimek, T. and Rybak, J.I. (1988). Long-term changes in littoral habitats and communities in Lake Mikolajskie (Poland). *Internationale Revue der Gesamten Hydrobiologie*, **73**, 361–78

Ploskey, G.R. (1985). Impact of terrestrial vegetation and preimpoundment clearing on reservoir ecology and fisheries in the USA and Canada. *Fisheries Technical Paper 258*, FAO, Rome

Reeders, H.H. and Bij de Vaate, A. (1990). Zebra mussels (*Dreissena polymorpha*): a new perspective for water quality management. *Hydrobiologia*, **200/201**, 437–50

Renman, G. (1989). Distribution of littoral macrophytes in a north Swedish riverside lagoon in relation to bottom freezing. *Aquatic Botany*, **33**, 243–56

Reynolds, C.S. (1987). The reponse of phytoplankton communities to changing lake environments. *Schweizerische Zeitschrift für Hydrologie*, **49**, 220–36

Rybak, J.I. (1980). The structure of littoral bottom deposits in several Masurian lakes. *Bulletin de l'Academie Polonaise des Sciences*, Cl. II, **28**, 389–94

Ryder, R.A., Kerr, S.R., Loftus, K.H. and Regier, H.A. (1974). The morphoedaphic index, a fish yield estimator – review and evaluation. *Journal of Fisheries Research Board of Canada*, **31**, 663–88

Salo, J. (1990). External processes influencing origin and maintenance of inland water–land ecotones. In Naiman, R.J. and Décamps, H. (eds.) *The Ecology and Management of Aquatic–Terrestrial Ecotones*, pp. 37–64. UNESCO/Parthenon, Paris/Carnforth

Sand-Jensen, K. and Borum, J. (1991). Interactions among phytoplankton, periphyton and macrophytes in temperate freshwater and estuaries. *Aquatic Botany*, **41**, 137–75

Schiemer, F. and Zalewski, M. (1992). The importance of riparian ecotones for diversity and productivity of riverine fish communities. *Netherlands Journal of Zoology*, **42**, 323–35

Schlosser, I.J. and Karr, J.R. (1981). Riparian vegetation and channel morphology impact on spatial patterns of water quality in agriculture watersheds. *Environmental Management*, **5**, 232–43

Shapiro, J. (1977). Biomanipulation – a neglected approach? Plenary Session of the 40th Annual Meeting of the American Society of Limnology and Oceanography, Michigan State University

Stanczykowska, A. (1984). Role of bivalves in the phosphorus and nitrogen budget in lakes. *Verhandlungen Internationale Vereinigung für Theoretische und Angewandte Limnologie*, **22**, 982–5

Straskraba, M. and Pieczyńska, E. (1970). Field experiments on shading effect by emergents on littoral phytoplankton production. *Rozpravy Ceskoslovenske Akademie Ved*, **80**, 7–32

Szilagyi, F., Somlyody, L., Herodek, S. and Istvanovics, V. (1990). The Kis–Balaton reservoir system as a means of controlling eutrophication of Lake Balaton, Hungary. In Jørgensen, S.E. and Lottler, H. (eds.) *Guidelines of Lake Management*. Volume 3. *Lake Shore Management*, pp. 127–51. International Lake Environment Committee Foundation and United Nations Environment Programme, Otsu, Japan

Thorpe, J.E. (1991). Review of the introductory session: general considerations. In Zalewski, M., Thorpe, J.E. and Gaudin, P. (eds.) *Fish and Land/Inland Water Ecotones*, pp. 11–15. UNESCO–MAB, Paris

Van der Valk, A.G. and Davis, C.B. (1979). A reconstruction of the recent vegetational history of a prairie marsh, Eagle Lake, Iowa, from its seed bank. *Aquatic Botany*, **6**, 29–51

Wetzel, R.G. (1983). *Limnology*. 2nd edn. Saunders College Publishing, Philadelphia

Wetzel, R.G. (1990). Land–water interfaces: metabolic and limnological regulators. *Verhandlungen Internationale Vereinigung für Theoretische und Angewandte Limnologie*, **24**, 6–24

Zalewski, M. (1991). The utilization of the shoreline ecotones by fish fry in reservoirs from the point of view of biomanipulation. In Zalewski, M., Thorpe, J.E. and Gaudin, P. (eds.) *Fish and Land/Inland Water Ecotones*, pp. 69–72. UNESCO-MAB, Paris

Zalewski, M., Brewinska-Zaras, B., Frankiewicz, P. and Kalinowski, S. (1990a). The potential for biomanipulation using communities in a lowland reservoir: concordance between water quality and optimal recruitment. *Hydrobiologia*, **200/201**, 549–56

Zalewski, M., Brewinska-Zaras, B. and Frankiewicz, P. (1990b). Fry communities as a biomanipulating tool in a temperate lowland reservoir. *Archiv für Hydrobiologie Beih. Ergebn. Limnologie*, **33**, 763–74

Zalewski, M. and Naiman, R.J. (1985). The regulation of riverine fish community by a continuum of abiotic–biotic factors. In Alabaster, J.S. (ed.) *Habitat Modification and Freshwater Fisheries*, pp. 3–9. Butterworths Scientific, London

Zalewski, M., Puchalski, W., Frankiewicz, P. and Bis, B. (1994). Riparian ecotones and fish communities in rivers – intermediate complexity hypothesis. In Cowx, I. (ed.) *Rehabilitation of Freshwater Fisheries*, pp. 152–60. Blackwell, Oxford

Zalewski, M., Puchalski, W., Frankiewicz, P. and Nowak, M. (1991). The relation between primary production and fish biomass distribution in an upland river system. *Verhandlungen Internationale Vereinigung für Theoretische und Angewandte Limnologie*, **24**, 2493–6

CHAPTER 4

MICROBIAL DIVERSITY AND FUNCTIONS IN LAND–INLAND WATER ECOTONES

Michel Aragno and Blanka Ulehlova

LAND–INLAND WATER ECOTONES GENERATE UNIQUE MICROBIAL HABITATS

Defined as the transition zones between purely aquatic (e.g. pelagic and/or benthic) ecosystems and purely terrestrial ones, the land–inland water ecotones can be described using scales ranging from 0.1 to several hundred metres. Plant and macrofaunal diversities can be described at this scale. Microbial, particularly bacterial habitats, the most important being transition zones (e.g. oxic/anoxic interfaces, the rhizosphere, surface of decaying particles) need often to be described at scales ranging from 1 to 1000 µm, that is several orders of magnitude below the scales mentioned above. Indeed, ecotones at macroscopical scale (we will call them macroecotones) are not directly relevant for microbial ecology and functional diversity of microorganisms. The microbial ecologist will therefore consider microecotones of micrometric to centimetric size. The conditions peculiar to macroecotones will generate a variety of unique microecotones.

In the past, a lot of research has been published dealing with microbial ecology, either concerning purely aquatic ecosystems (along water columns in holo- or meromictic lakes, in sediment or at the sediment/water interface, see e.g. Porter *et al.*, 1988) or concerning purely terrestrial ones, particularly in soils. These studies were often directed towards potential applications and/or environmental studies (microbially catalysed reactions in artificial lakes and reservoirs or in agricultural soils). In one case, at least, these studies concerned land–inland water ecotones: the microbial ecology of rice swamps. So far, natural ecotones have received much less attention from microbial ecologists (see however Ulehlova, 1976, 1990; Austin and Findlay, 1989). Therefore, the present paper has to be considered mainly as a prospective work. Indeed, the function of a macroecotone is the systemic combination of the functions of the microecotones it contains. These functions are largely catalysed by microorganisms. Thus, the microbiology of land–inland water ecotones would bring a major contribution to the understanding of the macroscale functions of such ecosystems, as well as of their biodiversity.

Following the second principle of thermodynamics, life needs a continuous supply of energy to be maintained, while the chemical elements constitutive of living material can (or should) be recycled. Either light (through phototrophy) or the energy provided by a chemical reaction (chemotrophy) can be used for sustaining living processes. The primary source of energy is almost exclusively light, with the exception of geothermal ecosystems, where all or part of the energy-giving chemical reactants have an abiotic underground origin (Aragno, 1991). In terrestrial ecosystems, the primary production is mainly due to plants, whereas algal and bacterial production (oxygenic and in some cases anoxygenic photosynthesis) dominate in offshore freshwaters. Ecotonal zones present both types of primary production.

Mainly conditioned by this primary production, chemotrophic microbial life will play an important role in gradients or at interfaces, that is in microecotones where energy fluxes occur, first from the producers to the primary consumers, then from consumer to consumer through the transfer of secretions, of metabolites or of dead organic materials, or by the exchange of redox inorganic compounds involved in the energetic metabolism as electron donors or acceptors. Thus these microecotones will give rise to a variety of ecological niches harbouring a diversity of microbial species responsible for the functional diversity of the system.

The presence, or absence, of oxygen in a biotope governs most of the functional characteristics of microbial life. Not only oxidative and reductive processes, but also precipitation and solubilization reactions are dependent, directly and indirectly, on the presence or absence of oxygen. A few organisms, the facultative anaerobes, can shift from aerobic to anaerobic way of life, whereas others (the aerotolerants) can pursue their anaerobic (fermentative) metabolism even in the presence of oxygen. Other organisms are either strictly aerobic or strictly anaerobic, and require or are excluded from oxic environments, respectively. Some strict aerobes are however sensitive to oxygen, and will grow only under reduced oxygen partial pressure. They are microaerophilic.

In shallow water environments of the littoral zone, typical of land–inland water ecotones, there is normally no stable water stratification, and dissolved oxygen reaches the surface of the sediment. The diffusion of oxygen within the sediment, as in water-submersed soils, is however severely restricted, its rate being $c\ 10^{-4}$ times that in a normally aerated porous soil. Therefore, the respiratory activity of aerobes in the first millimeters or centimeters of the sediment will in most cases remove all the diffusing oxygen, resulting in completely anoxic conditions underneath. This situation can also occur at the surface of offshore sediments, but, because of the higher primary production, will be more intensive in land–inland water ecotones. The presence of plant litter and its aerobic degradation at the surface of ecotone soils or sediments also increases this effect. Another unique feature of land–inland water ecotones is the presence of roots of aquatic plants in anoxic sediments or submersed soils. Oxygen is conducted to the roots by air channels and may eventually diffuse, generating a (micro) oxic sheath around the root (Armstrong and Armstrong, 1988).

Also important in ecotones, specially in the supralittoral zone, is the water table fluctuation. This delimits a zone where the soil is alternately drained (and therefore oxygenated) and submersed (and in consequence could become anoxic more or less rapidly).

SAMPLING, BIOMASS AND BIODIVERSITY EVALUATIONS

The main reasons for the scarcity of studies on microbial biodiversity in microheterogeneous habitats are probably the difficulty of obtaining representative samples and the unreliability of most methods for assessing microbial biomass and biodiversity.

Sampling

As stated above, most microbial habitats in land–inland water ecotones are of submillimetric size, or even can best be described as transitions with steep physicochemical gradients. However, most analytical methods require relatively important sample masses, from one to several hundred grams. Therefore, in the best case, biomass and biodiversity evaluation will give an overall rough image of much more finely structured biocoenosis. Although it is sometimes tedious, the direct epifluorescence observation of samples allows one to obtain a more direct image of the bacteria in their microenvironments. The development of fluorescent nucleic acid probes (see below) appears to be a very promising field.

Biomass measurement

Several parameters have been proposed to measure microbial biomass in soils, rhizosphere and other natural environments. 'Classical' biomass parameters include plate counts, epifluorescence microscope counts, and chloroform fumigation methods (Jenkinson and Powlson, 1976; Voroney and Paul, 1984). Even when performed using non-selective media, plate counts strongly under-estimate the actual biomass: only 0.01 to 10 % of the actual bacterial population will form colonies on such media (Atlas, 1983; Jones, 1977; Pickup, 1991). Three reasons could explain this:

(1) The organisms forming colonies on current, rich laboratory media are typically *r*-strategists ('zymogenous' according to Winogradski) which are not characteristic of the nutrient-limited, natural habitats. Autochthonous bacteria, adapted to very low nutrient concentrations

('*K*-strategists'), may have very low growth rates and will not form colonies during the incubation periods routinely used for plate counts (Roszak and Colwell, 1987).

(2) Under starvation conditions, bacterial cells may enter a viable, but non-culturable stage (Roszak and Colwell, 1987), which represents a form of dormancy, reversible or not.

(3) Microbial cells are often grouped, either adsorbed on particles or forming microcolonies. Such cells are often difficult to disperse, due to the cohesive properties of the mucilages embedding them. In such cases, each 'colony forming unit' may well represent a number of individual cells, either of the same species or of different ones.

A number of staining methods allow differentiation between living cells and dead particles in direct microscopical observations (Roszak and Colwell, 1987; Byrd and Colwell, 1992).

Other biomass parameters were proposed. Maire (1984, 1987) optimized ATP biomass determination in soils. Tunlid *et al.* (1985) and Tunlid and White (1990) utilized phospholipids and their fatty acids (PLFA) as biomass indicators. Both ATP and PLFA biomass determination take into account the whole biomass, regardless of the cultivability of the organisms (contrary to plate counts and MPN determinations). They exclusively measure the living biomass, because both phospholipids and ATP will rapidly be hydrolysed and destroyed after cell death.

ATP biomass determination is relatively simple. The ATP concentration in cells may vary according to the metabolic status of the cell, but only between certain limits, because it is submitted to well-known regulation mechanisms. Moreover, there is always a mixture of actively growing and of stationary cells in a soil microbial population, which makes ATP determination a reliable biomass parameter, in spite of its variability. However, it does not allow a distinction between taxonomic groups, such as between microbial and root biomass, for example.

PLFA determination is more difficult. The cost of a single analysis is high, which limits the number of samples which can be analysed. Precise analyses require relatively large samples. However, apart from those it shares with PLFA measurements, ATP analysis has its own substantial advantages: phospholipids are structural membrane compounds, and are therefore biomass components in relatively good balance with the whole 'living fraction' of the cell. Moreover, this technique can be used simultaneously to assess biodiversity (see below).

Microbial biodiversity

Prokaryotes show the widest genetic diversity. They encompass two of the three actually recognized 'domains' of life: the Bacteria *sensu stricto* and the

Archaea (Woese, 1987; Woese *et al.*, 1990). This diversity is accompanied by a considerable functional diversity, particularly in cell metabolism. However, there is not necessarily a close relationship between taxonomical and functional diversity. For example, aerobes and anaerobes, phototrophs and chemotrophs can show very close phylogenetical relationships, while similar functions are found in widely divergent phylogenetical groups. So, when speaking of bacterial diversity, one must distinguish between taxonomical (or genetical) diversity and the ecologically more relevant functional diversity.

Methods for assessing taxonomical diversity. Cultural methods are definitively not applicable for this purpose. Most of the strains characterized are common bacteria in clinical samples, or well-known fast growing *r*-strategists, representative of only 0.01 to 10% of the actual populations. Even if several attempts were made in the past to obtain diversity indices based upon such procedures, methods allowing one to take into account non-cultivable organisms should be preferred at present. They include fatty acids determination and molecular approaches.

The phospholipid fatty acids (PLFA) determination method appears promising (Korner and Laczkó, 1992). Indeed, a given species harbours a limited number of PLFAs. Some of them are group-specific at different taxonomical levels, and thus can be taken as a measure of the relative abundance of a given group. The whole spectrum of PLFAs extracted from a natural sample gives a finger-print image of the actual biodiversity in this sample, without previous knowledge of the individual organisms in it (e.g. by cultivation). A biodiversity index can be drawn from these profiles. Qualitative differences between profiles can also be established. The presence/absence of a given fatty acid in phospholipids is probably genetically controlled. However, the relative amounts of PLFAs can vary according to environmental parameters like temperature (more saturated FA at higher temperatures) and phosphate limitation (lower phospholipids/total lipids ratio). This variation is at least partly compensated by the heterogeneity of an environmental sample, meaning that an average population does not show variations of its eco-physiological status of the same amplitude as those which occur in pure culture systems.

There has been an increased use of molecular biology methods in microbial ecology during the last years (Pickup, 1991; Knight *et al.*, 1992). Nucleic acids do not normally respond to environmental parameters like PLFA or ATP. rDNA probes would allow detection and quantification of known taxa in a sample, without the filter of cultural methods. As the sequence variability along an rRNA is heterogeneous, it is possible to obtain rDNA probes which are characteristic of the whole bacterial domain, or of a large phylum (e.g. the 'Proteobacteria'), a distinct group (family, genus) or even of a single species. The use of rDNA fluorescent probes reacting *in situ* with ribosomes allows one to stain the cells belonging to the target group of the probe specifically, which will then be recognized and counted by epifluorescence microscopy.

Other methods include extraction of total DNA from soils samples, followed by amplification of selected genes using polymerase chain reaction (PCR) with specific probes.

Methods for assessing functional diversity. Microbial biomass and production data give no information on the metabolic capabilities or activities of the microbial populations. Much research deals with the characterization and measurement of the microbial activities *in situ*. However, most studies measure the transformations performed by the whole microbial community without distinguishing the role played by any individual population.

Assessing functional diversity often requires cultural methods, i.e. enrichment, most probable number (MPN) determination, isolation and functional characterization of the strains. Cultural methods have several drawbacks in the study of microbial populations: they often strongly underestimate the real population size (see above), and will frequently reveal potential activities which are not necessarily those actually used *in situ* by the microorganisms. However, it is also important to consider this potential. Most probable numbers, even if they give an underestimate of population sizes, can be used for comparative purposes. Combined with elective enrichment techniques, they will give a relative image of the populations of a given functional group. Finally, and perhaps most importantly, they allow one to isolate (from the last positive dilutions) the most abundant cultivable organisms representative of a given function. The use of microtiter plates for MPN determination makes it possible to obtain relatively precise MPN determinations in a small volume, and to subsequently isolate the organisms. Only a relatively small soil sample is necessary (1 g or less).

Elective media for enrichment, MPN determination and isolation of functional groups like chemolithoautotrophs, diazotrophs and denitrifiers are relatively easy to develop, such as the media for aerobic hydrogen-oxidizing bacteria (Aragno and Schlegel, 1991).

There is actually a need for molecular methods (such as gene probes) that make it possible to detect genes responsible for a given function in natural environments. For example, there is an important homology between the genes responsible for dinitrogen fixation (*nif* genes) of a wide span of diazotrophic bacteria. Combining reverse transcriptase and polymerase chain reaction (PCR) methods would allow detection and amplification of mDNA sequences from mRNAs in environmental samples (R.U. Meckenstock, personal communication), and so detect and quantify the genes that are actually induced in communities.

BACTERIAL FUNCTIONS ASSOCIATED WITH LAND–INLAND WATER ECOTONE MICROENVIRONMENTS

There are probably no bacterial functions exclusively associated with ecotones, but the variety of conditions and of interfaces means that most of the 'classical' geochemical functions of bacteria will be expressed in such ecosystems. We will just summarize here the main functions, the conditions for their occurrence and the types of organisms involved.

Primary production: phototrophic bacteria

Two types of bacterial photosynthesis can occur in freshwater environments:

(1) Oxygenic photosynthesis, which utilizes water as an electron source and releases molecular oxygen as oxidation product. This is almost identical to plant and algal photosynthesis, and is performed by the Cyanobacteria (formerly known as the 'blue-green algae'). Typical representatives are: *Oscillatoria, Aphanizomenon, Nostoc, Anabaena...* Many of them are also dinitrogen fixers (see below).

(2) Anoxygenic photosynthesis, which utilizes other donors (particularly reduced sulfur compounds), and does not release molecular oxygen. Anoxygenic photosynthesis is normally repressed in oxic conditions, therefore, the occurrence of anoxygenic photosynthetic bacteria is normally restricted to anoxic environments exposed to light. However, some otherwise anoxygenic phototrophs can grow aerobically as chemotrophs. Anoxygenic photosynthesis is probably of minor importance in land–inland water ecotones, except for S-dominated habitats. Typical representatives are *Chlorobium, Chromatium, Rhodospirillum.*

Chemotrophic metabolism

Chemotrophic organisms utilize organic and/or inorganic compounds in an energy-giving (exothermic) reaction. Part of the energy is converted to 'cellular energy', mainly adenosine-triphosphate (ATP) and trans-membrane potentials. The chemoheterotrophs need an organic source for carbon assimilation, whereas chemolithoautotrophs draw all of their cellular carbon from carbon dioxide. In this case, a reduced inorganic compound is both the source of reducing power for CO_2 reduction and of energy, through respiratory electron transfer to an inorganic electron acceptor.

Extracellular hydrolysis of biopolymers. Most of the primary production of organic compounds consists of biopolymers, which need to be hydrolysed extracellularly before the absorption of the relative monomers by chemoheterotrophic bacteria and fungi. Among unicellular organisms, only protozoa are able to engulf prey, particles and macromolecules in their cytoplasm through phagocytosis and pinocytosis. Extracellular depolymerases are produced by bacteria (aerobic and anaerobic) and fungi, and become part of the free enzyme pools of natural environments. In some cases, the depolymerases (particularly cellulases) are attached to the bacterial cell wall, and therefore the hydrolysis products are directly absorbed by the cell, without reaching the surrounding environment. Otherwise, the extracellular enzyme activities are beneficial, not only to the enzyme producers, but also to other microorganisms in the surroundings, which can then be considered as commensals. Extracellular polymer hydrolysis occurs in both aerobic and anaerobic conditions.

Chemotrophic functions expressed in anaerobic conditions. Two chemotrophic 'ways of life' can occur in the absence of oxygen: fermentations, and anaerobic respirations.

(1) *Fermentations.* Fermentations are essentially organotrophic ways of life: the transformation of organic compounds, particularly sugars, organic and amino acids, is coupled to ATP synthesis by substrate-level phosphorylation; the overall oxidation degree of the fermentation products is the same as that of the fermentation substrate. The fermentation products (alcohols, volatile fatty acids, molecular hydrogen, carbon dioxide) are released into the environment. Most of these products are not fermentable any more (there are exceptions, such as lactate), but they can be oxidized by organisms with respiratory (aerobic or anaerobic) metabolism. A great variety of fermentative processes occur in nature, performed by many different bacteria, either facultatively or obligate anaerobic, or aerotolerant. Several fermentative bacteria (particularly *Clostridium* spp.) secrete depolymerases (e.g. cellulases, pectinases) and therefore perform the first steps in the anaerobic decay of plant litter.

(2) *Anaerobic respirations.* Respiration consists of electron transport from a low-potential electron donor to a higher-potential electron acceptor. This transport occurs through a membrane-linked chain of redox components (flavoproteins, iron-sulfur proteins, quinones, porphyroproteins) and generates a transmembrane proton gradient, which in turn allows ATP synthesis and other functions to occur. In Eucarya (Eucaryotes), there is just one type of respiration, using metabolites from the tricarboxylic cycle as electron donors (NADH, succinate) and molecular oxygen as final acceptor. In Bacteria and Archaea, other types of respiration exist. Respirations which use compounds different from oxygen as acceptor are called 'anaerobic respirations'. Nitrate, sulfate, elemental sulfur, carbon dioxide, and even oxidized metallic ions can serve as acceptors (Figure 4.1).

Nitrate respiration (or denitrification) is an alternative pathway under O_2 limitation for several otherwise aerobic bacteria. It leads to the production of N_2 and sometimes of limited amounts of N_2O and of NO. It provokes a loss of combined nitrogen. Nitrammonification is the reduction of nitrate to ammonia. It combines respiratory and fermentative mechanisms, and is performed by otherwise anaerobic

Microbial diversity and functions

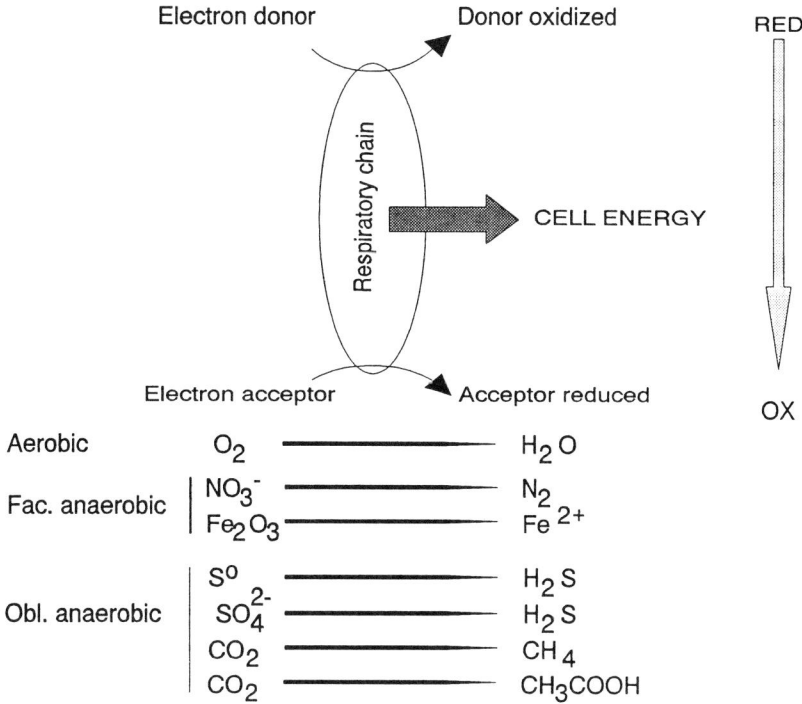

Figure 4.1 Principle of respiration with different external electron acceptors

(strict or facultative) organisms. This is the main nitrate-reduction pathway in permanently anoxic biotopes; it ensures the conservation of combined nitrogen and its conversion to a form assimilable by most anaerobic bacteria (e.g. methanogens).

The other anaerobic respirations are performed by strict anaerobes. They are therefore not substitutes for aerobic respiration.

Sulfate and elemental sulfur respirations (sulfatoreduction and sulforeduction) are important processes in anoxic environments. Together with anoxygenic, sulfur oxidizing phototrophs, they play a key role in 'sulfureta', associations of anaerobes based on the sulfur cycle. They are responsible for most non-geothermal sulfur springs. In the eutrophic waters of estuaries and sea-ponds, they produce high amounts of hydrogen sulfide, which is harmful to fishes and mussels. Sulfureta can also occur in heavily eutrophic inland waters. The well-known 'Winogradski columns' (Figure 4.2) are microcosms reproducing the functions of such sulfureta.

There are two different types of carbonate respiration: methanogenesis and acetogenesis. In the first, carbon dioxide is reduced to methane. The external electron source is mainly molecular hydrogen,

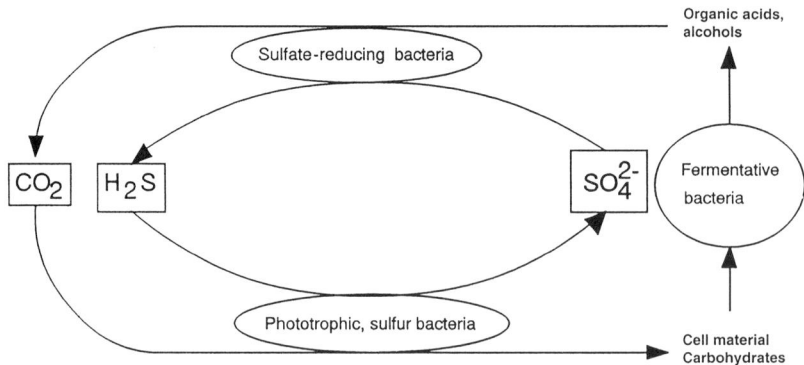

Figure 4.2 Combination of sulfur and carbon cycles in a 'sulfuretum'

sometimes formate or methylated compounds. All methanogens are Archaea. Their metabolism constitutes the final step in the anaerobic degradation of organic matter. Acetogens reduce carbon dioxide to acetate, the electron source being either molecular hydrogen (in *Clostridium aceticum* and *Acetobacterium woodii)* or electrons from sugars, *via* the Embden-Meyerhoff glycolytic pathway (in the 'homoacetogenic fermentations', e.g. in *Clostridium thermoaceticum*). Another group of methanogens split acetate into methane + carbon dioxide (acetoclastic methanogenesis).

The use of ferric oxides (as well as of other metallic oxides) as electron acceptors for anaerobic respiration is much less well documented. However, anaerobic mixed cultures have been shown to reduce and solubilize iron(III) oxides.

Chemotrophic functions expressed in aerobic conditions. In oxic environments, aerobic respiration is the most widespread way of life among chemotrophs. The external electron source for respiration may be an organic compound (organotrophy) or a reduced inorganic one (lithotrophy).

(1) *(Organo) heterotrophic mineralization.* The organic substrate is generally used as electron (and energy) source as well as carbon source for assimilation (heterotrophy). In natural environments, the substrates used in the respiratory process are often completely oxidized into water and carbon dioxide, with partial assimilation of the other elements combined in the organic substrates. The excess is mineralized, that is, released into the environment, generally as reduced, inorganic compounds (ammonia, hydrogen sulfide) (Ulehlova, 1990).

(2) *Chemolithoautotrophic oxidations.* Several inorganic compounds are used by specialized bacteria as an external electron source for aerobic respiration (Figure 4.3). This process is called lithotrophy. Often, the

Microbial diversity and functions

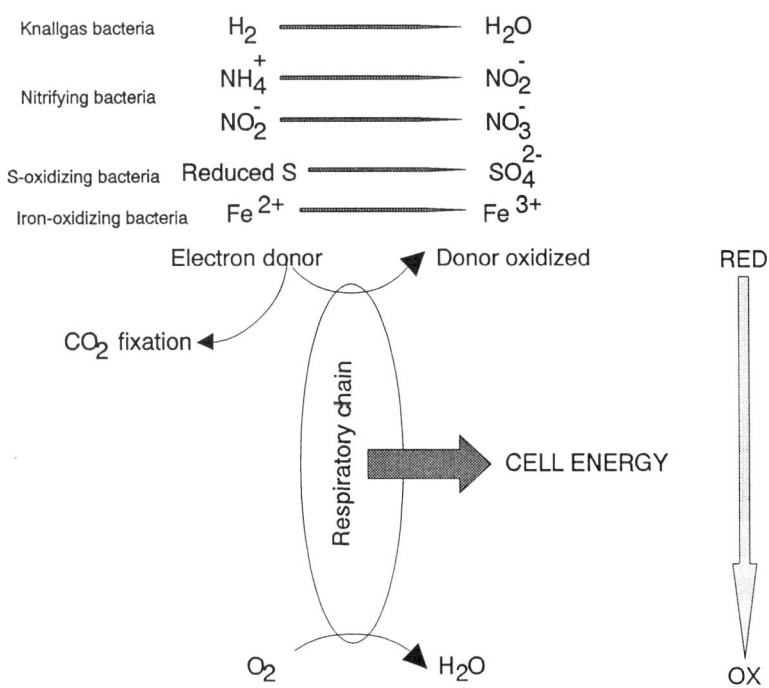

Figure 4.3 Principle of aerobic, chemolithoautotrophy with different external electron donors

reducing power from the inorganic compound and the ATP generated in the respiratory process allow the utilization of carbon dioxide as the main carbon source (autotrophy). Oxidation of nitrogen or sulfur compounds results in the production of strong acids (nitric or sulfuric), and may therefore lead to the acidification of the biotope. The chemolithoautotrophs are divided into several groups, according to the electron source:

(a) the *nitrifying bacteria*, which oxidize reduced nitrogen compounds. In fact, two functional groups are needed to perform the global oxidation of ammonia to nitrate: the nitrous bacteria, which oxidize ammonia to nitrite (the prefix *Nitroso-* is used for the generic names) and the nitric bacteria, which oxidize nitrite to nitrate (prefix *Nitro-*). No chemolithoautotrophic bacterium directly oxidizes ammonia to nitrate. Some are obligately autotrophic, whereas others can slowly assimilate a limited amount of organic compounds. Heterotrophic nitrification also exists, by some fungi and bacteria. However, compared to the lithoautotrophic nitrification, it probably has a negligible importance in natural cycles.

(b) the *sulfo-oxidizing bacteria*, which oxidize reduced sulfur compounds to sulfuric acid. Hydrogen sulfide, elemental sulfur, thiosulfate and other polythionates can serve as electron donors. Some of these bacteria are strict acidophiles (e.g. *Thiobacillus thiooxidans*) and can grow at pH values as low as 0.5, whereas other are neutrophilic. Several genera accumulate elemental sulfur globules in their cytoplasm as intermediate oxidation products. There are obligate autotrophs, as well as facultative ones.

(c) the aerobic *hydrogen-oxidizing bacteria*, (or Knall gas bacteria) which oxidize molecular hydrogen. In fact, molecular hydrogen is the only inorganic electron donor whose low redox potential allows the transfer of electrons to anaerobic respiratory chains, as in sulfate, sulfur and carbonate respirations (see above). Aerobic hydrogen-oxidizing bacteria are in general facultative autotrophs which, under organic carbon limitation, can lithotrophically draw a supplement of energy and of carbon from H_2 and from CO_2 (mixotrophy).

(d) the *iron-oxidizing bacteria*, which oxidize bivalent iron into iron(III) oxides or hydroxides. Two types were described, one occurring in low pH habitats, where Fe(III) ions are soluble (e.g. *Thiobacillus ferrooxidans*), the other (*Gallionella ferruginea*) in neutral pH conditions, and at low oxygen concentrations, where Fe(II) ions are not spontaneously oxidized. *Gallionella* cells form twisted mucilaginous ribbons, which become impregnated with iron(III) oxides and are easy to identify by microscopic examination. Other bacteria are often referred to as 'iron bacteria', like *Leptothrix* spp., *Crenothrix* spp. and *Siderocapsa* spp. But these are heterotrophs which accumulate iron(III) oxides in extracellular sheaths or capsules. Most probably, they catabolize naturally occurring iron(III) chelates and the iron moieties accumulate as oxides in the secreted mucilages.

Other functions to be considered

Other microbial functions, not related to energy metabolism, are also highly significant in biogeochemical cycles. The most important are dinitrogen fixation and exopolymer synthesis.

Biological dinitrogen fixation. Biological dinitrogen fixation is the ability to reduce elemental nitrogen N_2 to ammonia, which is an assimilable form. The overall reaction:

$$N_2 + 6H^+ + 6e^- \rightarrow 2NH_3$$

is catalysed by an enzyme complex, nitrogenase, comprising iron–sulfur and iron–molybdenum–sulfur proteins (the 'azoferredoxin' and the 'molybdoferredoxin', respectively). This complex brings electrons from ferredoxin to a very low redox potential, with ATP consumption. This low potential allows the reduction of dinitrogen (in three steps) to ammonia, but also other reductions. Indeed, the nitrogenase complex is specific to triple bound compounds, and will reduce cyanide to ammonia + methane, and acetylene to ethylene. This reaction is often used as a test for nitrogenase activity, in natural samples as well as in pure cultures. Moreover, the complex also has a hydrogenase activity, that is, it catalyses the reduction of protons to molecular hydrogen:

$$2 H^+ + 2 e^- \rightarrow H_2$$

This reaction could be considered a 'parasitic' side reaction, it is however unavoidable, and about one third of the electrons activated by the nitrogenase system are eventually 'lost' as molecular hydrogen. However, many dinitrogen-fixing bacteria possess a membrane-bound uptake hydrogenase, which activates molecular hydrogen and gives its electrons to the respiratory chain. This is both a way to recapture part of the energy lost, and to provide respiratory protection for the extremely sensitive nitrogenase complex by increasing the oxygen consumption. Otherwise, the hydrogen will diffuse into the environment.

Energetically speaking, dinitrogen fixation is an expensive pathway, at least 16 ATP being required for the reduction of one N_2 molecule. In the presence of sufficient concentrations of combined nitrogen (e.g. ammonia), nitrogenase is repressed, and assimilation of ammonia occurs through the 'normal', non-energy-consuming glutamate synthetase (dehydrogenase) or L-alanine dehydrogenase pathways which have a relatively low affinity for ammonia. Only under combined nitrogen limitation will nitrogenase be induced, as well as an ATP-consuming ammonia-assimilation system with much higher affinity, the glutamine–synthetase/glutamate–oxoglutarate aminotransferase cycle. This assimilation system will allow maintenance of a very low external ammonia concentration, and therefore prevent nitrogenase repression.

This need for energy explains why dinitrogen fixation is particularly favoured in conditions where both an important energy supply and a combined nitrogen deficiency occur.

Under illumination, a number of phototrophic bacteria, including Cyanobacteria (e.g. *Anabaena, Nostoc*), will fix dinitrogen. They have a particular importance in rice paddy, either as free-living organisms or as symbionts of the floating ferns *Azolla* spp. Their importance in the nitrogen cycle in land–inland water ecotones should also be investigated.

Chemotrophic, dinitrogen fixers can also benefit from light energy for dinitrogen fixation in an indirect way: by the utilization of sugars and other organic products resulting from plant photosynthesis. This is particularly the case at the root level, where mutual benefit associations between nitrogen-fixing bacteria and plants can occur, the plant bringing organic carbon and energy to the bacterium, which in turn provides the plant with combined nitrogen. This can occur at different levels of 'intimacy' (Aragno, 1984), in the external rhizosphere, on the rhizoplane, or in closer symbiotic associations, at histospheric (intratissular, extracellular location of the bacterial symbiont) or cytospheric (intracellular location) levels. The best-known dinitrogen-fixing symbioses are the cytospheric *Rhizobium/Fabaceae* and *Frankia/Alnus* root nodules.

Exopolymer biosynthesis. A number of bacteria synthesize and excrete polymers into their surroundings. These polymers are mainly polysaccharides, sometimes also unusual polypeptides. Depending on their hydration properties, these polymers either form well-defined capsules around the bacterial cells, or much looser, hydrated mucilaginous layers which spread into the environment and include other bacterial cells and particles. Often these polymers are very resistant to enzymatic attack, and have a long life-span in the environment. They are far more stable than most exopolymers secreted by plants. The bacterial exopolymers form a significant part of the organic fraction of soils, the so-called 'hydrolyzable fraction' of humus.

RELATIONSHIP OF BACTERIA WITH OTHER ORGANISMS IN THE BIOCOENOSES

An important parameter in soil microbial biomass and nutrient cycling control, often underestimated by microbial ecologists, is the predation of microbial cells by protozoa or microinvertebrates (Fenchel, 1984; Kuikman and van Veen, 1989; Kuikman *et al.*, 1990). This may considerably increase the carbon and nitrogen turnover. In that sense, heterotrophic bacteria or fungi can often be considered secondary producers, when they feed on dissolved or particulate organic matter (Moran and Hodson, 1989). Other bacteria, either phototrophs or chemolithoautotrophs, are even primary producers.

Microorganisms also often secrete growth factors (vitamins, essential amino or fatty acids, purine or pyrimidine precursors) into their environment, allowing the growth of auxotrophic organisms in their vicinity. Other microorganisms produce antibiotics, and may therefore exert an inhibitory effect on other organisms. Therefore, microbial secretions can exert positive as well as negative effects on other members of the the biocoenose.

Several microorganisms, bacteria and fungi, participate in mutualistic associations with other organisms. These associations are either of syntrophic nature (association at the trophic level only) or symbiotic (trophic and physical association). Well-known examples of symbioses are the mycorrhizae (fungal/root associations), the association between diazotrophic bacteria and plants, either rhizospheric or symbiotic (e.g. root nodules), and the lichens.

Several bacteria and fungi can also unilaterally benefit from other organisms: depending on their action on the other organism, they behave as commensals, parasites or predators.

As commensals, they benefit from products of the activity of other organisms, without significantly affecting their growth and differentiation.

As parasites, they live in tight association with the host organism, feeding on it and often modifying its growth and/or differentiation. However, true parasites normally do not kill their host. They are therefore biotrophic.

As predators, they literally eat (and so kill) their victim. Some are predators of other bacteria, like the *Myxobacteria* which destroy the cell walls of other bacteria or fungi and feed on the cell contents, or the *Bdellovibrios* which penetrate bacterial cells and multiply in them, until the cell envelope breaks. Let us also mention the nematode-trapping fungi, which have developed special organs (mycelial loops or vesicles) with which they capture the worms before sending nutritive hyphae into their internal cavity. The worm is then 'digested' by these internal hyphae.

Predator or prey, activator or inhibitor of other organisms, the microorganisms, bacteria and fungi not only act as 'decomposers' and 'biogeochemical agents', they are also involved in complex relationships with the other living members of the biocoenosis. These relationships are particularly important in ecosystems like land–inland water ecotones, where a high variety and large numbers of animals (unicellular and pluricellular), plants (algae and higher plants), fungi and bacteria (phototrophic and chemotrophic) coexist.

CONSIDERATIONS AND HYPOTHESES ON THE MICROBIAL ECOLOGY OF SOME MICROECOTONES (GRADIENTS, TRANSITIONS, INTERFACES) EXISTING IN LAND–INLAND WATER ECOTONES

Land–inland water ecotones show a superposition of characteristics from terrestrial and freshwater ecosystems. They house a rich diversity of higher plants and algae, and often present a very high biomass productivity, equalled only by the most fertilized cultivated fields. All intermediates between a permanently submersed sediment and an aerated soil exist, with spatial and temporal variations in the water table. This has important consequences on the decomposition of plant litter, as well as on root biology and the microbial ecology of the rhizosphere. We will try to discuss here some examples of the microecotones generated by land–inland water macroecotones, and the importance and integration of microbial functions in these microecotones that so far have been little investigated.

Carbon flux and related functions associated with litter decomposition in submersed conditions

On permanently emerged soils of land–inland water ecotones, the pattern of litter decomposition should be similar to that of terrestrial ecosystems, where aerobic humification processes dominate. However, a situation unique to land–inland water ecotones occurs where the ground is permanently submersed (swamp situation). The litter surface is in contact with water almost saturated with oxygen, and will be actively degraded aerobically, and the mineralization products like ammonia and hydrogen sulfide will be oxidized by chemolithoautotrophs to nitrate and sulfate respectively. But, as stated above, the diffusion of oxygen in submersed materials is very limited, either because of the physical properties of the mineral substrate, often very compact and impermeable, and/or because the attack on the fibrous materials will decrease the porosity of the litter layer. Therefore the elevated oxygen consumption by aerobes and its limited diffusion imply that the 'compensation point' where oxygen diffusion from water equals its consumption will be reached very near the litter or top soil surface.

Below, the conditions become anoxic. Because of the absence of molecular oxygen, a number of organic structures, particularly of lignified and suberified plant material, will not be decomposed and will therefore accumulate. This is one of the main causes of peat accumulation, which in the long term will transform swamps into terrestrial ecosystems (terrestrialization).

A tentative representation of the carbon flow in a submersed soil is shown in Figure 4.4. If several potential electron acceptors are present simultaneously, they will normally be reduced by mixed bacterial populations in a sequence going from the highest to the lowest redox potential. First oxygen will be utilized by aerobes (1). Then nitrates will be reduced to molecular nitrogen or ammonia (2), then iron(III) to iron(II) (3), then sulfate to hydrogen sulfide (4). Eventually the remaining metabolizable carbon will be transformed by the 'climax' methanogenic association (5 and 6), producing 'swamp gas' (biogas).

In this association, degradable organic compounds will be globally dismuted into methane and carbon dioxide by a complex association of anaerobic bacteria and functions, comprising enzyme hydrolysis, fermentative bacteria, and a tight association between obligately syntrophic, proton-reducing acetogenic bacteria (5) and methanogenic bacteria (6). Volatile fatty acids with more than two carbons, and alcohols, are oxidized by the acetogenic bacteria to acetate and CO_2. This oxidation requires the simultaneous reduction of protons to H_2. This is only possible thermodynamically, if H_2 and acetate concentrations are maintained at very low values, which implies the immediate uptake of these metabolites by other bacteria. This is the role of the H_2-consuming/CO_2-reducing methanogens and of the acetoclastic methanogens. Figure 4.5 shows the transformation of ethanol to methane and CO_2 by a

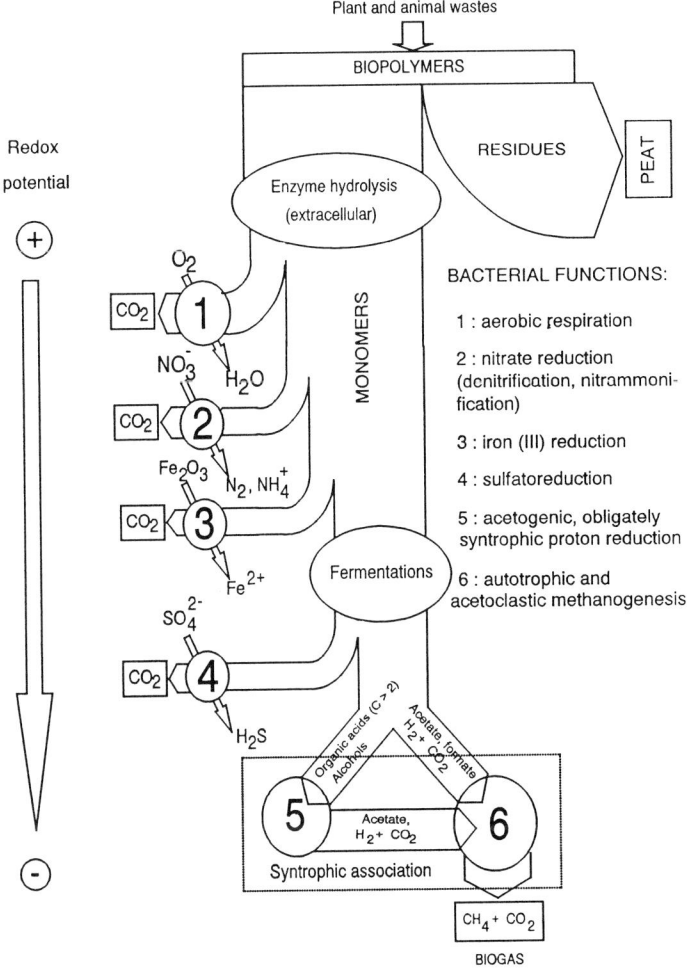

Figure 4.4 Carbon flow and the succession of bacterial functions in a swamp litter and sediment

syntrophic association between three bacteria, a proton-reducing acetogen (1), a H_2-oxidizing, CO_2-reducing methanogen (2) and an acetoclastic methanogen (3), the global reaction being:

$$\text{Ethanol} \rightarrow 1.5\,CH_4 + 0.5\,CO_2$$

This is a good illustration of the importance of the functional diversity of micro-organisms to achieve 'simple' transformations.

Interestingly, part of the methane produced in the sediment of swamps may diffuse to the atmosphere through the air channels of swamp plant roots.

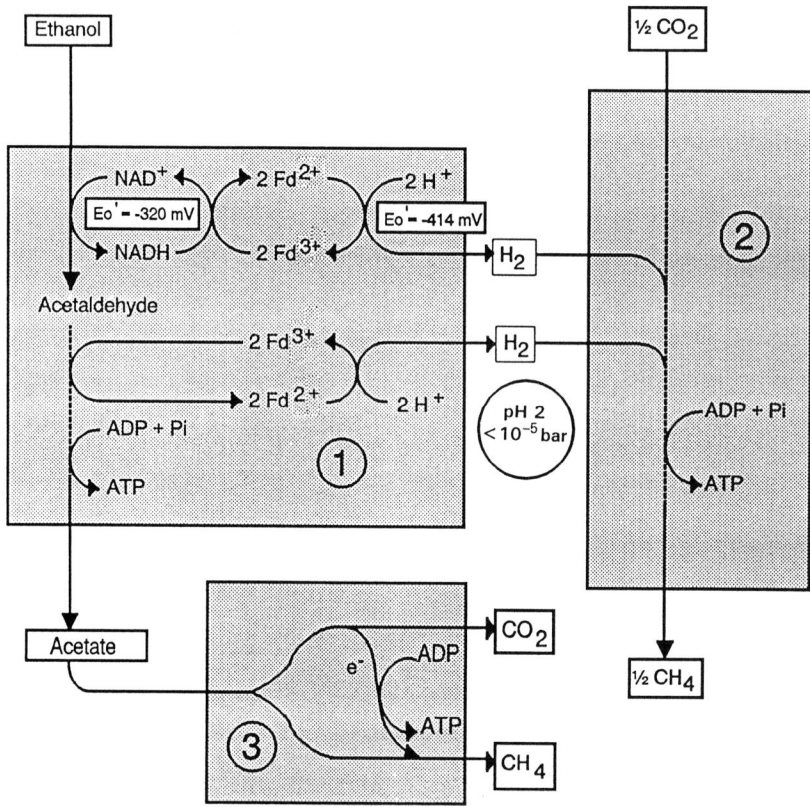

Figure 4.5 Structure of the syntrophic association between (1) an ethanol-oxidizing, proton-reducing obligately syntrophic acetogen, (2) a H_2-oxidizing methanogen and (3) an acetoclastic methanogen

Therefore, swamp plants, e.g. rice, contribute to increase the transfer of methane from anoxic sediments to the air (Conrad, 1993; Nouchi and Mariko, 1993).

Rhizospheric environment of aquatic plants

The rhizosphere, or the environment under the influence of the roots, is an extremely important interface influenced by the secretions, ion uptake and respiration of the roots. The rhizodeposition, all the organic material (exudates, secreted polymers, exfoliated cells, lysates) produced by a root in its environment is an important part of the total photosynthesized biomass. This therefore represents a considerable energy flux that benefits the rhizosphere microorganisms. Together with the root's own activities, the metabolism of rhizospheric microorganisms will therefore considerably modify the physical and chemical characteristics of the soil in the root vicinity.

We shall not discuss here in detail the 'normal' rhizospheric environment in an aerated soil. Let us just mention that root and microbial respiration will decrease the oxygen concentration in the root vicinity. If the root secretion has a relatively high C/N ratio, microbial growth from this secretion will immobilize combined nitrogen in the biomass. Both conditions (low combined nitrogen and low oxygen) will allow the nitrogen fixation system to function in N_2-fixing bacteria. Therefore, along with an important supply of energy, the rhizosphere will provide favourable conditions for dinitrogen-fixing organisms.

Typical of land–inland water ecotones is the occurrence of plant roots in anoxic, submersed soils or sediments. This implies unique characteristics of the root structure and functions, the most obvious of which are the air channels, which ensure an oxygen supply for the roots. Oxygen will then in most cases diffuse from the root into the rhizosphere, and form an oxic 'sheath' around the root (Armstrong and Armstrong, 1988).

So far, microbial ecology of the rhizosphere of swamp plants is an almost virgin field, with the exception of some studies on the rice rhizosphere. So we will just mention here several hypotheses on the microbial functions which could occur in this microenvironment (Figure 4.6). Whereas the plants preferably take up nitrogen and sulfur as oxidized ions (nitrate and sulfate respectively), these elements are present as reduced compounds (ammonia and hydrogen sulfide) in anoxic environments. The occurrence of aerobic nitrifying and sulfide-oxidizing bacteria in the oxic sheath around the root would then supply the root with the oxidized forms of these elements. The sulfide-oxidizers would also have a detoxifying effect, H_2S being a powerful respiratory inhibitor. Other bacteria characteristic of oxic/anoxic interfaces could also occur in the rhizosphere of swamp plants, like methane-oxidizers (methanotrophs), ethylene oxidizers, iron- and manganese oxidizing bacteria, etc.

Dinitrogen fixation could also be of importance in swamp plants. In a study of the swamps of the south shore of Lake Neuchâtel (Switzerland), Buttler (1987) has shown that the submersed soil in the vegetation dominated by *Phragmites communis* was extremely poor in soluble, combined nitrogen. However, the productivity of these fields was very high, comparable only to the richest cereal cultures. Therefore we suspected the occurrence of dinitrogen fixation in the *Phragmites* rhizosphere. In effect we measured a clear dinitrogen-fixing activity (by the acetylene reduction test) linked to *Phragmites* roots, an activity which was not detectable in anoxic sediments not in contact with roots. This activity was not removed by washing the roots, so the responsible bacteria were either attached to the rhizoplane or even embedded within the root tissues (the so-called 'endorhizosphere'). Several strains of dinitrogen-fixing bacteria were isolated from these roots (Arreguit, 1990). They form two distinct but related clusters of *Enterobacteriaceae* (DeLuca, 1991), which are at present under

Biodiversity in land–inland water ecotones

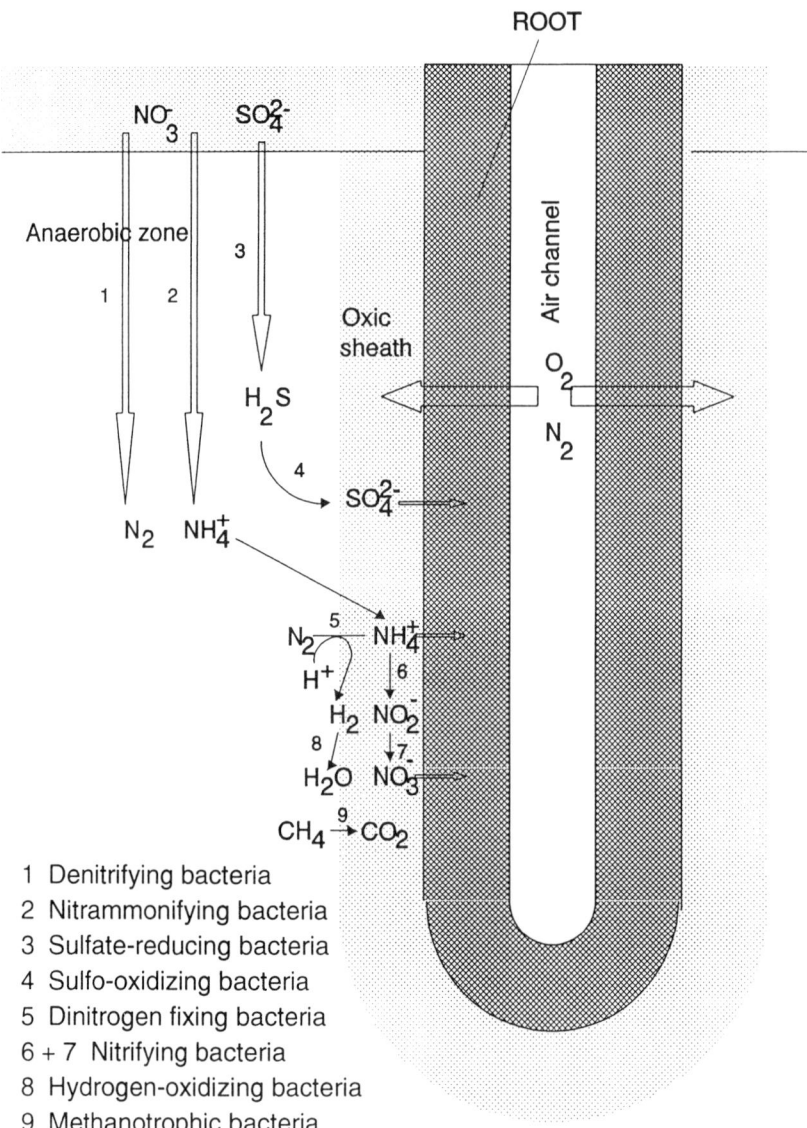

Figure 4.6 Oxidation and reduction of inorganic compounds by bacteria in the rhizosphere of a swamp plant

investigation in our laboratory. Indeed, the rhizosphere of *Phragmites* and probably of other swamp plants appears to be a good habitat for dinitrogen-fixing bacteria. Energy in the form of root exudates and oxygen at low concentrations are supplied by the plant in an extremely nitrogen-poor environment.

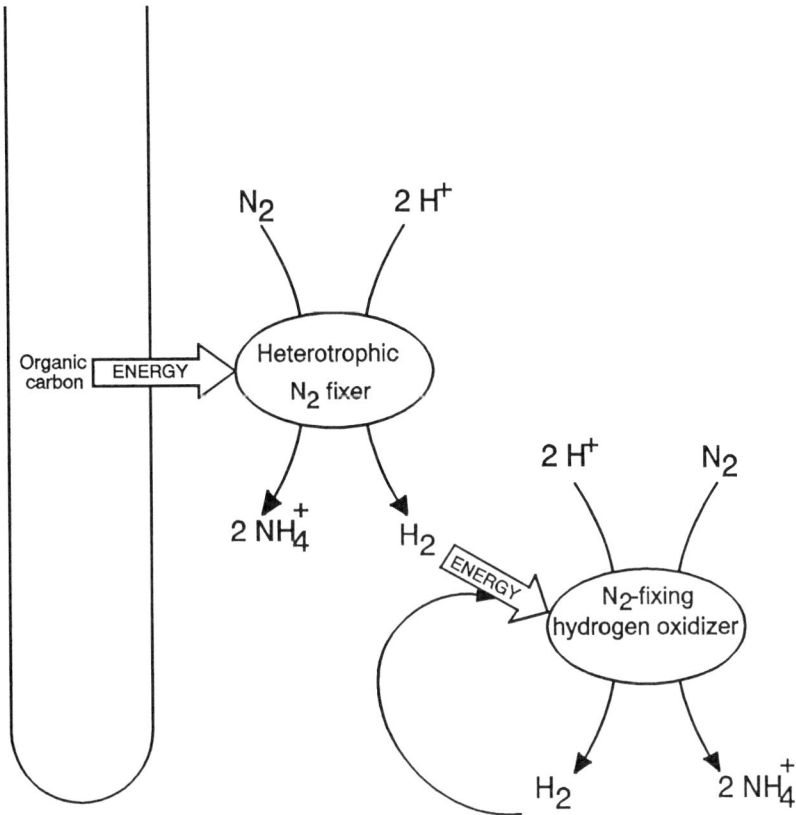

Figure 4.7 Energy flow from a root to a heterotrophic dinitrogen-fixing bacterium and an autotrophic, dinitrogen-fixing hydrogen-oxidizing bacterium

Interestingly, high numbers of aerobic hydrogen-oxidizing, chemolithoautotrophic bacteria ('knallgas' bacteria) were observed in the rhizosphere of swamp plants (Arreguit, 1990; Reding et al., 1991; T. Beffa, personal communication). Most isolates were dinitrogen fixers related to *Xanthobacter* spp. Due to the presence of hydrogen-consuming sulfate-reducing bacteria, hydrogen concentrations are kept very low in anoxic environments (Conrad et al., 1985). However, as stated above, significant hydrogen production always accompanies dinitrogen fixation and is not completely compensated by the uptake-hydrogenase activity (Conrad and Seiler, 1979). The occurrence of this otherwise unexpected population of 'knallgas' bacteria could therefore be a consequence of the presence of a heterotrophic, dinitrogen-fixing population on (or in) the roots (Figure 4.7). A similar enrichment in 'knallgas' bacteria was shown in the vicinity of legume root nodules (La Favre and Focht, 1983).

The possible consequences of a changing water table

A changing water table is a characteristic of land–inland water ecotones which could affect the soil microflora in the layers where a succession of emersion and submersion induces the alternation of oxic and anoxic conditions. Indeed, very little is known at present of the consequences of such a situation on microbial biodiversity. The relative duration of submersed/emersed periods, as well as the frequency of this variation (seasonal, daily), as well as amplitude should be taken into account.

As a rule, a variable environment would favour versatile organisms, possessing the ability to adapt rapidly to different growth conditions and substrates, and would restrict the development of specialized organisms, which would be favoured in more stable conditions.

As stated above, bacteria, in their response to oxygen, range from organisms absolutely requiring oxygen (strict aerobes) to organisms inhibited by oxygen (strict anaerobes). With these, one should distinguish between organisms which would be killed by even a short exposure to oxygen and organisms which could survive a rather long exposure, oxygen having then a mainly bacteriostatic effect. Endospore formation would also allow strict anaerobes to survive oxic conditions, even if the vegetative cells would rapidly be killed. Such oxygen-resistant organisms would be particularly adapted to environments subjected to an alternation of oxic and anoxic conditions. Facultative anaerobes, shifting from a respiratory to a fermentative metabolism and back when submitted to such alternations would be still better adapted, since they could be active in both conditions. Similarly, aerotolerant anaerobes would not be affected by the presence or absence of oxygen, and would maintain their fermentative metabolism in both situations.

Contrary to permanently anoxic biotopes, where nitrate reduction occurs almost exclusively through nitrammonification, biotopes alternating between oxic and anoxic conditions would favour denitrification in the strict sense, because it is performed by otherwise aerobic bacteria. This could explain the scarcity of combined nitrogen often noticed in anoxic sediments in the ecotone.

The often complex associations of bacteria such as discussed above would be more sensitive than individual species. Methanogenesis is extremely sensitive to changes in redox potential. Several days may be necessary for the association to recover from an important increase in redox potential, and the accumulation of volatile fatty acids during the transition would decrease the pH and inhibit the strictly neutrophilic methanogens. Therefore, a changing water table could result in a change from stable methanogenesis to a situation where intermediate fermentation products accumulate, products which would be rapidly oxidized after the establishment of oxic conditions.

THE ROLE OF MICROORGANISMS AS DEPOLLUTING AGENTS

Natural land–inland water ecotones are often considered as potential waste water purification systems. However, as in other autoepurative natural ecosystems, there are limits to the load and the nature of pollutants which are sustainable. If these limits are overshot, these complex ecosystems may be severely damaged and may not be able to fulfil their natural functions. The role of microorganisms in such waste purification appears essential and is multiple. Depending on their nature, the pollutants will be:

- partially or totally degraded;
- modified in their chemical or physico-chemical properties, e.g. precipitated or solubilized, oxidized or reduced;
- taken up and concentrated in the biomass.

In some cases, microorganisms occurring in masses can be pollutants in themselves, like the toxin-producing Cyanobacteria. Pollutants can tentatively be ranged into three categories:

- organic pollutants, of biological origin;
- synthetic organic pollutants, generally referred to as xenobiotic compounds, like pesticides, solvents and other compounds of industrial interest;
- inorganic pollutants.

As a rule, organic pollutants of biological origin will be completely degraded in aerobic conditions, giving rise to carbon dioxide, water and mineral salts, provided the biological activity is not restricted by some other factor (lack of a nutrient, low pH, presence of toxic compounds, etc.). However, these degradations are accompanied by oxygen consumption. A high load will therefore lead to anoxic conditions. This will considerably change the biological conditions, and in turn prevent the ulterior degradation of some of these organic compounds.

Because the chemical structures of xenobiotic organic pollutants are completely different from naturally occurring ones, many of them will not be biologically degraded, or only partially, giving rise to other end-compounds which can even be more toxic than the original ones. This is the case with perchloroethylene, for example, which is reduced by anaerobic bacteria to the much more toxic tri-, di- and monochlorinated ethylenes (Vogel and McCarthy, 1985). In some cases, aromatic xenobiotics will undergo chemical or enzymatical transformation into chemical radicals, which will in turn react and incorporate into complex humic compounds (Bollag, 1983). It is not clear if this is a long-term detoxification pathway or not,

Some inorganic compounds (e.g. sodium chloride) are toxic only at relatively high concentrations. Other elements exist in toxic and non-toxic forms and can be made toxic or, on the contrary, detoxified by reduction or oxidation performed by microorganisms (e.g. hydrogen sulfide and sulfate). Finally, some elements (essentially the 'heavy metals') are toxic at relatively low concentrations. Some (e.g. Co, Cu, Mn, Ni, W, Zn...) are necessary (at very low concentrations) to living processes, whereas others (e.g. Ag, Cd, Pb...) are not. They may be inactivated by precipitation. Several microbial activities, however, will help to solubilize them, by acidification, reduction, or synthesis of chelating compounds.

Provided the load is kept within certain limits, ecotone microorganisms will assume the role of degraders of organic pollutants of natural origin efficiently. This will be particularly favoured by the diffusion of oxygen in the rhizosphere of aquatic plants and by the water table movement allowing air or water containing dissolved oxygen to penetrate permeable soils and sediments. Due to their limited biodegradability, the fate of xenobiotics in naturally occurring ecotones is much more problematic. They will be better eliminated by treating them in systems using selected or even modified microorganisms, or better by avoiding their dispersal into the environment altogether.

The fate of inorganic pollutants in ecotones is much more difficult to estimate. Nitrate elimination by denitrification, for example, could be favoured as stated above by the alternation of aerobiosis and anaerobiosis typical of many 'microecotones'. However, heavy metals would be submitted to continuous precipitation/solubilization cycles in such interface conditions.

Immobilization of certain toxic compounds in the microbial biomass is conceivable: microorganisms are very efficient at concentrating several compounds, such as heavy metals, due to their active ion transport systems. However, microbial biomass can by no means be considered as a 'final deposit' for such pollutants, because it is itself submitted to a high turnover, due to spontaneous cell death and probably still more to grazing by protozoa and microinvertebrates. It seems moreover illusory to imagine harvesting this pollutant-containing microbial biomass in an ecotone. Consider only the efficiency of collecting all the mushrooms in contaminated areas, where they would have accumulated heavy metals and radionucleides...

CONCLUSIONS

Bacteria and other microorganisms play an essential role in land–inland water ecotones, as in most other ecosystems. However, knowledge of the microbial part of these ecosystems is almost non-existent. Two main reasons can be advanced for this:

(1) The severe methodological limits which hindered microbial ecologists from obtaining data comparable to those of plant or animal ecologists. With the development of modern analytical and molecular methods, there is hope that such limitations will disappear, and that microbial ecology will progress considerably, as did microbial taxonomy over the last 20 years.

(2) Only a few microbial ecologists actually study 'real life' situations in the field. Most restrict themselves to simplified laboratory ecosystems, like chemostats or plug-flow systems with immobilized biomass. Field microbiologists then deal mainly with ecosystems which are more homogeneous and/or have potential applications, like pelagic waters or agricultural soils.

However, it is clear that transient ecosystems like microecotones provide a multiplicity of ecological niches for a variety of functional groups of bacteria and other microorganisms, and that macroecotones harbour a variety of different microecotones. Many of these niches (dinitrogen fixation for example) are occupied exclusively by bacteria and are ideal for understanding the flow of elements through the system. Bacteria are almost the only participants in anaerobic ecosystems like 'sulfureta', which are most likely to appear in underwater environments subjected to a high input of energetic substrates, as are some parts of the land–inland water ecotones.

Apart from the development of basic knowledge on microbial ecology, future research in the microbial ecology of land–inland water ecotones should try to answer the following questions:

- how do the ecotone microorganisms influence the quality of offshore waters, and conversely, how do the respective properties of offshore, pluvial and 'terrestrial' waters affect the ecotone microflora?
- how will the seasonal variations affect the microbial activities in an ecotone, particularly the presence and activity of vegetation?
- how do ecotone microorganisms respond to pollutants?
- in what measure will they participate in the reduction and/or immobilization of water contaminants and constitute a screen between pollutants of terrestrial origin and the free waters?
- what is the fate of potentially pathogenic microorganisms in ecotone ecosystems?

ACKNOWLEDGEMENTS

We would like to thank Alexandre Buttler and Jean-Bernard Lachavanne for their critical reading of the manuscript, and Catherine Fischer who corrected the English text. The research performed in the first author's laboratory was supported in part by grant 31-28597.90 from the Swiss National Science Foundation.

REFERENCES

Aragno, M. (1984). Interactions à bénéfice mutuel entre le monde bactérien et les plantes vasculaires. *Botanica Helvetica*, **94**, 235–48

Aragno, M. (1991). Thermophilic, aerobic, hydrogen-oxidizing (knallgas) bacteria. In Balows, A., Trüper, H.G., Dworkin, M., Harder, W. and Schleifer, K.H. (eds.) *The Prokaryotes, a Handbook on Biology of Bacteria*, pp. 3917–33. 2nd edn, vol. 4. Springer Verlag, New York

Aragno, M. and Schlegel, H.G. (1991). The hydrogen-oxidizing (knallgas) bacteria. In Balows, A., Trüper, H.G., Dworkin, M., Harder, W. and Schleifer, K.H. (eds.) *The Prokaryotes, a Handbook on Biology of Bacteria*, pp. 344–84. 2nd edn, vol. 1. Springer Verlag, New York

Armstrong, J. and Armstrong, W. (1988). *Phragmites australis*, a preliminary study of soil-oxidizing sites and internal gas transport pathways. *New Phytologist*, **108**, 373–89

Arreguit, M. (1990). Etude de la fixation de l'azote atmosphérique dans la rhizosphère du roseau (*Phragmites communis*). I. Diploma work, Institute of Botany, University of Neuchâtel

Atlas, R.M. (1983). Use of microbial diversity measurements to assess environmental stress. In Klug, M.J. and Reddy, C.A. (eds.) *Current Perspectives in Microbial Ecology*, pp. 540–5. American Society of Microbiology, Washington DC

Austin, H.K. and Findlay, S.E.G. (1989). Benthic bacterial biomass and production in the Hudson River estuary New York USA. *Microbial Ecology*, **18**, 105–16

Bollag, J.M. (1983). Cross-coupling of humus constituents and xenobiotic substances. In Christman, F.R. and Gjessing, E.T. (eds.) *Aquatic and Terrestrial Humic Materials*, pp. 127–41. Ann Harbor Scientific Publications, Ann Harbor, MT

Buttler, A. (1987). Etude écosystémique des bas-marais non boisés de la rive sud du lac de Neuchâtel (Suisse): phytosociologie, pédologie, hydrodynamique et hydrochimie, production végétale, cycles biochimiques et influence du fauchage sur la végétation. Ph.D. thesis, University of Neuchâtel

Byrd, J.J. and Colwell, R.R. (1992). Microscopy applications for analysis of environmental samples. In Levin, M.A., Seidler, R.J. and Rogul, M. (eds.) *Microbial Ecology*, pp. 93–112. McGraw-Hill, New York

Conrad, R. (1993). Mechanisms controlling methane emission from wetland rice fields. In Oremland, R.S. (ed.) *Biogeochemistry of Global Change: Radiatively Active Trace Gases*, pp. 336–52. Chapman and Hall, New York

Conrad, R., Phelps, T.J. and Zeikus, J.G. (1985). Gas metabolism evidence in support of juxtapositioning between hydrogen producing and methanogenic bacteria in sewage sludge and lake sediments. *Applied Environmental Microbiology*, **50**, 595–601

Conrad, R. and Seiler, W. (1979). Field measurement of hydrogen evolution by nitrogen-fixing legumes. *Soil Biology and Biochemistry*, **11**, 689–90

DeLuca, A. (1991). Etude de la fixation de l'azote atmosphérique dans la rhizosphère du roseau (*Phragmites communis*). II. Diploma work, Institute of Botany, University of Neuchâtel

Fenchel, T. (1984). Suspended marine bacteria as a food source. In Fasham, M.J.R. (ed.) *Flows of Energy and Materials in Marine Ecosystems*, pp. 301–15. Plenum Press, New York

Jenkinson, D.S. and Powlson, D.S. (1976). The effect of biocidal treatment on metabolism in soil. V A method for measuring soil biomass. *Soil Biology and Biochemistry*, **8**, 209–13

Jones, J.G. (1977). The effect of environmental factors on estimated viable and total populations of planktonic bacteria in lakes and experimental enclosures. *Freshwater Biology*, **7**, 61–97

Knight, I.T., Holben, W.E., Tiedje, J.M. and Colwell, R.R. (1992). Nucleic acid hybridization techniques for detection, identification and enumeration of microorganisms in the environment. In Levin, M.A., Seidler, R.J. and Rogul, M. (eds.) *Microbial Ecology*, pp. 65–91. McGraw-Hill, New York

Korner, J. and Laczkó, E. (1992). A new method for assessing soil microorganism diversity and evidence of vitamin deficiency in low diversity communities. *Biology and Fertility of Soils*, **13**, 58–60

Kuikman, P.J., Jansen, A.G., van Veen, J.A. and Zehnder, A.J.B. (1990). Protozoan predation and the turnover of soil organic carbon and nitrogen in the presence of plants. *Biology and Fertility of Soils*, **10**, 22–8

Kuikman, P.J. and van Veen, J.A. (1989). The impact of protozoa on the availability of bacterial nitrogen to plants. *Biology and Fertility of Soils*, **8**, 13–18

La Favre, J.S. and Focht, D.D. (1983). Conservation in soil of H_2 liberated from N_2 fixation by Hup^- nodules. *Applied Environmental Microbiology*, **46**, 304–11

Maire, N. (1984). Extraction de l'adénosine triphosphate dans les sols: une nouvelle méthode de calcul des pertes en ATP. *Soil Biology and Biochemistry*, **16**, 361–6

Maire, N. (1987). Evaluation de la vie microbienne dans les sols par un système d'analyses biochimiques standardisé. *Soil Biology and Biochemistry*, **19**, 491–500

Moran, M.A. and Hodson, R.E. (1989). Bacterial secondary production on vascular plant detritus: relationships to detritus composition and degradation rate. *Applied Environmental Microbiology*, **55**, 2178–89

Nouchi, I. and Mariko, S. (1993). Mechanism of methane transport by rice plants. In Oremland, R.S. (ed.) *Biogeochemistry of Global Change: Radiatively Active Trace Gases*, pp. 336–52. Chapman and Hall, New York

Pickup, R.W. (1991). Development of molecular methods for the detection of specific bacteria in the environment. *Journal of General Microbiology*, **137**, 1009–19

Porter, K.G., Paerl, H., Hodson, R., Pace, M., Priscu, J., Riemann, B., Scavia, D. and Stockner, J. (1988). Miocrobial interactions in lake food webs. In Carpenter, S.R. (ed.) *Complex Interactions in Lake Communities*, pp. 209–27. Springer Verlag, New York

Reding, H.K., Hartel, P.G. and Wiegel, J. (1991). *Xanthobacter* as a rhizosphere organism of rice. *Plant and Soil*, **138**, 221–30

Roszak, D.B. and Colwell, R.R. (1987). Survival strategies of bacteria in the natural environment. *Microbiology Review*, **51**, 365–79

Tunlid, A., Baird, B.H., Trexler, M.B., Olsson, S., Findlay, R.H., Odham, G. and White, D.C. (1985). Determination of phospholipid ester-linked fatty acids and poly beta hydroxybutyrate for the estimation of bacterial biomass and activity in the rhizosphere of the rape plant *Brassica napus*. *Canadian Journal of Microbiology*, **31**, 1113–19

Tunlid, A. and White, D.C. (1990). Use of lipid biomarkers in environmental samples. In Fox, A. *et al. Analytical Microbiology Methods*. Plenum Press, New York

Ulehlova, B. (1976). *Microbial Decomposers and Decomposition Processes in Wetlands*. Studie CSAV, ACADEMIA, Praha, 17

Ulehlova, B. (1990). Release and uptake of minerals during decomposition of plant litter in fishpond littoral. *Folia Geobotanica Phytotaxanomia, Praha.*, **25**, 303–8

Vogel, T.M. and McCarthy, P.L. (1985). Biotransformation of tetrachloroethylene to trichloroethylene, dichloroethylene, vinyl chloride and carbon dioxide under methanogenic conditions. *Applied Environmental Microbiology*, **49**, 1080–3

Voroney, R. and Paul, E.A. (1984). Determination of K_c and K_n *in situ* for calibration of the chloroform fumigation–incubation method. *Soil Biology and Biochemistry*, **16**, 9–14

Woese, C.R. (1987). Bacterial Evolution. *Microbiological Review*, **51**, 221

Woese, C.R., Kandler, O. and Wheelis, M.L. (1990). Towards a natural system of organisms: proposal for the domains Archaea, Bacteria and Eucarya. *Proceedings of the National Academy of Science, USA*, **87**, 4576–9

CHAPTER 5

PATTERNS AND REGULATION OF PLANT DIVERSITY IN LACUSTRINE ECOTONES

Raphaëlle Juge and Jean-Bernard Lachavanne

INTRODUCTION

Land–inland water ecotones, which are considered to be particularly productive and to be sites of very active structural and functional dynamics, exhibit a great heterogeneity of life conditions at various spatio-temporal scales (Hansen *et al.*, 1988; Décamps and Naiman, 1990) and are therefore capable of sheltering a great diversity of life-forms.

Higher plants constitute a biological response that is particularly demonstrative of the diversity of the habitat (Sculthorpe, 1967; Hutchinson, 1975; Pieczyńska, 1976; Hejny and Hroudová, 1987; Duarte *et al.*, 1994, etc.). Indeed, various development and adaptation strategies (ecological plasticity) enable them to exploit the environment's resources in a highly differentiated manner (Sculthorpe, 1967). The result is the well-known structural characteristics (physiognomy of communities) of vegetation in land–inland water ecotones: zonation of the various growth forms in relation to hygrometry and hydrometry (Luther, 1949, 1951; Hejny, 1960; Den Hartog and Segal, 1964; Segal, 1971; Den Hartog and Van der Velde, 1988). On a finer scale, the mode of plant distribution (features, architecture) within large formations is modulated by the variability of the micro-environment (microtopography, nutrient availability of substrate and water, etc.) under the influence of the seasonal climate (temperature, light) and by the biotic relations (especially competition).

As the expression of the diversity of environmental conditions, vegetation is also a factor of diversity, since it is the biotic component that contributes the most – through its physiognomy, its composition in terms of species and biological types, its biomass, the surface it occupies, its location and distribution (patchiness) – to the diversification of habitats for microflora (Wetzel, 1975) and fauna (Cody, 1975; Pieczyńska, 1976, 1990; Tonn and Magnuson, 1982).

The 'ecotone' and 'biodiversity' concepts bring immediately to mind what Leopold (1933) called the 'edge effect', namely the tendency of plant communities in the ecotone to be more diversified and denser than in the adjacent ecosystems (Odum, 1971). The whole issue of the edge effect is very much debated, particularly in the case of land–inland water ecotones, which are systems that are neither stable nor mature and where very important and sometimes sudden changes in water level can produce the opposite effect (Van der Maarel, 1976).

Another phenomenon which is related to the ecotone concept and its plant diversity is the plant communities succeeding under the effect of allogenous and autogenous forces (Clements, 1916; Connell and Slatyer, 1977; Van der Valk, 1981, 1982; Mitsch and Gosselink, 1986, etc.). However, the mechanisms involved here have neither been quantified nor fully understood.

Furthermore, at the regional scale, it is accepted that maximal diversity occurs at a certain level of heterogeneity of habitat (Noss, 1983) and that the viability of populations is closely linked to the dynamics of all populations in the mosaic of the environment (Pulliam, 1988).

Thus the special dynamics of ecotones, and that of wetlands in particular, have provided a great source of interest among scientists for several decades and have given rise to many concepts of fundamental ecology, especially in connection with the various expressions of biodiversity (Clements, 1916; Leopold, 1933; Margalef, 1968; Odum, 1971, 1979; Greeson *et al.*, 1979; Sousa, 1979; Wilcox, 1984; Mitsch and Gosselink, 1986; Wilson, 1988; Di Castri and Younès, 1990; Gosselink and Maltby, 1990; Odum, 1990, etc.). Our examples and concepts refer mainly to circumlacustrine ecotones in temperate regions, therefore any transposition to other types of ecotone requires adaptation. They also describe only the diversity of higher plants (Spermatophyta), whose variability between terrestrial and aquatic ecosystems, as well as within the ecotonal zone, reflects particularly well the effects of ecological gradients and fluctuations of life conditions. It should be noted, however, that inclusion of the Bryophyta and Thallophyta (especially algae) would demonstrate that the aquatic system offers an even greater potential for diversity.

Our aim is therefore to describe some relevant data and concepts susceptible to documenting the key issues of plant diversity in ecotones:

– What is the extension of vegetation diversity, in space and time?

– How are floristic and structural diversities of plant communities regulated in ecotone-specific life conditions?

– Under what conditions is diversity greater in ecotones than in related ecosystems (the 'edge effect' theory, Leopold, 1933)?

– What ecological roles does plant diversity play within the ecotone?

VEGETATION: A KEY TO ECOTONE COMPLEXITY

A multi-scale and multi-criteria approach to the vegetation permits one to account for the complexity and heterogeneity of ecotones because the composition and distribution of plant communities present in and around lakes and ponds constitute particularly relevant global criteria for characterizing and delimiting the circumlacustrine ecotone (Lachavanne and Juge, in preparation).

Indeed, the structure, the diversity and the speed of reaction of plant communities can be considered as highly significant indicators of ecotone spatio-temporal organization, functioning and dynamics. In fact, these plant community features result from the combined action in space and time (on the scale of seasons or decades, etc.) of the ecological factors (especially the hydric one) and of the perturbation regimes.

Vegetation, being a particularly sensitive indicator of the degree of integrity of the structure and functioning of an ecosystem as well as of the direction of its evolution, can also reveal past dysfunctions and anomalies in ecotone dynamics in the short and long terms. It can hence serve as a memory of prior climatic or topographic accidents.

PATTERNS OF PLANT DIVERSITY IN THE CIRCUMLACUSTRINE ECOTONE

The pattern of plant diversity in an ecosystem is the product of variable interactive forces (Solbrig, 1991); it differs from one type of ecosystem to another and among ecosystems of the same type, depending on their age, degrees of stability and functional integrity. The same is true of ecotones whose vegetation exhibits diversities in terms of physiognomic structure and specific composition that are dependent upon autogenic and allogenic forces dominated primarily by the impact of variations in hydrology.

Criteria for determination of plant diversity

The perception of the structure and diversity of plant communities and of their variability depends on the criteria under consideration and on the spatial and temporal scales within which the analysis is conducted. These are determined by the scope of the study and the questions asked or the hypotheses to be tested.

Thus in the case of land–inland water ecotones, one can observe significant differences in floristic diversity, depending on the functional and spatial definitions given to the term ecotone (see Lachavanne, Chapter 1). We can either consider the entire transition zone between terrestrial and limnetic ecosystems, in which case biological diversity tends to be high owing particularly to the great variety of abiotic conditions generated by the hydrological gradient and its variations; or, on the other hand, we can limit the ecotone system to the interface zone (eulittoral according to Hutchinson, 1975), as advocated by Van der Maarel (1990), which would correspond at least functionally to the definition that Di Castri and Hansen (1992) give to the ecotone with regard to the ecocline: 'abrupt change in spatial patterns leading to non-linear behaviour'. In that case diversity is rather low in respect to the rigour of the environmental conditions (strong hydraulic constraints). Thus it is mandatory that a clear spatial or ecosystemic delimitation should be stated before conducting any comparative analysis.

Diversity can be described at all the organizational levels, from the chemical, molecular or genetic level to the individual organisms and to the levels of populations, life-forms, communities, ecosystems, landscapes and the biosphere (Di Castri and Younès, 1990; Solbrig, 1991; Hansen et al., 1992). The analysis of diversity is related to genetical and phenotypical aspects within species and populations and with floristic (α and β) or architectural aspects within plant communities (Solbrig, 1991). In this way, by applying the hierarchy theory to the vegetation (see Van der Maarel, 1976; Allen and Starr, 1982; O'Neill et al., 1986; Urban et al., 1987; Farmer and Adams, 1991), each of the integration levels of the biological systems reveals distinct properties (e.g. patterns of diversity) that vary differently in space and time. What may appear to be homogeneous at a given spatial and temporal level of observation is generally not the case on a more sensitive scale.

It is worthwhile noticing that the floristic richness observed at a given time on a site represents only an apparent diversity compared to all the potential developments that are not expressed but exist in the form of longstanding organs. Indeed, many plants subsist in the soil in various dormancy forms (seeds, rhizomes, etc.). There are species present in the past (e.g. pioneering plants) which are capable of developing when the environmental conditions become favourable; they represent a potential diversity that should be taken into consideration when evaluating the actual diversity.

Types of diversity

The criteria selected to measure diversity depend on the particular function one wishes to highlight. In this way, three types of diversity can be globally distinguished.

Taxonomic diversity. Floristic richness and diversity reveal both the coexistence of competing species reflecting the degree of availability of, and accessibility to, resources and the coexistence of species belonging to more or less overlapping ecological niches reducing the degree of variety and quality of life conditions.

Diversity of life strategies. The diversity of life-forms and growth forms reveals plants' abilities to develop diversified strategies to exploit the resources (space, matter, energy) available for their development, reproduction and propagation under different regimes of environmental conditions (Sculthorpe, 1967; Grime, 1979). Raunkiaer's classification (1934, 1937) highlights, for all the higher plants, species diversity within life-forms as a function of the environmental constraints, e.g. the poorness of hydrophyte and helophyte species compared to terrestrial life-forms. Among the hydrophytes, Den Hartog and Van der Velde (1988) distinguish 14 growth forms of rhizophytes and four of pleustophytes (Figure 5.1). This plant variety in terms of plants' space occupation strategy shows that, although difficult life conditions connected with hydric constraints tend to reduce diversity of species, they can, on the other hand, express themselves through a great variety of growth forms (Sculthorpe, 1967; Murphy et al., 1990; Duarte and Roff, 1991) and modes of propagation (Bartley and Spence, 1987).

Diversity is enhanced by the specially high adaptation ability of plants which exhibit great ecological plasticity (MacCreary et al., 1983; MacCreary and Carpenter, 1987; MacCreary, 1991): there are adaptations that are morphological (Duarte et al., 1994), anatomical (macro- or microphytization), physiological (growth rate and production I, dormancy) (Howard-Williams et al., 1995) or ontogenic.

Many authors have demonstrated that changing and constraining life conditions of palustrine and especially aquatic environments lead to a great diversity of development strategies and even to the ability to change them during the life cycle (Hejny, 1957, 1971; Spence et al., 1973; Hutchinson, 1975; Carignan and Kalff, 1980; Hroudová, 1980; Spence, 1982; Luther, 1983; Chambers and Kalff, 1985, 1987; Mitsch and Gosselink, 1986; Chambers, 1987; Duarte and Kalff, 1987; Hejny and Hroudová, 1987; Duarte and Roff, 1991; Madsen, 1991). According to Farmer and Adams (1989), any species exhibiting great genetic diversity points to the existence of different physiological races within this species and differential tolerance of chemical or biological regulators. This appears to be true for certain aquatic and palustrine plants. Yet, the reproduction function of aquatic plants is predominantly vegetative (Sculthorpe, 1967; Kadlec and Wentz, 1974; Van der Valk, 1987), and could have implications for these plants' adaptation potential by weakening genetic diversity (Wain et al., 1985; Farmer and Adams, 1989).

It is likely that the variety of morphological types observable in certain species corresponds more often to accommodations than to ecotypes. By developing very different phenotypes depending on the hydric conditions of its implantation site (whether terrestrial, palustrine or aquatic). *Sagittaria sagittifolia*, for example, illustrates well the ecological plasticity that characterizes certain species of plants in land–inland water ecotones.

Structural diversity: physiognomic or architectural (communities). Diversity and structural complexity (mosaic) of communities result from the variety of shape and spatial pattern of individual plant assemblages, and hence from the variety of their growth strategies (Murphy et al., 1990) linked to the plants' growth forms (Sculthorpe, 1967; Hutchinson, 1975; Spence, 1982; Luther, 1983; Chambers and Kalff, 1985, 1987; Chambers, 1987; Duarte and Kalff,

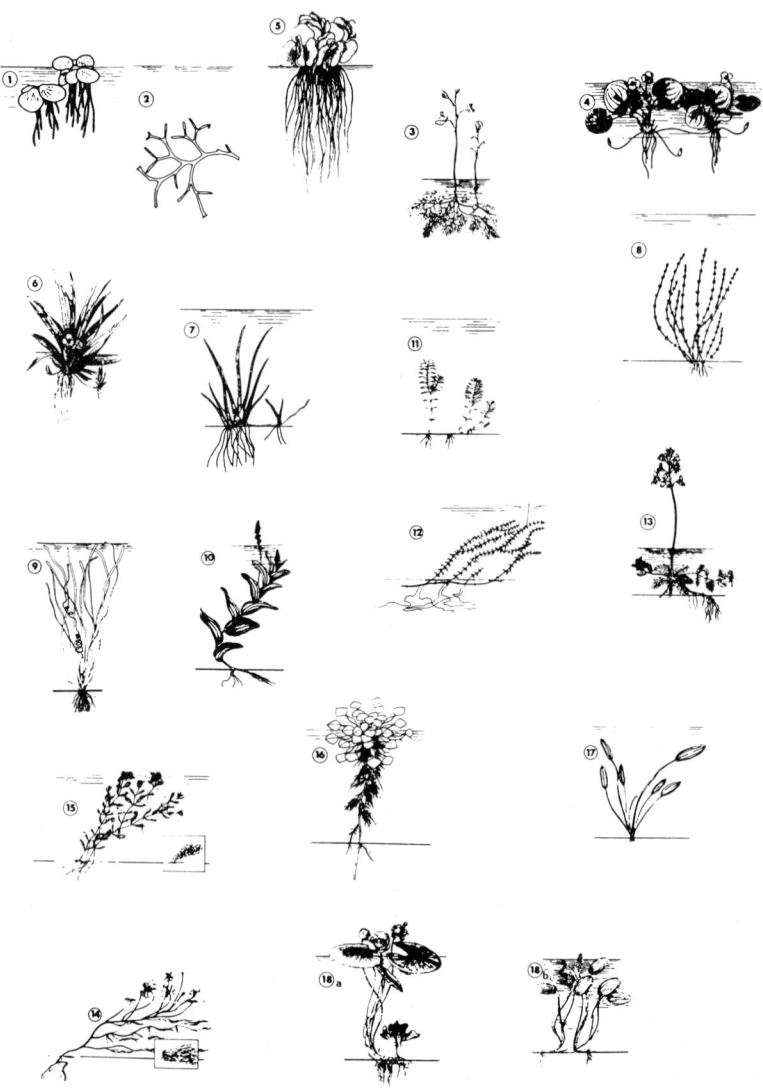

Figure 5.1 Growth forms of hydrophytes and modes of implantation in and colonization of the habitat (according to the classification of Den Hartog and Van der Velde, 1988). Drawings (some redrawn) taken from: Cook *et al.* (1974) (1,2,5,9,11,16), Dethioux (1989) (12,14,15), Montegut (1987) (7,17), Bursche (1971) (3,4,6,10,13,18a,18b), Fasset (1968) (8). (1) *Spirodella polyrhiza* (L.) Schleid, (2) *Riccia fluitans* L., (3) *Utricularia vulgaris* L., (4) *Hydrocharis morsus-ranae* L., (5) *Salvinia molesta* Mitchell, (6) *Stratiotes aloïdes* L., (7) *Littorella uniflora* (L.) Ascherson, (8) *Chara aspera* L., (9) *Vallisneria spiralis* L., (10) *Potamogeton perfoliatus* L., (11) *Groenlandia densa* (L.) Fourr., (12) *Elodea canadensis* Michaux, (13) *Hottonia palustris* L., (14) *Ranunculus penicillatus* (Dum.) Bab., (15) *Callitriche stagnalis* Scop., (16) *Trapa natans* L., (17) *Aponogeton distyachos* L. fil., (18a) *Nuphar lutea* (L.) Smith, (18b) *Potamogeton natans* L. (from Lachavanne *et al.*, 1995)

Plant diversity in lacustrine ecotones

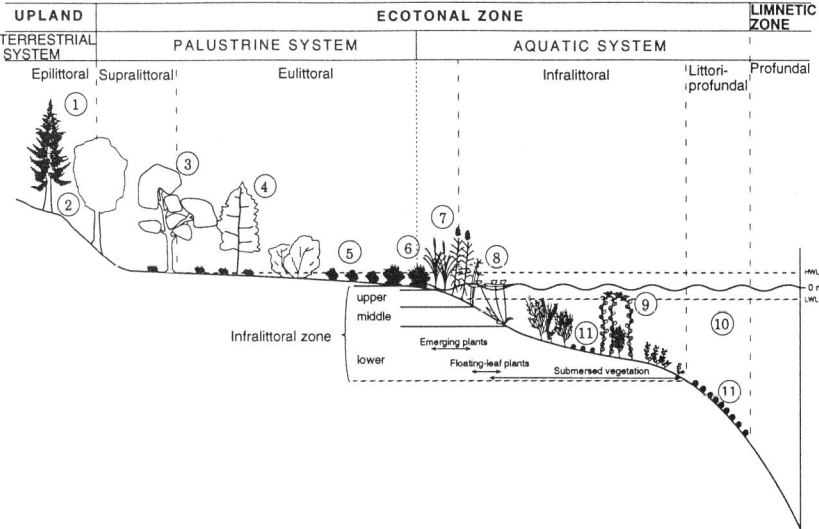

Figure 5.2 Zonation of circumlacustrine ecotone vegetation. Formations: (1) terrestrial ligneous, (2) terrestrial herbaceous, (3) wetland hardwood (e.g. *Fraxinus* sp. group), (4) wetland softwood (birch, poplar, alder and willow groups), (5) short sedges (*Orchio-Schoenetum nigricans*, various *Molinion, Ranunculo-Caricetum hostianae* groups, etc.), (6) tall sedges (e.g. *Carex elata, Cladium mariscus, Phalaris arundinacea* groups), (7) emerging plants (*Phragmites australis, Typha* sp., *Scirpus lacustris* groups), (8) floating-leaf plants (*Nuphar, Nymphaea* groups), (9) submersed plants (*Potamogeton, Myriophyllum* groups, etc.), (10) phyto- and bacterioplankton, (11) epibenthic phytobiota (based on Hutchinson, 1975, modified). The littoral zone (ecotone) is composed of the supralittoral, the eulittoral, the infralittoral and the littori-profundal

1987, 1990; Kautsky, 1988; Den Hartog and Van der Velde, 1988, quoted by Duarte and Roff, 1991).

Zonation and patchiness of vegetation in ecotones

Conventionally, the vegetation of circumlacustrine ecotones is represented as more or less spread-out concentric belts parallel to the shoreline and reflecting the biological types adapted to gradients (gradients of increasing humidity and further off gradients of flooding) that characterize the land–water transition zone (Luther, 1951; Hejny, 1960; Den Hartog and Segal, 1964; Segal, 1971, 1982; Den Hartog and Van der Velde, 1988; see also Daubenmire, 1968 for the different modes of transition between plant communities). The zonation of the circumlacustrine ecotone presented in Figure 5.2 (Lachavanne and Juge, in preparation) corresponds in part to that of the littoral zone described by Hutchinson (1975) on the basis of criteria related to hydrology and vegetation.

Infralittoral. The higher intra-aquatic plants of the infralittoral, with which charophytes are generally associated, include plants that are rooted and free, entirely submersed, with submersed and floating organs or emerged with floating leaves. These various communities exhibit a generally low floristic richness because their dynamics are conditioned predominantly by abiotic factors independent of population density and also because the pool of species likely to colonize the lake beds is very modest, due to the relatively homogeneous (isotropy of the aqueous environment) and very constraining life conditions. Higher plant organisms are indeed morphologically and physiologically ill-adapted to aquatic life conditions because of constraints of a physical nature like hydrostatic pressure, hydraulic forces deriving from currents or waves and texture of the substrate, and because of constraints of a metabolic nature like reduction of the quantity of light energy available and modification of the quality of the spectrum permitting photosynthesis, etc.

Eulittoral and land–water interface. The eulittoral can be defined as the interface or contact zone between aquatic and terrestrial environments, alternately flooded or emerged and confined between the limits of high and low waters. Its extent depends on the shore slope and on the amplitude of water-level variations.

Despite the presence of a hydric gradient, a clear geographical and ecological discontinuity (shoreline) separates the aquatic and terrestrial ecosystems within the eulittoral. Here, the mechanical action of waves is most strongly exerted. In that way the land–water interface is often characterized by a set of environmental conditions with strong and frequent fluctuations which constitute the culminating point of the interactions between neighbouring ecosystems. Few species colonize this part of the eulittoral in an extensive manner, and to a lesser extent if the water-level fluctuations are very frequent, have a too high amplitude or take place during crucial moments of plant development, and if other environmental constraints occur (erosion, sediment instability, see Mitsch and Gosselink, 1986).

The stands of emergent plants which colonize the shoreline tend to be monospecific and are dominated by one or several species, such as: *Phragmites australis* (Cav.) Trin., *Scirpus lacustris* L., *Typha latifolia* L. or *Carex elata* All. They are generally characterized by a dominance of vegetative propagation, leading to a clear competitive advantage.

Supralittoral. Diversity and richness tend to increase in the more emerged parts (Juge *et al.*, in preparation). Zonation is more irregular and the patchiness of communities reflects very active dynamics and a more pronounced spatio-temporal heterogeneity of environmental conditions. In addition, a clear vertical stratification appears with bushes being added to the herb layer. These features constitute elements of diversification of habitats rather favourable to great floristic richness and diversity (α and β).

The limit of the epilittoral zone is indicated by tree formation, where the root system reaches the groundwater. This limit given by the trees at the boundary of low marshes is often modified by local hydrological conditions such as surface runoff, groundwater circulation, springs, or by geological features such as molasse outcroppings (Buttler *et al.,* 1985; Buttler and Gobat, 1991).

The low marshes of the land–water ecotones constitute a highly productive environment which therefore generally presents a very high diversity of species and life-forms (Odum, 1971; Gosselink and Maltby, 1990).

The example presented in Figure 5.3 (Juge *et al.*, in preparation) demonstrates the existence of an edge effect at the level of the wetland as a whole. In addition, diversity tends, in overall terms, to increase along the land–water ecological gradient, but in a non-linear manner. It should be noted that there are also in natural (cliffs, screes, zones of intense erosion) or in man-modified conditions, incomplete plant zonations presenting a potential physiognomic, phytosociological and taxonomic diversity lower than that observed in natural flat shores (as shown in Figure 5.2 and Figure 5.3).

Vegetation dynamics: ecotone instability and plant succession. Lacustrine ecotones permanently exhibit progressive and regressive plant successions under the combined influence of allogenic and autogenic forces (Watt, 1947; Mitsch and Gosselink, 1986; Odum, 1993). These dynamics give rise to a large variety of habitats which form the basis of great plant and animal diversity (Pieczyńska, 1975, 1990; Naiman and Décamps, 1991; Pieczyńska and Zalewski, Chapter 3).

The natural process of lake infilling by sediments takes place in particular at the land–water level which presents in that way a certain instability. According to Wilson's predictive model (1969), there is an intensification in species interactions. Those species that are capable of coexisting and adapting to the system see the numbers of their populations increase as the system evolves (Zwölfer, 1987). This is true, for example, in wetlands and ripisylvas with communities exhibiting intermediate stability between aquatic and terrestrial systems.

According to Clements (1916), the zoning of plant groups along the hydric gradient also corresponds to the different stages of the succession of plant formations in time (terrestrialization). The successions are therefore partly consecutive to the changes in hydrological conditions, particularly to their timing (Margalef, 1968; Gosselink and Turner, 1978; Von Schwaar, 1989). Clements' interpretation is contested by Tutin (1941), Odum (1971), Mitsch and Gosselink (1986), Van der Valk (1987), Rickleffs (1990) and others, for whom zoning does not necessarily indicate succession; the vegetation of wetlands is adapted to the environmental dynamics specific to these systems. This means that palustrine vegetation can form climax communities (Tutin, 1941; Van der Valk, 1987). Often the water-level variations, as a periodic perturbation, keep vegetation in a state of 'pulse stabilized subclimax' (Odum, 1971), prevent plant communities from reaching an equilibrium and permit

Biodiversity in land–inland water ecotones

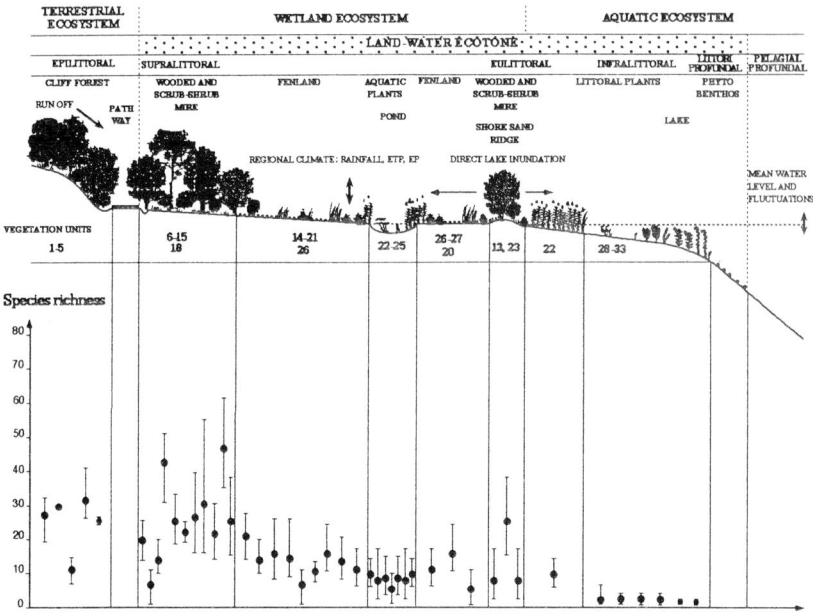

Figure 5.3 Variation of richness in species and plant associations along the hydrological gradient exemplified by the southern shore of Lake Neuchâtel (from Juge *et al.*, in preparation, modified)

the coexistence of antagonistic species (Connell, 1978; Sousa, 1979; Miller, 1982; Silvertown and Law, 1987). Odum (1993) points out, however, that a stable intermediate climax state can only be maintained if the entire community (plants, animals and microorganisms) is adapted to the fluctuating conditions that are imposed.

For Van der Valk (1987), changes in specific composition (Gleasionian succession, Van der Valk, 1981), in size (maturing and accumulation of biomass that are typical of forested wetland) and in age-structure/population-size (reversible, due to fluctuations in environmental conditions) depend on species life histories (i.e. demographic strategies). The various environmental factors modulate the conditions of seed germination, propagule dispersion and clonal growth which govern these changes of vegetation. According to Connell and Slatyer (1977) and to Rickleffs (1990), facilitation is the more common feature of secondary succession and may be expressed in priority effects, conferring competitive dominance on the first arrival.

Conventional theory attributes a greater stability to the most diversified systems (Margalef, 1969; Odum, 1969) since, according to Odum (1971), spatial heterogeneity, specific diversity, biomass and complexity in the organization of communities increase simultaneously during successions. On the other hand, May (1973) and Orians (1975) observe that introducing an increase in diversity in mathematical models of ecosystems causes a reduction of the systems' stability.

Whittaker (1975) suggests that at intermediate stages of succession, vegetation may be more diversified than in the advanced stages. Our observations on the southern shore of Lake Neuchâtel seem to confirm this (Juge *et al.*, in preparation). According to Rickleffs (1990), we do not really know whether the increase in diversity along ecological successions depends more on an increase in production, on environmental stability or on habitat heterogeneity (although it appears obvious that the three phenomena are difficult to separate under natural conditions).

Therefore, plant diversity does not necessarily increase during successions, rather certain stages of this evolutive process may represent a relative poverty in plant communities, particularly in relation to unfavourable or greatly fluctuating environmental conditions. This is the case, for example, of aquatic reed beds: they are situated in the zone of high tension between terrestrial and aquatic systems, a zone which does not exhibit a real tendency to evolve towards a climax.

The processes of sedimentation and terrestrialization which make up the succession of plant communities to form part of an evolutive series are not the only factors responsible for the transformation of the plant landscape of the ecotone. Conversely, the phenomenon of shore erosion due to the action of currents, waves and wind, draws plant communities into a regressive series, the speed of which can be amplified by other factors such as eutrophication.

Can the balance of the antagonistic action of equivalent physical forces lead to a stasis of the environmental conditions whose consequence would be the maintenance of adapted communities? This is what Odum (1971) refers to as the 'pulse stabilized subclimax'. Although the author refutes the possibility of this phenomenon creating violent interactions, it appears that very exposed aquatic reed beds do conform to such characteristics. It is merely a question of identifying the time scale during which this situation can last.

Indeed, in view of the imbalance between the inputs and outputs of matter in lake systems (tendency to trap the matter brought by tributaries), one may expect the situation to evolve endlessly at the geological scale, while the lake is infilled by sediments.

REGULATION OF PLANT DIVERSITY IN ECOTONES

Plant diversity results from: (i) the interaction of all internal and external ecological factors to the system under consideration, (ii) the pool size of the available species, as well as (iii) the development capabilities and (iv) the functional and adaptative peculiarities specific to each species. The action and effect of the ecological factors may themselves be modulated by human activities.

According to Gosselink and Maltby (1990), the wetland must be considered as an open system severely exposed to the effects of upstream inputs (nutrients, sediments, water, organisms) and influencing in turn the downstream aquatic system, after having functioned as a semi-permeable barrier for ecological flows of matter, energy and organisms between the terrestrial and aquatic ecosystems (Naiman et al., 1988).

However, Whittaker (1977) stresses the difficulty of determining the regulating factors specific to plant diversity as well as their respective roles. He attributes the reasons for the difficulty to the fact that the size of the populations and the reproduction rate are not determined precisely in plants. The chances of colonizing a habitat and the seed dispersal remain hazardous. In addition, it is difficult to distinguish the modes of utilization and distribution of the resources within the different species, partly because all plants share similar light, water and nutrient requirements and also because of the difficulty of evaluating some processes such as the uptake of underground resources.

Modes of regulation of plant diversity more or less specifically related to land–water ecotones

Dependency on abiotic factors is very important in pioneering and unstable ecosystems and diminishes in stable ones in favour of a dependency on the variety, relative intensity and temporal variability of biotic relations (Horn, 1976). It is therefore mainly allogenic factors which regulate the changes of communities in systems that undergo great variations in water level and mechanical wave action, whereas autogenic succession is more significant in those where the water level is more or less stable (Gopal and Masing, 1990). Table 5.1 shows the factors that regulate plant diversity. The most important of them are presented below.

The key role of hydrology. Numerous authors emphasize the effect of hydrology on the composition, structure and distribution of plant communities in wetlands, alluvial zones and lacustrine littoral zones (Hejny, 1960, 1971; Robel, 1962; Walker and Coupland, 1968; Ager and Kerce, 1970; Heinselman, 1970; Segal, 1971; Pieczyńska, 1975, 1990; Gosselink and Turner, 1978; Robertson et al., 1978; Weller, 1978; Bedinger, 1979; Mitsch et al., 1979; Mitsch and Ewel, 1979; Van der Valk and Davis, 1980; Brown, 1981; MacKnight et al., 1981; Clayton, 1982; Leveque and Quensiere, 1983; Mitsch and Rust, 1984; Shugart, 1984; Mitchell and Rogers, 1985; Mitsch and Gosselink, 1986; Brock et al., 1987; Bren, 1988; Wetzel et al., 1989; Burgess et al., 1990; Gasith and Gaafny, 1990; Crowder and Painter, 1991; Wilcox and Meeker, 1991; Bueche et al., personal communication).

The degree of humidity and flooding of the soil and, farther off, the depth of water, vary as a function of the overall slope of the site and of the local

Table 5.1 Main factors regulating aquatic and palustrine plant diversity

NATURAL REGULATING FACTORS (AUTOGENIC AND ALLOGENIC)

- **Abiotic factors (density-independent)**

Climatology:
- type and variability of the climate (sunshine hours, annual and daily variations of air temperature, degree of air humidity, wind, precipitation) depending on latitude and altitude.

The substrate

Geological and pedological characteristics of the lake bed and catchment basin:
- nature (inorganic and organic chemical contents), quality (humus, salinity, degree of anoxia) and texture (granulometry, degree of aeration) of the soil/sediments;
- accumulation of organic matter (plant biomass, litter);
- degree of spatial heterogeneity and mode of distribution of this heterogeneity (substrate patchiness);
- erosion, sedimentation and aggradation (allogenic inputs of material by surface water, alluvia, rock and by subterranean water).

Morphology and topography:
- slope, orientation, exposure, form and configuration of the shore and the littoral bank: therefore extent of the ecotone (surface and volume);
- development of shores, organization and configuration of lake beds, degree of morphological and topographical heterogeneity (transverse and longitudinal) of the soil/sediments.

The water

Hydrology:
- frequency, duration and amplitude of water-level variations (dependency on the water regime of the catchment basin and of the shore slope);
- degree of humidity of the soil;
- fetch, wave action, exposure to prevailing winds, currents, seiches;
- hydrostatic pressure, water depth.

Physico-chemical characteristics of the water:

Quality and quantity of resources and degree of heterogeneity of their spatial distribution:
- mineral chemical composition and organic content (nutrients, trophic level, carbon dioxide, etc.);
- dissolved oxygen content;
- alkalinity/acidity;
- water and soil temperatures;
- availability of light (spectrum and intensity), therefore transparency/turbidity of the water (trophic level), degree of shading;
- allogenic inputs of nutrients *via* air and water.

- **Biotic factors (density-dependent)**

Vital attributes (genotype and phenotype) of species (influencing their competitiveness, i.e. their solutions for survival and exploitation of resources):
- speed and intensity of metabolic processes (photosynthesis, translocation of nutrients, growth and development);
- length of life cycle;

continued

Table 5.1 continued

- ecological tolerance spectrum;
- modes and speed of reproduction (vegetative, pollination [helps dissemination and reproduction], dormancy, forms of seed resistance, etc.);
- space occupancy strategies (growth forms, mode and capability of development and dispersion);
- capability and strategies of adaptation and accommodation (biochemical, physiological, histological and morphological);
- colonization capability of the ecotone and related ecosystems (species' ability to spill over from the geographical zone of the ecotone);
- degree of proneness to mutations, speciation, extinction.

Intraspecific relations:

- effects of density and competition.

Interspecific relations:

- trophic relations (grazing, predation, decomposition, detritivorous activity);
- competition for vital resources (space, energy, material);
- mutualism, co-operation;
- supporting epiphytes and microbiological activity in water and sediments/soil, particularly that of detritivorous (micro)organisms (decomposition of plant organic matter and bioavailability of nutritive elements);
- pathologic relations (allelopathy, antibiosis, parasitism, bacterial and viral infections);
- activity of living organisms as vector of inputs and exports (dissemination *via* air, water) of pollen, seeds, propagules, pathogenic elements and participation in reproduction.

Perturbations: natural cataclysms (fortuitous events that intervene by modifying the characteristics of the environment and/or the biological characteristics):

- strong change in water level, floods, swelling, fires, avalanches, landslides, earthquakes, eruptions, cyclones, invasions, epidemics.

MAN-MADE REGULATING FACTORS

- **Perturbations (acting on plants directly and/or *via* modifications of the environment's characteristics):**

- eutrophication of water (*versus* oligotrophication);
- pollution (pesticides, mutagenic substances, etc.);
- acidification;
- water-level regulation;
- development (constructions, embankments, harbours, etc.);
- exploitation of littoral bank (gravel dredging/ditch digging);
- trampling (man, domestic and breeding animals);
- floating wood and waste (mechanical action);
- anchoring of boats (mooring zones);
- parking of boats, constructions, access roads, etc. (in the reed beds, marshes, for example);
- exploitation of resources (reed belts, fibres, cellulose, plantations, rice fields, etc.);
- introduction of plant and animal species;
- marsh drainage for agriculture.

These perturbations lead to fragmentation, impairment and/or elimination of habitats.

- **Management**

- creation, enlargement and maintenance of biotopes;
- restoration of impaired sites;

Continued

Table 5.1 continued
- returning built-up land to nature;
- clearing of weeds;
- isolation of sites (natural protection, buffer zones with exploitation regulations, planning);
- passive and active species protection (feeding, building of shelters or protective sites, care);
- population regulation and fight against invading species (mechanical or chemical control of undesirable species);
- re-introduction of species that had died out;
- exploitation of resources;
- cleaning up of water;
- etc.

The allogenic factors derive from the related ecosystems (surface and subterranean terrestrial, aerial, aquatic, benthic), the catchment basin and various origins at the scale of the region, landscape or planet.

---

topography. The irregularity of these two parameters, in addition to water quality (nutrient content and suspended matter) and the variability of the substrate determine the degree of heterogeneity and the mosaic of the habitat which condition the distribution and availability of resources (energy and matter). Therefore the greater is the complexity of the spatial heterogeneity, the greater is the number of different habitats and the greater is the opportunity, for each species, to complete a successful colonization (Jacobs, 1975).

Hydrodynamics and the hydrological gradient appear to be the key factors in the regulation of plant diversity in the ecotonal zone, as they condition all the other abiotic and biotic parameters. The frequency, the duration, the moment and, to a certain extent, the amplitude of water-level variations are determinants and produce a temporal heterogeneity of life conditions added to the basic spatial heterogeneity.

Hydrology conditions, primarily the distribution and diversity of plant communities, either by directly influencing plant metabolism or through the modifications produced to all the characteristics of the habitat which its dynamics inevitably causes (Wilson and Keddy, 1985, 1988; Wisheu and Keddy, 1989).

The various patterns of species and life-form diversity found in different parts of the ecotonal zone clearly illustrate the relative importance of each regulation factor, depending partly on the ecotone and partly on the community under consideration. For example, submersed plant communities are primarily sensitive to the availability of light, whose variable intensity and radiation spectrum determine the depth of colonization (Chambers and Kalff, 1987; Chambers and Prepas, 1988; Duarte and Kalff, 1988) and floristic composition. It should be stressed that the degree of illumination may also be impaired by water-level variations (Howard-Williams *et al.*, 1995) and by the particles in suspension (inorganic particles, phytoplankton). Hydrostatic pressure and the quality of the substrate (composition, texture, fluidity) also exert a strong influence on subaquatic plant populations. On the other hand, reed belts are mainly dependent on water-level variations and on the mechanical action of waves.

The hydrological regime may lead either to uniformity or to diversity, depending on the regime of a specific wetland. Whether or not diversity increases with time is often determined by whether or not the biocoenosis that develops has a reciprocal effect on the hydrological regime (Gorham, 1957).

On one hand, flood waters provide a vehicle for moving dissolved or suspended matter. This may minimize spatial diversity and so produce a uniform mixture. Therefore wetlands subject to sheet flow by flood waters tend to be rather uniform and have large monospecific or poor species composition. On the other hand, by modifying the microtopography and the conditions of oxygenation and by contributing to the quality of the substrate and the (also temporal) variability of its texture and chemical composition (content of nutrients and organic matter), the hydrological regime is the prime source of specific diversity in wetlands (Hinde, 1954). Heinselman's data (1970) indicate for example, that plant species richness increases as the water flow rate (and probably the rate of renewal) increases in the peat bogs of northern Minnesota.

Wilson and Keddy (1988) and Wilcox and Meeker (1991) note that natural water-level fluctuations can be considered as an 'intermediate disturbance' in Connell's sense (1978) and can thus promote floristic diversity (Keddy, 1983). Waterlogged soils that are perpetually being moved around impose very harsh and constraining conditions on plants. A greatly reduced substrate and anoxic conditions particularly modify the conditions of nutrient uptake by plants. This requires morphological, anatomical and physiological adaptations from the plants, or even changes in their whole life cycle (Hejny and Hroudová, 1987). Only a limited number of species are capable of accomplishing such changes (Wilson and Keddy, 1985). Indeed, an anoxic environment constitutes a stress factor and growth is generally reduced by markedly anaerobic conditions (Patrick and Mikkelsen, 1971; Armstrong, 1975).

The most selective life conditions prevail at the land–water interface (eulittoral) characterized by a fluctuation between terrestrial and aquatic conditions, and in the infralittoral, where life conditions deteriorate with increasing depth of water. On the other hand, hydrological dynamics tend to be rather favourable to specific diversity within all eu- and supra-littoral palustrine zones. Under such conditions, we observe a certain variation in degrees and forms of constraints from the supralittoral to the infralittoral.

Gosselink and Turner (1978) point out that the characteristics of the hydrological regime determinant for phytocoenosis are: (i) the origin of the water, which determines the ionic composition, oxygen saturation and toxin load, (ii) the speed of currents which modifies turbulence and the water's ability to carry particles in suspension, (iii) the rate of water renewal which depends on the depth and the speed of flooding and, lastly (iv) time – i.e. the frequency of flooding (daily, seasonal) and its regularity or predictability – which influences the system's potential for succession and maturation (Margalef, 1968).

Biotic interrelations as factors of regulation for the structure of plant communities and their diversity. A great variety of mechanisms intervene in the regulation of population sizes – thus in their biodiversity – within the ecosystem when they are not uniquely dependent on overly strong environmental constraints (density-dependent regulation).

In land–inland water ecosystems let us mention, for example, grazing on aquatic plants by crayfishes, fishes, birds and musk rats, whose effects, which are minimized by Gregory (1983) and Wetzel (1983), turn out to be non-negligible for other authors (Kvet and Hudec, 1971; Pelikan *et al.*, 1971; Anderson and Low, 1976; Danell, 1977, Jupp and Spence, 1977; Carpenter, 1980; Kiorboe, 1980; Prejs, 1984; Sheldon, 1984, 1987; Smith and Kadlec, 1985; Carpenter and Lodge, 1986; Lodge and Lorman, 1987; Lodge *et al.*, 1988; Chambers *et al.*, 1990; Chambers *et al.*, 1991; Lodge, 1991; Olsen *et al.*, 1991).

Interspecific competition between plants for light, nutrient and space resources, etc. is also mentioned by many authors (Daubenmire, 1968; Harman, 1974; Titus and Adams, 1979; Grace and Wetzel, 1981; MacCreary *et al.*, 1983; Carpenter and Titus, 1984; Keddy, 1984; Moss, 1988; Chambers and Prepas, 1990; MacCreary, 1991; Sand-Jensen and Borum, 1991). The relative contribution of competition and allelopathy to the regulation of aquatic and palustrine plant communities is still a matter of debate and Bazzaz (1979) and Wetzel (1983) express doubts concerning their importance. Other authors tend to think that such a mechanism is possible or even operative (MacNaughton, 1968; Szczepanska, 1971, 1977, 1987; Szczepanski, 1971; Wium-Andersen *et al.*, 1982, 1983; Grace, 1983; Wium-Andersen, 1987; Jeffries and Mills, 1990).

Eutrophication, a perturbation factor of vegetation. Eutrophication of wetlands and of the water of the lacustrine littoral zone is one of the main factors of vegetation perturbation in land–inland water ecotones. The consequences of the process for aquatic vegetation, for instance, provide a good illustration of the theory of intermediate perturbation which states that a periodic or a gradual, continuous perturbation (as is the case with the water eutrophication phenomenon) prevents communities from reaching a balance and allows the coexistence of antagonistic species (Connell, 1978; Sousa, 1979; Miller, 1982; Silvertown and Law, 1987). Many authors (Seddon, 1972; Best, 1981; Lachavanne, 1982, 1985; Roelofs, 1983; Best *et al.*, 1984; Crowder and Bristow, 1986; Lachavanne *et al.*, 1986a, 1986b; Hough *et al.*, 1989; Arts *et al.*, 1990; Crowder and Painter, 1991; Mason, 1991; Van Groenendael *et al.*, 1993; Papastergiadou and Babalonas, 1993a, 1993b, etc.) have indeed observed or experimented on the deleterious effect of water accelerated enrichment in nutrients on plants, particularly on the specific diversity and structural complexity of communities. They tend initially to increase then to diminish when the waters become eutrophic. The reversibility of the phenomenon (oligotrophication) appears to be confirmed in the case of aquatic plant communities (Forest, 1987; Crowder and Painter, 1991; Hough et al., 1991; Lachavanne et al., 1991) when efficient control measures for nutritional pollution are installed in watersheds.

ROLES OF PLANT DIVERSITY

We do not know much about the functional role of biodiversity in ecosystems, which depends on the degree of redundancy of species' functions and on the magnitude of the role of key species (Di Castri and Younès, 1990). It appears, however, that high diversity and complexity would confer a definite ability to resist and absorb certain perturbations on any system (Di Castri and Younès, 1990; Putman, 1994).

Furthermore, no one seems to contest that plant diversity plays a leading role in the (spatio-temporal) dynamics of the ecosystem. In land–water ecotones, the diversity of floristic composition and of the structural complexity of aquatic and palustrine vegetation contribute to many essential functions for the ecosystem and in connection with related ecosystems. Plants determine, with the abiotic factors, the structural and functional conditions permitting the occurrence of other living organisms (consumers and decomposers). Indeed, vegetation structures the space, diversifies habitats and engenders microclimates affecting the distribution of these organisms (Pieczyńska, 1976, 1990; Lodge et al., 1988; Pieczyńska and Zalewski, Chapter 3). Vegetation also contributes to the filtering role played by ecotones between terrestrial and aquatic ecosystems (Peterjohn and Correll, 1984; Wiens et al., 1985; Naiman et al., 1989).

Vegetation also constitutes a structuring element of the ecotone by virtue of: (i) physiognomy, (ii) composition – both in species and in biological types (relative importance, at the scale of microbiotopes, of aerial, aquatic and subterranean-root vegetative and reproductive systems), (iii) biomass, (iv) the surface occupied, (v) location and distribution (patchiness). This structuring function can increase the variety of life conditions and of resources for other organisms. In connection with the physico-chemical quality of water and soil, for instance, several authors stress the roles played by vegetation and decomposing plant material in the nutrient cycle (De Marte and Hartman, 1974; Wetzel, 1975; Carignan and Kalff, 1980; Carignan, 1982; Pieczyńska et al., 1984; Pieczyńska, 1986). The rhizosphere also conditions the chemical and structural characteristics of soils. For example, oxygenation of sediments by rooted aquatic plants can cause an increase in the redox potential (Tessenow and Baynes, 1978; Carpenter *et al.*, 1983; Jaynes and Carpenter, 1986), but this ability varies greatly from species to species (Sand-Jensen *et al.*, 1982).

We have seen above that the degree of spatial or architectural heterogeneity generated by plant populations engenders a potential for the development of diversity of other living organisms that are more or less vegetation-dependent. Thus aquatic vegetation (submersed green parts and rhizosphere) offers at the same time a considerable colonization surface for the associated microflora (Wetzel, 1975), a substrate, a shelter or even food for an increased diversity of invertebrates (Anderson and Day, 1986; Shiel, 1986; Giudicelli and Bournaud, Chapter 6) as well as a place for numerous species of fish to achieve all or parts of their vital functions (Tonn and Magnuson, 1982).

In palustrine and terrestrial ecosystems, structural diversity of vegetation, more than specific diversity itself, correlates with diversity in terms of insects (Murdoch *et al.*, 1972; Heithaus, 1974; Southwood *et al.*, 1979), lizards (Pianka, 1967, 1973) and birds (MacArthur and MacArthur, 1961; Cody, 1975; Roth, 1976; Dobrowolski, Chapter 9). However, it is difficult to distinguish the contribution of vegetation to this animal diversity from that of other ecological (abiotic) factors, also responsible for plant diversity (Begon *et al.*, 1990).

CONCLUSIONS

Far from being exhaustive, this Chapter attempts to review the plant diversity patterns at the different organizational levels of living matter. It tries to outline their ecological functions and, above all, to emphasize the main ecological factors (abiotic and biotic, autogenic and allogenic) which specifically determine plant diversity in land–inland water ecotones at different spatial scales, as well as their mode and variability of intervention in time. However, a considerable amount of research remains to be completed to obtain a better understanding of the mechanisms of their actions and interactions.

There are many threats to ecotones and biodiversity. Therefore it is a matter of urgency to focus scientific research towards the acquisition of a concrete knowledge concerning the mode of action of the factors of plant diversity regulation in the ecotone, which appear to be the main key to understanding the origin and causes of plant diversity and to help managers involved in conservation measures.

Future studies to be developed on vegetation should therefore focus primarily on the issues related to its various functions (qualitative and quantitative), both within the ecotone itself and in ecological flows (sediments, nutrients, organisms) between adjacent ecosystems. Some of the key questions connected with the study of vegetation in ecotones benefit from partial knowledge, while others are to be addressed *de novo*:

- How and over what time spans does vegetation react to the changes in the environment's conditions?
- What is the productivity of the various vegetation groups in the ecotone and how is the produced organic matter integrated into the trophic networks of producers and decomposers?
- What are the tolerance spectra of plant species/groups to natural and man-made factors?
- What are the impacts of the various man-made modifications of ecotones (qualitative and quantitative approach) on vegetation?
- How and at what point does vegetation help to condition the structure of habitats (spatial organization, degree of heterogeneity) and from then, what influence does it exert on animal diversity in the ecotone itself and in adjacent ecosystems: importance of (i) nature of plant species and groups/associations, (ii) their size, (iii) their physiognomic structure, (iv) their organization, (v) their spatial distribution (patchiness), (vi) their seasonal or annual cycles, etc.?
- How, in a systemic approach, can one exploit the numerous existing but often dissimilar data that have been collected in the past on vegetation in ecotones (particularly qualitative and often partial quantitative data)?

Knowledge of ecotone vegetation can contribute:
- to delimit the geographic limits of the ecotone, to help establish a functional definition of ecotones;
- to draw up a typology of ecotones centred on various classification criteria: conservation of biotopes and threatened species, maintenance of biodiversity, utilization of pollution retention and filtration capabilities, stabilization of related ecosystems, etc.;
- to determine the diversity of the life conditions and habitats of ecotones both in relation to the specific variety of vegetation itself and in relation to the fundamental role played by the structural organization of this vegetation in the construction of habitats for fauna and microorganisms;
- to evaluate the role and importance of vegetation in matter and energy transfer and in transformation functions within the ecotone and between the ecosystems which it separates; and more precisely;
- to quantify the vegetation's abilities to retain, capture, filter, transform or release nutrients and pollutants on the one hand, and living organisms on the other (maintenance and renewal of genetic heritage).

ACKNOWLEDGEMENTS

We thank E. Castella for his stimulating remarks, K. Whiteley for translating the text and both O. Rossier and J. Mueller for preparing the figures.

REFERENCES

Ager, L.A. and Kerce, K.E. (1970.) Vegetational changes associated with water level stabilization in Lake Okeechobee, Florida. In *Proceedings of the 24th Annual Conference of the Southeastern Association of Game and Fish Commissioners*, pp. 338–51.

Allen, T.F.H. and Starr, T.B. (1982). *Hierarchy: Perspectives for Ecological Complexity*. University of Chicago Press, Chicago, IL

Anderson R.V. and Day, D.M. (1986). Predictive quality of macroinvertebrate–habitat associations in lower navigation pools of the Mississippi River. *Hydrobiologia*, **136**, 101–12

Anderson, M.G. and Low, J.P. (1976). Use of sago pondweed by waterfowl on the Delta Marsh, Manitoba. *Journal of Wildlife Management*, **40**, 233–42

Armstrong, W. (1975). Waterlogged soils. In Etherington, J.R. (ed.) *Environment and Plant Ecology*, pp. 181–218. John Wiley, London

Arts, G.H.P., Van der Velde, G., Roelofs, J.G.M. and Van Swaay, C.A.M. (1990). Successional changes in the soft-water macrophyte vegetation of (sub)atlantic, sandy, lowland regions during this century. *Freshwater Biology*, **24**, 287–94

Bartley, M.R. and Spence, D.H.N. (1987). Dormancy and propagation in helophytes and hydrophytes. *Archiv für Hydrobiologie, Beiheft Ergebnisse der Limnologie*, **27**, 139–55

Bazzaz, F.A. (1979). The physiological ecology of plant succession. *Annual Review of Ecological Systematics*, **10**, 351–71

Bedinger, M.S. (1979). Relation between forest species and flooding. In Greeson, P.E., Clark, J.R. and Clark, J.E. (eds.) *Wetland Functions and Values: the State of Our Understanding*, pp. 427–35. American Water Resource Association, Minneapolis, Minnesota

Begon, M., Harper, J.L. and Townsend, C.R. (eds.) (1990). *Ecology: Individuals, Populations and Communities*. 2nd edn. Blackwell, Cambridge, USA

Best, E.P.H. (1981). A preliminary model for growth of *Ceratophyllum demersum*. L. *Verhandlungen Internationale Vereinigung für Theoretische und Angewandte Limnologie*, **21**, 1484–91

Best, E.P.H., De Vries, D. and Reins, A. (1984). The macrophytes in the Loosdrecht Lakes: a story of their decline in the course of eutrophication. *Verhandlungen Internationale Vereinigung für Theoretische und Angewandte Limnologie*, **22**, 868–75

Bren, L.J. (1988). Effects of river regulaton on flooding of a riparian red gum forest on the River Murray, Australia. *Regulated Rivers*, **2**, 65–78

Brock, Th. C.M., Van der Velde, G. and Van de Steeg, H.M. (1987). The effects of extreme water level fluctuations on the wetland vegetation of a nymphaeid-dominated oxbow lake. *Archiv für Hydrobiologie, Beiheft Ergebnisse der Limnologie*, **27**, 57–73

Brown, J.H. (1981). Two decades of homage to Santa Rosalia: toward a general theory of diversity. *American Zoologist*, **21**, 877–88

Burgess, N.D., Evans, C.E. and Thomas, G.J. (1990). Vegetation change on the Ouse Washes Wetland, England, 1972–88 and effects on their conservation importance. *Biological Conservation*, **53**, 173–89

Bursche, E.M. (1971). *A Handbook of Water Plants*. Warne, London

Buttler, A., Bueche, M., Cornali, Ph. and Gobat, J.M. (1985). Historischer und ökologischer Ueberblick über das Südostufer des Neuenburger Sees. *Telma (Hannover)*, **15**, 31–42

Buttler, A. and Gobat, J.M. (1991). Les sols hydromorphes des prairies humides de la rive sud du lac de Neuchâtel (Suisse). *Bulletin of Ecology*, **22**(3–4), 405–18

Carignan, R. (1982). An empirical model to estimate the relative importance of roots in phosphorus uptake by aquatic macrophytes. *Canadian Journal of Fisheries and Aquatic Sciences*, **39**, 243–7

Carignan, R. and Kalff, J. (1980). Phosphorus sources for aquatic weeds: water or sediments. *Science*, **207**, 987–9

Carpenter, S.R. (1980). Enrichment of Lake Wingra, Winsconsin, by submersed macrophyte decay. *Ecology*, **61**, 1145–55

Carpenter, S.R., Elser, J.J. and Olson, K.M. (1983). Effects of roots of *Myriophyllum verticillatum* L. on sediment redox conditions. *Aquatic Botany*, **17**, 243–9

Carpenter, S.R. and Lodge, D.M. (1986). Effects of submersed macrophytes on ecosystem processes. *Aquatic Botany*, **26**(3–4), 341–70

Carpenter, S.R. and Titus, J.E. (1984). Composition and spatial heterogeneity of submersed vegetation in a softwater lake in Wisconsin. *Vegetatio*, **57**, 153–65

Chambers, P.A. (1987). Light and nutrients in the control of aquatic plant community structure. II. *In situ* observations. *Journal of Ecology*, **75**, 621–8

Chambers, P.A., Hanson, J.M., Burke, J.M. and Prepas, E.E. (1990). The impact of the crayfish *Orconectes virilis* on aquatic macrophytes. *Freshwater Biology*, **24**, 81–91

Chambers, P.A. and Kalff, J. (1985). Depth distribution and biomass of submersed aquatic macrophyte communities in relation to Secchi depth. *Canadian Journal of Fisheries and Aquatic Sciences*, **42**, 701–9

Chambers, P.A. and Kalff, J. (1987). Light and nutrients in the control of aquatic plant community structure. I. *In situ* experiments. *Journal of Ecology*, **75**, 611–19

Chambers, P.A. and Prepas, E.E. (1988). Underwater spectral attenuation and its effect on the maximum depth of angiosperm colonization. *Canadian Journal of Fisheries and Aquatic Sciences*, **45**, 1010–17

Chambers, P.A. and Prepas, E.E. (1990). Competition and coexistence in submerged aquatic plant communities: the effects of species interactions *versus* abiotic factors. *Freshwater Biology*, **23**, 541–50

Chambers, P.A., Prepas, E.E., Hamilton, H.R. and Bothwelt, M.L. (1991). Current velocity and its effect on aquatic macrophytes in flowing waters. *Ecological Applications*, **1**, 249–57

Clayton, J.S. (1982). Effect of fluctuation in water level and growth of *Lagarosiphon major* on aquatic vascular plants in Lake Rotoma, 1973–80. *New Zealand Journal of Marine and Freshwater Research*, **16**, 89–94

Clements, F.E. (1916). *Plant Succession: an Analysis of the Development of Vegetation*. Publication 242, Carnegie Institution of Washington, Washington, DC

Cody, M.L. (1975). Towards a theory of continental species diversities: bird distributions over Mediterranean habitat gradients. In Cody, M.L. and Diamond, J.M. (eds.) *Ecology and Evolution of Communities,* pp. 214–57. Belknap Press of Harvard University Press, Cambridge, MA

Connell, J.H. (1978). Diversity in tropical rain forests and coral reefs. *Science,* **199,** 1302–10

Connell, J.H. and Slatyer, R.O. (1977). Mechanisms of succession in natural communities and their role in community stability and organization. *American Naturalist,* **111,** 1119–44

Cook, C.D.K., Gut, B.J., Rix, E.M., Schneller, J. and Seitz, M. (1974). *Water Plants of the World: a Manual for Identification of the Genera of Freshwater Macrophytes.* Dr. W. Junk b.v., The Hague

Crowder, A.A. and Bristow, J.M. (1986). Aquatic macrophytes of the Bay of Quinte: 1972–82. In Minns, C.K., Hurley, D.A. and Nicholls, K.H. (eds.) *Project Quinte: Point Source Phosphorus Control and Ecosystem Response in the Bay of Quinte, Lake Ontario,* pp. 114–27. Canadian Journal of Fisheries and Aquatic Sciences, Special Publication 86

Crowder, A. and Painter, D.S. (1991). Submerged macrophytes in Lake Ontario: current knowledge, importance, threats to stability, and needed studies. *Canadian Journal of Fisheries and Aquatic Sciences,* **48**(8), 1539–45

Danell, K. (1977). Short-term plant successions following the colonization of a northern Swedish lake by the muskrat, *Ondatra zibethica. Journal of Applied Ecology,* **14,** 933–47

Daubenmire, R. (1968). *Plant Communities. A Textbook of Plant Synecology.* Harper and Row, New York

Décamps, H. and Naiman, R.J. (1990). Towards an ecotone perspective. In Naiman, R.J. and Décamps, H. (eds.) *The Ecology and Management of Aquatic–Terrestrial Ecotones,* pp. 1–5. MAB Book Series 4. UNESCO/Parthenon, Paris/Carnforth

De Marte, J.A. and Hartman, R.T. (1974). Studies on absorption of ^{32}P, ^{59}Fe and ^{45}Ca by water-milfoil (*Myriophyllum exalbescens* Fernald). *Ecology,* **55,** 188–94

Den Hartog, C. and Segal, S. (1964). A new classification of the water-plant communities. *Acta Botanica Neerlandica,* **13,** 367–93

Den Hartog, C. and Van der Velde, G. (1988). Structural aspects of aquatic plant communities. In Symoens, J.J. (ed.) *Handbook of Vegetation Science.* Vol. 15. *Vegetation of Inland Waters,* pp. 113–53. Kluwer Academic Publishers, Dordrecht

Dethioux, M. (1989). Espèces aquatiques des eaux courantes. Aménagement des cours d'eau. Ministère de la Région Wallonne, Service Promotion et Communication

Di Castri, F. and Hansen, A.J. (1992). The environment and development crises as determinants of landscape dynamics. In Hansen, A.J. and di Castri, F. (eds.) *Landscape boundaries: consequences for biotic diversity and ecological flows,* pp. 3–18. Ecological Studies 92, Springer Verlag, New York

Di Castri, F. and Younès T. (eds.) (1990). Ecosystem Function of Biological Diversity. *Biology International,* Special Issue, **22,** 1–20

Dobrowolski, K.A. (1996). Bird diversity in ecotonal habitats. In Lachavanne, J.-B. and Juge, R. (eds.) *Biodiversity in Land–Inland water Ecotones,* pp. 205–221. UNESCO/Parthenon, Paris/Carnforth

Duarte, C.M. and Kalff, J. (1987). Weight–density relationships of submersed plants: the importance of light and growth form. *Oecologia*, **72**, 612–17

Duarte, C.M. and Kalff, J. (1988). Influence of lake morphometry on the response of submerged macrophytes to sediment fertilization. *Canadian Journal of Fisheries and Aquatic Sciences*, **45**, 216–21

Duarte, C.M. and Kalff, J. (1990). Biomass density and the relationship between submersed macrophyte biomass and plant growth form. *Hydrobiologia*, **196**, 17–23

Duarte, C.M., Planas, D. and Peñuelas, J. (1994). Macrophytes, taking control of an ancestral home. In Margalef, R. (ed.) *Limnology Now: A Paradigm of Planetary Problems*, pp. 59–79. Elsevier Science B.V

Duarte, C.M. and Roff, D.A. (1991). Architectural and life history constraints to submersed macrophyte community structure: a simulation study. *Aquatic Botany*, **42**(1), 15–29

Farmer, A.M. and Adams, M.S. (1989). A consideration of the problems of scale in the study of the ecology of aquatic macrophytes. *Aquatic Botany*, **33**, 177–89

Farmer, A.M. and Adams, M.S. (1991). The nature of scale and the use of hierarchy theory in understanding the ecology of aquatic macrophytes. *Aquatic Botany*, **41**, 253–61

Fasset, N.C. (1968). *A Manual of Aquatic Plants*. The University of Wisconsin Press, Madison, WI

Forest, H. (1987). The remarkable recovery of submerged vegetation in Iron-dequoit Bay. *Bulletin 296. Rochester Committee for Scientific Information*. Rochester, New York

Gasith, A. and Gaafny, S. (1990). Effects of water level fluctuation on the structure and function of the littoral zone. In Tilzer, M. and Serruya, C. (eds.) *Large Lakes. Ecological Structure and Function*, pp. 156–71. Springer Verlag, New York

Giudicelli, J. and Bournaud, M. (1996). Invertebrate biodiversity in land–inland water ecotonal habitats. In Lachavanne, J.-B. and Juge, R. (eds.) *Biodiversity in Land–Inland water Ecotones*, pp. 143–160. UNESCO/Parthenon, Paris/Carnforth

Gopal, B. and Masing, V. (1990). Biology and Ecology. In Patten, B.C. (ed.) *Wetlands and Shallow Continental Water Bodies*. Vol. 1. *Natural and Human Relationships*, pp. 91–239. SPB Academic Publishing, The Hague

Gorham, E. (1957). The development of peatlands. *Quarterly Review of Biology*, **32**, 145–66

Gosselink, J.G. and Maltby, E. (1990). Wetland losses and gains. In Williams, M. (ed.) *Wetlands: A Threatened Landscape*, pp. 296–322. Basil Blackwell, Oxford

Gosselink, J.G. and Turner, R.E. (1978). The role of hydrology in freshwater wetland ecosystems. In Good, R.E., Whigham, D.F. and Simpson, R.L. (eds.) *Freshwater Wetlands: Ecological Processes and Management Potential*, pp. 63–78. Academic Press, New York

Grace, J.D. (1983). Autotoxic inhibition of seed germination by *Typha latifolia*: an evaluation. *Oecologia*, **59**, 366–9

Grace, J.B. and Wetzel, R.G. (1981). Habitat partitioning and competitive displacement in cattails (*Typha*): experimental field studies. *American Naturalist*, **118**, 463–74

Greeson, P.E., Clark, J.R. and Clark, J.E. (eds.) (1979). *Wetland Functions and Values: The State of Our Understanding*. Proceedings of National Symposium on Wetlands,

Lake Buena Vista, Florida, American Water Resources Association, Technical Publication. TPS 79-2, Minneapolis, MN

Gregory, S.V. (1983). Plant–herbivore interactions in stream systems. In Barnes, J.R. and Minshall, G.W. (eds.) *Stream Ecology. Application and Testing of General Ecological Theory*, pp. 157–89. Plenum Press, New York

Grime, J.P. (1979). *Plant Strategies and Vegetation Processes*. John Wiley, Chichester

Hansen, A.J., di Castri, F. and Naiman, R.J. (1988). Ecotones: what and why? In di Castri, F., Hansen, A.J. and Holland, M.M. (eds.) A New Look at Ecotones: Emerging International Projects on Landscape Boundaries. *Biology International*, Special Issue, **17**, 9–46

Hansen, A.J., Risser, P.G. and di Castri, F. (1992). Epilogue: biodiversity and ecological flows across ecotones. In Hansen, A.J. and di Castri, F. (eds.) *Landscape boundaries: consequences for biotic diversity and ecological flows*, pp. 423–38. Ecological Studies 92, Springer Verlag, New York

Harman, W.N. (1974). Phenology and physiognomy of the hydrophyte community in Otsego Lake, New York. *Rhodora*, **76**, 497–508

Heinselman, M.L. (1970). Landscape evolution, peatland types and the environment in the Lake Agassiz Peatlands Natural Area, Minnesota. *Ecological Monographs*, **40**, 235–61

Heithaus, E.R. (1974). The role of plant–pollinator interactions in determining community structure. *Annals of the Missouri Botanic Garden*, **61**, 675–91

Hejny, S. (1957). Ein Beitrag zur ökologischen Gliederung der Makrophyten der tschechoslowakischen Niederungsgewässer. *Preslia*, **29**, 349–68

Hejny, S. (1960). *Ökologische Charakteristik der Wasser- und Sumpfpflanzen in den Slowakischen Tiefebenen (Donau- und Theissgebiet)*. Vydavatelstvo SAV, Bratislava

Hejny, S. (1971). The dynamic characteristics of littoral vegetation with respect to changes of water level. *Hydrobiologia*, **12**, 71–85

Hejny, S. and Hroudová, Z. (1987). Plant adaptations to shallow water habitats. *Archiv für Hydrobiologie, Beiheft Ergebnisse der Limnologie*, **27**, 157–66

Hinde, H.P. (1954). Vertical distribution of salt marsh phanerogams in relation to tide levels. *Ecological Monographs*, **24**, 209–25

Horn, H.S. (1976). Succession. In May, R.M. (ed.) *Theoretical Ecology*, pp. 187–204. Blackwell Scientific Publications, Oxford

Hough, R.A., Allenson, T.E. and Dion, D.D. (1991). The response of macrophyte communities to drought-induced reduction of nutrient loading in a chain of lakes. *Aquatic Botany*, **41**, 299–308

Hough, R.A., Fornwall, M.D., Negele, B.J., Thompson, R.L. and Putt, D.A. (1989). Plant community dynamics in a chain of lakes: principle factors in the decline of rooted macrophytes with eutrophication. *Hydrobiologia*, **173**, 199–217

Howard-Williams, C., Schwarz, A.M. and Vincent, W.F. (1995). Deep-water aquatic plant communities in an oligotrophic lake: physiological responses to variable light. *Freshwater Biology*, **33**, 91–102

Hroudová, Z. (1980). An ecological study of the species *Sagittaria sagittifolia* L., *Butomus umbellatus* L., *Bolboschoenus maritimus* (L.) Palla, *Oenanthe aquatica* (L.) Poir. (in Czech). Mcsr., CSc. Thesis, Institute of Botany CSAV, Pruhonice

Hutchinson, G.E. (1975). A *Treatise on Limnology*. III. *Limnological Botany*. John Wiley, New York

Jacobs, J. (1975). Diversity, stability and maturity in ecosystems influenced by human activities. In Van Dobben, W.H. and Lowe-McConnell, R.H. (eds.) *Unifying Concepts in Ecology,* pp. 187–207. Dr W. Junk b.v., The Hague

Jaynes, M.L. and Carpenter, S.R. (1986). Effects of vascular and nonvascular macrophytes on sediment redox and solute dynamics. *Ecology*, **67**, 875–82

Jeffries, M. and Mills, D. (1990). *Freshwater Ecology*. Bellhaven Press, London

Jupp, B. and Spence, D.H.N. (1977). Limitations on macrophytes in a eutrophic lake, Loch Leven. I. Effects of phytoplankton. *Journal of Ecology*, **65**, 175–86

Kadlec, J.A. and Wentz, W.A. (1974). *State-of-the-art survey and evaluation of marsh plant establishment techniques: induced and natural.* Dredged Materials Research Program. Contract report D-74-9, U.S. Army Coastal Engineering Research Center, Fort Belvoir, VA

Kautsky, L. (1988). Life strategies of aquatic soft bottom macrophytes. *Oïkos*, **53**, 126–35

Keddy, P.A. (1983). Shoreline vegetation in Axe Lake, Ontario: effects of exposure on zonation patterns. *Ecology*, **64**, 331–44

Keddy, P.A. (1984). Plant zonation on lakeshores in Nova Scotia: a test of the resource specialization hypothesis. *Journal of Ecology*, **72**, 797–808

Kiorboe, T. (1980). Distribution and production of submerged macrophytes in Tipper Grund (Ringköbing Fjord, Denmark) and the impact of waterfowl grazing. *Journal of Applied Ecology*, **17**, 675–87

Kvet, J. and Hudec, K. (1971). Effects of grazing by grey-lag geese on reedswamp plant communities. *Hydrobiologia*, **12**, 351–9

Lachavanne, J.B. (1982). Influence de l'eutrophisation des eaux sur les macrophytes des lacs suisses: résultats préliminaires. In Symoens, J.J., Hooper, S.S. and Compère, P. (eds.) *Studies on Aquatic Vascular Plants,* pp. 333–9. Royal Botanical Society of Belgium, Brussels

Lachavanne, J.B. (1985). The influence of accelerated eutrophication on the macrophytes of Swiss lakes: abundance and distribution. *Verhandlungen Internationale Vereinigung für Theoretische und Angewandte Limnologie*, **22**, 2950–5

Lachavanne, J.B. (1996). Why study biodiversity in land–inland water ecotones? In Lachavanne, J.-B. and Juge, R. (eds.) *Biodiversity in Land–Inland water Ecotones*, pp. 1–45. UNESCO/Parthenon, Paris/Carnforth

Lachavanne, J.B., Juge, R. and Noetzlin, A. (1986a). Evolution des macrophytes du Léman (rives genevoises – 1972–1984). *Sciences de l'Eau*, **5**, 419–33

Lachavanne, J.B., Juge, R., Perfetta, J., Noetzlin, A. and Lods-Crozet, B. (1986b). *Etude chorologique et écologique des macrophytes des lacs suisses en fonction de leur altitude et de leur niveau trophique*, 1976–1985. Rapport FNRS

Lachavanne, J.B., Juge, R. and Perfetta, J. (1991). The consequences of water oligotrophication on macrophytic vegetation of Swiss lakes. *Verhandlungen Internationale Vereinigung für Theoretische und Angewandte Limnologie*, **24**, 943–8

Lachavanne, J.B., Juge, R. and Perfetta, J. (1995). Structure des peuplements de macrophytes. In Pourriot, R. and Meybeck, M. (eds.) *Limnologie Générale,* pp. 473–93. Masson, Paris

Leopold, A. (1933). *Game Management*. Charles Scribner's Sons, New York

Leveque, C. and Quensiere, J. (1983). Un lac tropical sous climat semi-aride: le lac Tchad. In Lamotte, M. and Bourlière, F. (eds.) *Problèmes d'Écologie: Structure et Fonctionnement des Ecosystèmes Limniques*, pp. 161–241. Masson, Paris

Lodge, D.M. (1991). Herbivory on freshwater macrophytes. *Aquatic Botany*, **41**, 195–224

Lodge, D.M., Barko, J.W., Strayer, D., Melack, J.M., Mittelbach, G.G., Howarth, R.W., Menge, B. and Titus, J.E. (1988). Spatial heterogeneity and habitat interactions in lake communities. In Carpenter, S.R. (ed.) *Complex Interactions in Lake Communities*, pp. 181–208. Springer Verlag, New York

Lodge, D.M. and Lorman, J.G. (1987). Reductions in submersed macrophyte biomass and species richness by the crayfish *Orconectes rusticus*. *Canadian Journal of Fisheries and Aquatic Sciences*, **44**, 591–7

Luther, H. (1949). Vorschlag zu einer ökologischen Grundeinteilung der Hydrophyten. *Acta Botanica Fennica*, **44**, 1–15

Luther, H. (1951). Verbreitung und Ökologie der höheren Wasserpflanzen im Brackwasser der Ekenäs-Gegend in Südfinnland. II. Spezieller Teil. *Acta Botanica Fennica*, **50**, 1–370

Luther, H. (1983). On the life forms and aboveground and underground biomass of aquatic macrophytes: a review. *Acta Botanica Fennica*, **123**, 1–23

MacArthur, R.H. and MacArthur, J.W. (1961). On bird species diversity. *Ecology*, **42**, 594–8

MacCreary, N.J. (1991). Competition as a mechanism of submersed macrophyte community structure. *Aquatic Botany*, **41**, 177–93

MacCreary, N.J. and Carpenter, S.R. (1987). Density-dependent growth interactions between *Eleocharis acicularis* (L.) R. and S. and *Juncus pelocarpus* forma *submersus* Fassett. *Aquatic Botany*, **27**, 229–41

MacCreary, N.J., Carpenter, S.R. and Chaney, J.E. (1983). Coexistence and interference in two submersed freshwater perennial plants. *Oecologia*, **59**, 393–6

MacKnight, J.S., Hook, D.D., Langdon, O.G. and Johnson, R.L. (1981). Flood tolerance and related characteristics of trees of the bottomland forests of the southern United States. In Clark, J.R. and Benforado, J. (eds.) *Wetlands of Bottomland Hardwood Forests*, pp. 29–69. Elsevier, Amsterdam, Oxford and New York

MacNaughton, S.J. (1968). Autotoxic feedback in the regulation of *Typha* populations. *Ecology*, **49**(2), 367–9

Madsen, J.D. (1991). Resource allocation at the individual plant level. *Aquatic Botany*, **41**, 67–86

Margalef, R. (1968). *Perspectives in Ecological Theory*. University of Chicago Press, Chicago

Margalef, R. (1969). Diversity and stability: a practical proposal and a model of interdependence. In *Diversity and Stability in Ecological Systems*, pp. 25–37. Brookhaven Symposium in Biology No 22

Mason, C.F. (1991). *Biology of Freshwater Pollution*. 2nd edn. Longman Scientific and Technical; John Wiley, New York

May, R.M. (1973). *Stability and Complexity in Model Ecosystems*. Princeton University Press, Princeton

Miller, T.E. (1982). Community diversity and interactions between the size and frequency of disturbance. *American Naturalist*, **120**, 533–6

Mitchell, D.S. and Rogers, K.H. (1985). Seasonality/aseasonality of aquatic macrophytes in southern hemisphere inland waters. *Hydrobiologia*, **125**, 137–50

Mitsch, W.J., Dorge, C.L. and Wiemhoff, J.R. (1979). Ecosystem dynamics and a phosphorus budget of an alluvial cypress swamp in southern Illinois. *Ecology*, **60**, 1116–24

Mitsch, W.J. and Ewel, K.C. (1979). Comparative biomass and growth of cypress in Florida wetlands. *American Midlands Naturalist*, **101**, 417–26

Mitsch, W.J. and Gosselink, J.G. (1986). *Wetlands*. Van Nostrand Reinhold, New York

Mitsch, W.J. and Rust, W.G. (1984). Tree growth responses to flooding in a bottomland forest in northeastern Illinois. *Forest Science*, **30**, 499–510

Montegut, J. (1987). *Le Milieu Aquatique, Entretien, Désherbage*. 2è édition. ACTA, Paris

Moss, B. (1988). *Ecology of Fresh Waters: Man and Medium*. 2nd edn. Blackwell Scientific Publications, Oxford

Murdoch, W.W., Evans, F.C. and Peterson, C.H. (1972). Diversity and pattern in plants and insects. *Ecology*, **53**, 819–29

Murphy, K.J., Rørslett, B. and Springuel, I. (1990). Strategy analysis of submerged lake macrophyte communities: an international example. *Aquatic Botany*, **36**, 303–23

Naiman, R.J. and Décamps, H. (1991). Landscape boundaries in the management and restoration of changing environments: a summary. In Holland, M.M., Risser, P.G. and Naiman, R.J. (eds.) *Ecotones: the Role of Landscape Boundaries in the Management and Restoration of Changing Environments*, pp. 130–7. Chapman and Hall, New York and London

Naiman, R.J., Décamps, H. and Fournier, F. (eds.) (1989). Role of Land–Inland Water Ecotones in Landscape Management and Restoration: a Proposal for Collaborative Research. *MAB Digest 4*, UNESCO, Paris

Naiman, R.J., Décamps, H., Pastor, J. and Johnston, C.A. (1988). The potential importance of boundaries to fluvial ecosystems. *Journal of the North American Benthological Society*, **7**, 289–306

Noss, R.F. (1983). A regional landscape approach to maintain diversity. *BioScience*, **33**, 700–6

Odum, E.P. (1969). The strategy of ecosystem development. *Science*, **164**, 262–70

Odum, E.P. (1971). *Fundamentals of Ecology*. 3rd edn. W.B. Saunders, Philadelphia

Odum, E.P. (1979). Ecological importance of the riparian zone. In Johnson, R.R. and McCormick, J.F. (technical coordinators). *Strategies for Protection and Management of Floodplain Wetlands and other Riparian Ecosystems*, pp. 2–4. Proceedings of the Symposium, Callaway Gardens, Georgia, December 11–13, 1978, US Forest Service General Technical Report. WO–12, Washington, DC

Odum, E.P. (1993). *Ecology: and our Endangered Life-support Systems*. 2nd edn. Sinauer Associates, Inc., Sunderland, MA

Odum, W.E. (1990). Internal processes influencing the maintenance of ecotones: do they exist? In Naiman, R.J. and Décamps, H. (eds.) *The Ecology and Management of Aquatic–Terrestrial Ecotones,* pp. 91–102. MAB Book Series 4. UNESCO/Parthenon, Paris/Carnforth

Olsen, T.M., Lodge, D.M., Capelli, G.M. and Houlihan, R.J. (1991). Mechanisms of impact of an introduced crayfish (*Orconectes rusticus*) on littoral congeners, snails and macrophytes. *Canadian Journal of Fisheries and Aquatic Sciences,* **48**(10), 1853–61

O'Neill, R.V., DeAngelis, D.L., Waide, J.B. and Allen, T.F.H. (1986). A hierarchical concept of ecosystems. *Monographical Population Biology 23.* Princeton University Press, Princeton, NJ

Orians, G.H. (1975). Diversity, stability and maturity in natural ecosystems. In Van Dobben, W.H. and Lowe-McConnell, R.H. (eds.) *Unifying Concepts in Ecology,* pp. 139–50. Dr. W. Junk b.v., The Hague

Papastergiadou, E. and Babalonas, D. (1993a). The relationships between hydrochemical environmental factors and the aquatic macrophytic vegetation in stagnant and slow flowing waters. 1. Water quality and distribution of aquatic associations. *Archiv für Hydrobiologie,* **90**(4), 475–91

Papastergiadou, E. and Babalonas, D. (1993b). The relationships between hydrochemical environmental factors and the aquatic macrophytic vegetation in stagnant and slow flowing waters. 2. Evaluation of plant associations indicative value. *Archiv für Hydrobiologie.,* **90**(4), 493–506

Patrick, W.H. Jr. and Mikkelsen, D.S. (1971). Plant nutrient behavior in flooded soils. In *Fertilization Technology and Use.* 2nd edn., pp. 187–215. Soil Science Society of America, Madison, Wisconsin

Pelikán, J., Svoboda, J. and Kvet, J. (1971). Relationship between the population of muskrats (*Ondatra zibethica*) and the primary production of cattail (*Typha latifolia*). *Hydrobiologia,* **12**, 177–80

Peterjohn, W.T. and Correll, D.L. (1984). Nutrient dynamics on an agricultural watershed: observations on the role of a riparian forest. *Ecology,* **65**, 1466–75

Pianka, E.R. (1967). On lizard species diversity: North American flatland deserts. *Ecology,* **48**, 333–51

Pianka, E.R. (1973). The structure of lizard communities. *Annual Review of Ecological Systematics,* **4**, 53–74

Pieczyńska, E. (1975). Ecological interactions between land and the littoral zones of lakes (Masurian Lakeland, Poland). In Hasler, A.D. (ed.) *Coupling of Land and Water Systems,* pp. 263–76. Ecological Studies Vol. 10, Springer Verlag, Berlin

Pieczyńska, E (ed.) (1976). *Selected problems of lake littoral ecology.* Warsaw University, Warsaw.

Pieczyńska, E. (1986). Sources and fates of detritus in the shore zone of lakes. *Aquatic Botany,* **25**, 153–66

Pieczyńska, E. (1990). Lentic aquatic–terrestrial ecotones: their structure, functions and importance. In Naiman, R.J. and Décamps, H. (eds.) *The Ecology and Management of Aquatic–Terrestrial Ecotones,* pp. 103–40. MAB Book Series 4. UNESCO/Parthenon, Paris/Carnforth

Pieczyńska, E., Balcerzak, D., Kolodziejczyk, A., Olszewski, Z. and Rybak, J.I. (1984). Detritus in the littoral of several Masurian lakes (sources and fates). *Ekologia Polska,* **32**, 387–440

Pieczyńska, E. and Zalewski, M. (1996). Habitat complexity in land–inland water ecotones. In Lachavanne, J.-B. and Juge, R. (eds.) *Biodiversity in Land–Inland Water Ecotones*, pp. 61–79. UNESCO/Parthenon, Paris/Carnforth

Prejs, A. (1984). Herbivory by temperate freshwater fishes and its consequences. *Environmental Biology of Fishes*, **10**, 281–96

Pulliam, H.R. (1988). Sources, sinks and population regulation. *American Naturalist*, **132**, 652–61

Putman, R.J. (1994). *Community Ecology*. Chapman and Hall, London

Raunkiaer, C. (1934). *The Life Forms of Plants and Statistical Plant Geography*. Oxford University Press, Oxford

Raunkiaer, C. (1937). *Plant Life Forms*. Oxford University Press, Oxford

Rickleffs, R. (1990). *Ecology*. 3rd edn., Freeman and Company, New York

Robel, R.J. (1962). Changes in submersed vegetation following a change in water level. *Journal of Wildlife Management*, **26**, 221–4

Robertson, P.A., Weaver, G.T. and Cavanaugh, J.A. (1978). Vegetation and tree species patterns near the northern terminus of the southern floodplain forest. *Ecological Monographs*, **48**, 249–67

Roelofs, J.G.M. (1983). Impact of acidification and eutrophication on macrophyte communities in soft waters in The Netherlands. I. Field observations. *Aquatic Botany*, **17**, 139–55

Roth, R.R. (1976). Spatial heterogeneity and bird species diversity. *Ecology*, **57**, 773–82

Sand-Jensen, K. and Borum, J. (1991). Interactions among phytoplankton, periphyton and macrophytes in temperate freshwaters and estuaries. *Aquatic Botany*, **41**, 137–75

Sand-Jensen, K., Prahl, C. and Stockholm, H. (1982). Oxygen release from roots of submerged aquatic macrophytes. *Oïkos*, **38**, 349–54

Sculthorpe, C.D. (1967). *The Biology of Aquatic Vascular Plants*. Edward Arnold, London

Seddon, B. (1972). Aquatic macrophytes as limnological indicators. *Freshwater Biology*, **2**(2), 107–30

Segal, S. (1971). Principles on structure, zonation and succession of aquatic macrophytes. *Hydrobiologia*, **12**, 89–95

Segal, S. (1982). General trends in structure development during succession of aquatic macrophyte vegetation. In Symoens, J.J., Hooper, S.S. and Compère, P. (eds.) *Studies on Aquatic Vascular Plants*, pp. 249–56. Royal Botanical Society of Belgium, Brussels

Sheldon, S.P. (1984). The effects of herbivory and other factors on species abundance and the diversity of freshwater macrophyte communities. PhD Thesis, University of Minnesota

Sheldon, S.P. (1987). The effects of herbivorous snails on submerged macrophyte communities in Minnesota lakes. *Ecology*, **68**, 1920–31

Shiel, R.J. (1986). Zooplankton of the Murray–Darling system. In Davies, B.R. and Walker, K.F. (eds.) *The Ecology of River Systems*, pp. 661–677. Dr W. Junk, Dordrecht

Shugart, H.H. (1984). *A Theory of Forest Dynamics: The Ecological Implications of Forest Succession Models*. Springer Verlag, New York

Silvertown, J. and Law, R. (1987). Do plants need niches? Some recent developments in plant community ecology. *Trends in Ecology and Evolution*, **2**, 24–6

Smith, L.M. and Kadlec, J.A. (1985). Fire and herbivory in a Great Salt Lake marsh. *Ecology*, **66**, 259–65

Solbrig, O. T. (1991). Biodiversity: Scientific Issues and Collaborative Research Proposals. *MAB Digest 9*, UNESCO, Paris

Sousa, W.P. (1979). Disturbance in the marine intertidal boulder fields: the nonequilibrium maintenance of species diversity. *Ecology*, **60**, 1225–35

Southwood, T.R.E., Brown, V.K. and Reader, P.M. (1979). The relationships of plant and insect diversities in succession. *Biological Journal of the Linnean Society*, **12**, 327–48

Spence, D.H.N. (1982). The zonation of plants in freshwater lakes. *Advances in Ecological Research*, **12**, 37–125

Spence, D.H.N., Campbell, R.M. and Crystal, J. (1973). Specific leaf areas and zonation of freshwater macrophytes. *Journal of Ecology*, **61**, 317–28

Szczepanska, W. (1971). Allelopathy among the aquatic plants. *Polskie Archiwum Hydrobiologii*, **18**(1), 17–30

Szczepanska, W. (1977). The effect of remains of helophytes on the growth of *Phragmites communis* Trin. and *Typha latifolia* L. *Ekologia Polska*, **25**(3), 437–45

Szczepanska, W. (1987). Allelopathy in helophytes. *Archiv für Hydrobiologie, Beiheft Ergebnisse der Limnologi*, **27**, 173–9

Szczepanski, A. (1971). Allelopathy and other factors controlling the macrophytes production. *Hydrobiologia*, **12**, 193–7

Tessenow, U. and Baynes, Y. (1978). Experimental effects of *Isoetes lacustris* L. on the distribution of Eh, pH, Fe and Mn in lake sediment. *Verhandlungen Internationale Vereinigung für Theoretische und Angewandte Limnologie*, **20**, 2358–62

Titus, J.E. and Adams, M.S. (1979). Coexistence and the comparative light relations of the submersed macrophytes *Myriophyllum spicatum* L. and *Vallisneria americana* Michx. *Oecologia*, **40**, 272–86

Tonn, W.M. and Magnuson, J. (1982). Patterns in the species composition and richness of fish assemblages in northern Wisconsin lakes. *Ecology*, **63**, 1149–66

Tutin, T.G. (1941). The hydrosere and current concepts of the climax. *Journal of Ecology*, **29**, 268–79

Urban, D.L., O'Neill, R.V. and Shugart, H.H. (1987). Landscape ecology. *BioScience*, **37**, 119–27

Van der Maarel, E. (1976). On the establishment of plant community boundaries. *Berichte Deutschen Botanischen Gesesellschaft Bd.*, **89**, 415–43

Van der Maarel, E. (1990). Ecotones and ecoclines are different. *Journal of Vegetation Science*, **1**, 135–8

Van der Valk, A.G. (1981). Succession in wetlands: a Gleasonian approach. *Ecology*, **62**, 688–96

Van der Valk, A.G. (1982). Succession in temperate North American wetlands. In Gopal, B., Turner, R.E., Wetzel, R.G. and Whigham, D.F. (eds.) *Wetlands: Ecology and Management*, pp. 169–79. National Institute of Ecology and International Scientific Publications, Jaipur, India

Van der Valk, A.G. (1987). Vegetation dynamics of freshwater wetlands: a selection review of the literature. *Archiv für Hydrobiologie, Beiheft Ergebnisse der Limnologie*, **27**, 27–39

Van der Valk, A.G. and Davis, C.B. (1980). The impact of a natural drawdown on the growth of four emergent species in a prairie glacial marsh. *Aquatic Botany*, **9**, 301–22

Van Groenendael, J.M., Van Mansfeld, M.J.M., Roozen, A.J.M. and Westhoff, V. (1993). Vegetation succession in lakes in the coastal fringe of West Connemara, Ireland. *Aquatic Conservation: Marine and Freshwater Ecosystems*, **3**, 25–41

Von Schwaar, J. (1989). Succession of reeds, sedges, alder woodlands and other swamp communities. *Phytocoenologia*, **17**(4), 507–68

Wain, R.P., Haller, W.T. and Martin, D.F. (1985). Isozyme studies of aquatic plants. *Journal of Aquatic Plant Management*, **23**, 42–5

Walker, B.H. and Coupland, R.T. (1968). An analysis of vegetation–environment relationships in Saskatchewan sloughs. *Canadian Journal of Botany*, **46**, 509–22

Watt, A.S. (1947). Patterns and process in the plant community. *Journal of Ecology*, **35**, 1–22

Weller, M.W. (1978). Management of freshwater marshes for wildlife. In Good, R.E., Whigham, D.F. and Simpson, R.L. (eds.) *Freshwater Wetlands: Ecological Processes and Management Potential*, pp. 267–84. Academic Press, New York

Wetzel, R.G. (1975). *Limnology*. Saunders College Publishing, Philadelphia

Wetzel, R.G. (1983). *Limnology*. Saunders College Publishing, Philadelphia

Wetzel, R., Adams, M., Almassy, A., Dinka, M., Eiseltova, M., Fustec, E., Gibert, J., Gopal, B., Leichtfried, M., Livingstone, R.J., Mathieu, J., Parma, S., Pastor, J., Pieczyńska, E., Roberts, G., Sagova, M., Thompson, K. and Uerhoog, F. (1989). Nutrient, energy and water flows through land–water ecotones. In Naiman, R.J., Décamps, H. and Fournier, F. (eds.) Role of Land–Inland Water Ecotones in Landscape Management and Restoration: a Proposal for Collaborative Research, pp. 61–4. *MAB Digest No 4*, UNESCO, Paris

Whittaker, R.H. (1975). *Communities and Ecosystems*. 2nd edn. Macmillan, New York

Whittaker, R.H. (1977). Evolution of species diversity in land communities. *Evolutionary Biology*, **10**, 1–67

Wiens, J.A., Crawford, C.S. and Gosz, J.R. (1985). Boundary dynamics: a conceptual framework for studying landscape ecosystems. *Oïkos*, **45**, 421–7

Wilcox, B.A. (1984). *In situ* conservation of genetic resources: determinants of minimum area requirements. In Neeley, J.A. and Miller, K.R. (eds.) *National Parks: Conservation and Development*, pp. 639–47. Smithsonian Institution Press, Washington, DC

Wilcox, B.A. and Meeker, J.E. (1991). Disturbance effects on aquatic vegetation in regulated and unregulated lakes in northern Minnesota. *Canadian Journal of Botany*, **69**(7), 1542–51

Wilson, E.O. (1969). The species equilibrium. In Woodwell, G.M. and Smith, H.H. (eds.) *Diversity and Stability in Ecological Systems*. Brookhaven Symposium of Biology, **22**, 38–47

Wilson, E.O. (ed.) (1988). *Biodiversity*. National Academy Press, Washington, DC

Wilson, S.D. and Keddy, P.A. (1985). Plant zonation on the shoreline gradient: physiological response curves of component species. *Journal of Ecology*, **73**, 851–60

Wilson, S.D. and Keddy, P.A. (1988). Species richness, survivorship and biomass accumulation along an environmental gradient. *Oikos*, **53**, 375–80

Wisheu, I.C. and Keddy, P.A. (1989). Species richness–standing crop relationship along four lakeshore gradients: constraints on the general model. *Canadian Journal of Botany*, **67**, 1609–17

Wium-Andersen, S. (1987). Allelopathy among aquatic plants. *Archiv für Hydrobiologie, Beiheft Ergebnisse der Limnologie*, **27**, 167–72

Wium-Andersen, S., Anthoni, U., Christophersen, C. and Houen, G. (1982). Allelopathic effects on phytoplankton by substances isolated from aquatic macrophytes (Charales). *Oikos*, **39**, 187–90

Wium-Andersen, S., Anthoni, U. and Houen, G. (1983). Elemental sulphur, a possible allelopathic compound from *Ceratophyllum demersum*. *Phytochemistry*, **22**(11), 2613

Zwölfer, H. (1987). Species richness, species packing and evolution in insect–plant systems. In Schulze, E.D. and Zwölfer, H. (eds.) *Potentials and Limitations of Ecosystem Analysis*, pp. 301–19. Ecological Studies 61, Springer Verlag, Berlin

CHAPTER 6

INVERTEBRATE BIODIVERSITY IN LAND–INLAND WATER ECOTONAL HABITATS

Jean Giudicelli and Michel Bournaud

INTRODUCTION

The land–water ecotones are generally characterized by higher richness and density than adjacent patches.

Considerable information exists for fish and bird communities in ecotonal situations (Carothers *et al.*, 1974; Welcomme, 1979, 1986; Goulding, 1980). For instance many natural floodplains have diverse and abundant fish fauna, and fishery workers recognise a positive relationship between the abundance of fish stocks and flooding intensity.

Macroinvertebrate studies in the terrestrial–water ecotonal habitats were scarce but have increased in number over the last few years, particularly those on stream ecology. They have been mainly focused on the floodplain which is a wide ecotonal system made of different habitat units. For instance, a special issue of the journal *Regulated Rivers* (1991) was devoted to surface water invertebrates of European alluvial floodplains.

Most of the available studies cover only specific invertebrate groups and emphasize the use of characteristic taxa as environmental describers. Attempts to describe land–water ecotonal habitats in terms of spatial and trophic resources based on total invertebrate biocoenosis composition are fairly rare.

Throughout the present discussion we shall focus on biodiversity in the ecotonal habitats along river systems. These ecotones vary from narrow strips in headwaters – here we find hygropetric or madicolous habitats – to broad wetlands in the floodplain of the lower reaches.

EVALUATION OF BIODIVERSITY AND REPRESENTATIVITY OF RESULTS

Sampling macroinvertebrates in ecotonal habitats

The sampling techniques usually used for epigean water systems are generally insufficient in the case of terrestrial aquatic ecotones. Every ecotone being a complex of microhabitats, sampling has to be performed using tools adapted to every microhabitat type. For instance, sampling the twelve types of micro-habitats of submerged banks of the Rhône river made the use of small hand nets, hand grabs, scraping knives, shears and corers necessary; the sampled areas or volumes were as small as possible, the linear dimensions being lower or equal to 10 cm (Bournaud and Cogérino, 1986).

Evaluation of biodiversity

In most of the existing studies, the biodiversity in a certain sector is expressed by the total number of species found (total richness).

It is preferable to take into account two patterns of invertebrate diversity related to different spatial scales and their relative importance:

– alpha (i.e. within habitat) diversity

– beta (i.e. between habitat) diversity.

Castella *et al.* (1991) conducted their study of aquatic invertebrate assemblages of two contrasting floodplains in this way, those of the Rhône and Ain rivers (France). In the regulated Rhône floodplain there is a clear distinction between the diverse faunal assemblages together with a high taxa richness at the scale of the studied sector (the beta diversity being predominant). In the unregulated Ain river sector, the faunal assemblages are highly overlapping and the taxa richness is high at the scale of a sample (the alpha diversity being predominant). These patterns of faunal distribution and diversity in these ecotonal systems seem to be related to the geomorphology of the floodplains and to the human impacts on them.

ILLUSTRATION OF ECOTONAL BIODIVERSITY IN LAND–STREAM BOUNDARIES

In recent studies on the ecology of fluvial systems, the need is stressed to take into account the channel, the banks, and the alluvial plain at the same time (Jenkins *et al.*, 1984; Minshall *et al.*, 1985).

The banks of a watercourse represent a linear ecotonal system between epigean aquatic, terrestrial, and phreatic environments; a highly complex system owing to the variety of topographic, physico-chemical, and biotic influences exerted on it. The banks have their own abiotical peculiarities: low current velocity, emergence of phreatic water, riparian vegetation, shade.

The macroinvertebrate communities of aquatic banks of a large river in a potamic reach may be considered as an example of an ecotonal community between the aquatic community of the channel and the mainly terrestrial community of the terrestrial bank (Naiman *et al.*, 1988; Cummins *et al.*, 1984). The habitats of the aquatic bank are much more diversified than those of the main channel because the influence exerted by the substrate and by the riparian vegetation enhances the complexity of the riparian system. Several studies have clearly demonstrated the abundance and richness of invertebrate communities near the bank, as well as the diminution of the faunistic richness from the banks towards the channel (Mordukhai-Boltovskoi, 1979; Biggs and Malthus, 1982; Guill, 1985; Cellot and Bournaud, 1986; Anderson and Day, 1986; Moore, 1987).

We should emphasize here that the river bank, as an ecotone, is a habitat complex inhabited by several types of different communities, each having its own biotic characteristics, whereas they are all connected by multiple interactions. Inside the riparian ecotone we have to distinguish between aquatic banks and terrestrial banks.

The aquatic bank sensu lato. This is the zone stretching from the water line towards the channel, up to the zone where the grain-size of the substrate changes. When the slope is slight this change in the substrate granulometry is often scarcely apparent; in such cases the bank *sensu lato* is defined as being the zone whose inferior limit corresponds to that of the hydrophytes; it is the littoral zone of Dussart (1966).

The aquatic bank sensu stricto. This is the upper zone of the ecotone – the most characteristic and the most easily studied. Like Bournaud and Cogérino (1986) for the Upper Rhône river, we consider that its lower limit corresponds to the projection on the river's bottom of a horizontal distance of one meter from the water line. This arbitrary limit is based on the observation that outside this zone (towards the channel) the current velocity rapidly increases.

In other words, the bank *sensu lato* is clearly limited by a change in the grain size; whereas the bank *sensu stricto* may be distinguished from the remainder of the river bed by distinct microhabitats.

The terrestrial bank is represented by water-saturated emergent substrates through direct contact with the aquatic environment. The terrestrial riparian environment is inhabited by two macroinvertebrate communities quite distinct in their composition and structure:

- the *wet edaphic riparian community* represents a mixture of invertebrates belonging to the upper layer of the stream bottom and of species living buried in humid sand or silt of emergent banks;

- the *epigean riparian community* is represented by terrestrial hygrophilous arthropods.

For a biodiversity analysis of the communities of the riparian ecotone we can only rely on a small number of studies, this kind of community being seldom included in studies of lotic ecosystems. Moreover, some papers take into account only one type of riparian fauna (aquatic, terrestrial endogeic, terrestrial epigean); in other publications comparisons are made between riparian communities, and between riparian fauna and benthic fauna of the watercourse (Boumezzough, 1988; Playoust *et al.*, 1989; Prévot *et al.*, 1993). We

Table 6.1 Number of macroinvertebrate taxa recorded in five types of microhabitats of the aquatic banks of the Upper Rhône River (Cogérino, 1989)

Habitat classification	Total recorded	Mean no./microhabitat
Eroding	92	18.0
Submerged (detritus, wood)	78	15.4
Aquatic vegetation	56	26.7
Semi-aerial (roots, helophytes)	36	7.9
Depositing (sand, silt, clay)	32	7.9

shall refer here to results obtained in the study of the upper Rhône system, of a hydrographic basin in southern France, the Arc, and of two basins, the Reghaya and the N'Fis, in the High Atlas Mountains (Morocco).

In the Upper Rhône 170 taxa were identified on the aquatic banks (Cogérino, 1989), which are richer than the river channel (Berly, 1989; Tachet et al., 1991). The taxonomic richness (number of macroinvertebrate taxa) is not homogeneous along the banks; the reaches with hydrology close to that of the most natural river and with the highest variety of microhabitats are inhabited by a more diversified fauna (90–100 taxa) than sectors with artificial variations of the hydrologic regime (43–67 taxa in regulated sectors). Moreover, the taxonomic richness (number of taxa) varies also according to the microhabitat type (Table 6.1).

There is, also, variation in the composition of the riparian aquatic fauna depending on the microhabitat types. Invertebrates are characteristic of the different microhabitats: Leptocerid caddisflies in the eroding-type microhabitat, *Hydropsyche* (also Trichoptera) in the plant roots, and Oligochaeta in the depositing-type microhabitat.

If we turn now to the relative abundance of the functional feeding groups in the river bank zone, we find that:

- all the groups are present;
- the collectors are the prevailing group, as a result of the accumulation of fine debris; inside this group the rheophilic taxa (Simuliidae, Hydropsychidae) are poorly represented, taxa of lentic waters (Chironomini, Tanytarsini) being more abundant;
- the scrapers (i.e. most Molluscs and Coleoptera) represent nearly one third of the total fauna, this environment being characterized by a strong algal development (due to reduced depth, high degree of insolation, rich trophic resources);
- the remaining groups (shredders, especially predators) are not abundant, although their taxonomic richness is high (41 taxa).

We may conclude that the substrate type determines the diversity and the density of taxonomic and functional groups depending on the capacity of its microhabitats to support the development of a periphyton and to determine zones propitious to the accumulation of organic matter. The abundance of riparian fauna depends on the capacity of the invertebrates to make use of the substrate as 'life space' and of its trophic resources.

The wet edaphic riparian communities are poorly known; they are represented by two studies: on one watercourse of low altitude in southern France, the Arc, (Playoust, 1988) and on two mountain watercourses in the High Atlas (Boumezzough, 1988). The high taxonomic richness of these communities is notable: 158 taxa were sampled in the Arc (varying between 46 and 76 taxa in the different sampling stations), the figures for the High Atlas being 135 (between 11 and 58 taxa). It becomes apparent from these studies that this kind of community varies, but very slightly, in composition and structure from one region to another. In the two mentioned areas 93% and 91% of the sampled specimens belong to three taxonomic groups:

- the Diptera show a relative abundance of 56% in the Arc and of 55% in the High Atlas. Not less than 20 families belong to this assemblage, the best represented being Psychodidae, Chironomidae, Tipulidae, Stratiomyiidae, Limoniidae and Dolichopodidae. Their larvae are consumers of organic detritus abundantly present in most microhabitats. These Dipterans are thus responsible for a high biodiversity in the riparian ecotone of the watercourses, whereas their richness and densities are much lower in the epibenthic aquatic assemblage;
- the Oligochaetes have a relative abundance of 24% and 18% respectively;
- a third dominant group is represented by the Coleoptera (Elmidae, Dryopidae, Hydraenidae) – all common in sandy and gravelly margins of streams and also abundant in the aquatic epibenthic assemblage of the channel.

The riparian population (158 taxa) of the Arc was analysed using reciprocal averaging (Prévot et al., 1993). Results show that spatial variation of the riparian edaphic zoocoenosis is mainly determined by the nature of the substratum, by upstream–downstream eutrophication and, to a lesser extent, by the hydrological variations along the watercourse. Axis I plotted with axis III (Figure 6.1) illustrates a correlation between the gradient of enrichment of the environment and the nature and structure of the substratum. It isolates three faunal sets typical of a polluted sandy and silty environment, of a silty environment with vegetal fragments, and of a sedimentary substratum covered with bryophytes.

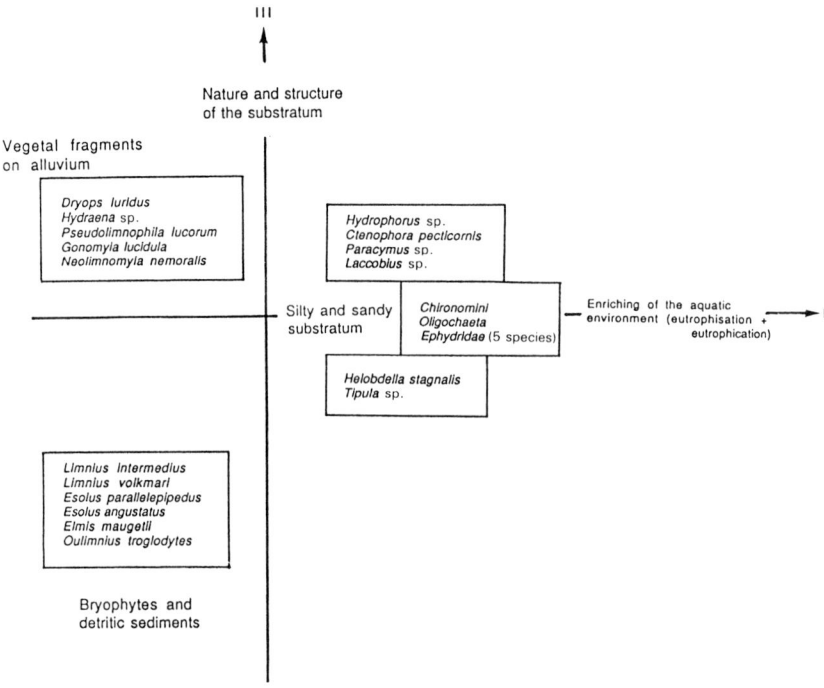

Figure 6.1 Distribution of the most abundant wet edaphic riparian invertebrates and of their typical microhabitats in the factorial plane I/III (from Prévot et al., 1993)

The riparian endogenous zoocoenosis of the Arc was compared with that of the benthal channel, inhabited by 216 taxa (Playoust et al., 1989). The two assemblages are clearly distinct in their composition and structure, despite the fact that 45% of species are shared by both of them. If reciprocal averaging is used, benthic and riparian samples cluster separately. On axis II the benthic/riparian samples pair from each site show the same value according to an upstream–downstream trophic gradient. Therefore, the wet edaphic riparian populations, because of their high richness, can be considered as bioindicators of the water quality in the same way as benthic populations.

The communities of epigean riparian invertebrates, although less rich than those previously analysed, are still rather diversified (some 100 taxa) in the studied Mediterranean hydrographic networks. Their faunal composition is clearly different. Two groups are prevalent in terms of species and individuals: Staphylinid and Carabid Coleopterans. For instance, 132 Carabid species were found in the riverine habitats of the Weser floodplain in Germany (Gerken et al., 1991). In this ecotonal environment species are spatially selected by some ecofactors; indeed, the Carabidae are fairly well known bioindicators of grain size, humidity, vegetation and flood dynamics.

Simultaneous analysis of epigean and endogeic riparian assemblages of 26 stations on two watercourses in the High Atlas, showed that these two community types are quite distinct in their faunistic composition, although sharing 41 taxa of 125 in total (Boumezzough, 1988). In this case, use of reciprocal averaging showed clear relations between the composition of the riparian communities and abiotic factors. It was possible to distinguish:

- an assemblage of psychrostenothermic epigean and endogeic species inhabiting environments of the 'Mittelgebirge' or of the high mountains;
- an assemblage of endogeic and thermophilic species characterizing watercourses of low altitudes;
- an assemblage of epigean species typical for environments with an unstable regime with periods of low water level, or even of total drying up;
- an assemblage of epigean and endogeic eurythermic species able to tolerate the strong variations in temperature and hydrological regime – two main factors in North African streams.

All these studies show that the communities of the riparian ecotone are endowed with high biodiversity within spatial (local, and upstream–downstream) and temporal variations.

ROLE OF TERRESTRIAL–AQUATIC BOUNDARY CHARACTERISTICS IN INFLUENCING BIODIVERSITY

The tendency of communities to possess an increase in abundance and richness in an ecotonal situation is called 'the edge effect' (Odum, 1971). The edge effect comes from two phenomena:

(1) The intra-ecotonal complexity and heterogeneity

The environment of terrestrial–aquatic ecotones is influenced by four types of natural disturbances: erosion, sedimentation, inundation, desiccation. The interaction of these processes creates environmental complexity and heterogeneity within ecotonal areas. Indeed, these zones represent highly differentiated environments with respect to spatial and temporal variability in hydrodynamics, in water chemistry, in structure of sediments, in detritus distribution, and in development of aquatic and semiaquatic vegetation. In most riparian ecotones the submerged vegetation, branches and roots provide additional feeding substrates and shelters. These characteristics which define the terrestrial–aquatic ecotonal habitats may affect composition, density and richness of animal communities. Therefore, patterns of distribution, abundance and richness of invertebrates in this kind of ecotone appear to be affected by the various habitats present and the amount of nutritional resources.

In their faunal approach to aquatic microhabitats of the banks of a large watercourse, the Upper Rhône, Bournaud and Cogérino (1986) explored the zone corresponding to the upper fringe of the submerged banks. They distinguished here 19 different microhabitats and defined three major habitats: erosional with rough stable substrate, depositional with fine substrate, and vegetation. As a result, in the fluvial systems the substrate diversity as well as the range of current speeds are higher at the banks, the channel being much more homogeneous in these respects. The new habitats that appear in the aquatic bank of the river induce a particular fauna, often with a high taxonomic richness.

Similar results were obtained in the floodplain ecotonal areas which are patchworks of habitats. Thus, Dumont (1986a) related the 'extreme diversity' of the limnetic zooplankton in a wide floodplain, the Internal Delta of the Niger river system (mean species per lake: 10, total of the delta: 22) to the presence of diverse biotopes. Of course, in this area during the flood season, Welcomme (1986) distinguished 2 major habitats and 3 microhabitats in the main channel, 5 major habitats and 14 microhabitats in the floodplain.

(2) The use of terrestrial–aquatic ecotonal systems by invertebrates

Riparian ecotones are connecting lines or areas for species flux. Terrestrial–aquatic boundaries may be permeable to aquatic insects. Many species occupy different habitats during their life and depend on the land–water interface to perform their life cycle and get their food.

Nearly all aquatic insects are terrestrial at some stage of their cycle. About 6,000 species in 8 families of beetles are aquatic in their adult and larval stages; in addition, about 3,500 more species in another 12 families of beetles are aquatic or terrestrial in either their adult or larval stages (Spangler, 1986). Many aquatic insects spend some portion of their life in the riparian zone. They can find in this ecotonal environment suitable areas for feeding, pupation, adult emergence and mating, egg-laying.

An important ecological aspect of semiaquatic habitats is the presence of such material as driftwood and various animal and plant detritus. These materials provide food sources, shelter and sites for breeding and development. Terrestrial leaf litter is an important part of the diet of aquatic insects (Cummins, 1973; Erman, 1983; Hawkins and Sedell, 1981). Most of them are omnivorous, food needs changing with instars (Anderson and Cummins, 1979; Chapman and Demory, 1963). Indeed, some species feed on epilithic algae produced within the stream in the first instars, later they shred decaying leaves from the riparian vegetation.

Many aquatic insects move towards the banks, into slower water, to pupate or to emerge. This applies to several genera of Ephemeroptera, e.g. *Ecdyonurus, Epeorus, Oligoneuriella, Baetis, Ephemerella* (Verrier, 1956). Nearly all aquatic Tipulidae leave the water to pupate in the nearby wet soil, moss or litter (Byers, 1978). Pupation also is terrestrial in almost all species of aquatic Coleopterans.

Egg laying in the riparian zone is also a source of increasing biodiversity in the riparian ecotone. Most imagoes of aquatic insects avoid entering the water to oviposit, they lay their eggs at the water's edge. Several species of Trichoptera Limnephilidae lay eggs away from water (Wiggins, 1973). This is an adaptation to drying conditions and to hydrologic variations of land–water ecotonal areas: eggs are laid along the edge of drying streams and ponds, larvae hatch but remain in the gelatinous egg mass until the stream or pond is reflooded (Wiggins, 1973). Many aquatic Coleopteran (Dytiscidae, Hydraenidae, Elmidae, Dryopidae, Helodidae) oviposit at the shoreline (Doyen and Ulrich, 1978; Holland, 1972). Oviposition at the water edge, followed by migration of the larvae to their open water habitat, has been observed in several Diptera, e.g. most species of *Simulium* (Peterson and Wolfe, 1958; Rivosecchi, 1978), Culicidae of genus *Aedes* (MacHaffey, 1972), and Blephariceridae (Hubault, 1927; Giudicelli, 1964). So riparian ecotonal habitats are important for invertebrates as nursery areas.

TAXONOMIC RICHNESS AND SPECIFICITY IN THE ECOTONAL COMMUNITIES

The ecotonal communities commonly contain many of the organisms of each overlapping community; this induces a high taxonomic richness. They also contain organisms which are characteristic and sometimes restricted to the ecotone (Odum, 1971); this induces a faunistic specificity. There follow a few examples. In the upper Mississippi river the highest invertebrate diversity was found in shallow channel border areas with macrophyte development (Anderson and Day, 1986), (Figure 6.2).

Out of more than 400 taxa of Rotifera and Crustacea identified in the Murray (the longest Australian river) about one third were recorded only from the vegetated littoral zone (Shiel, 1986). In the French Upper Rhône, upstream from Lyon, 170 taxa were identified on the aquatic banks (Cogérino, 1989). The banks of the Upper Rhône are richer in this respect than the river channel where 118 taxa have been identified (Berly, 1989). In the same river sector 178 taxa were encountered in the floodplain ecotonal system (Castella *et al.*, 1991). Considering the known facts on the riparian aquatic fauna of the Rhône river and of other watercourses, the faunal density in the banks is at least two times higher than in the channel (Cogérino, 1989), and two thirds of the species living in a large watercourse can be found in the river banks (Moore, 1987).

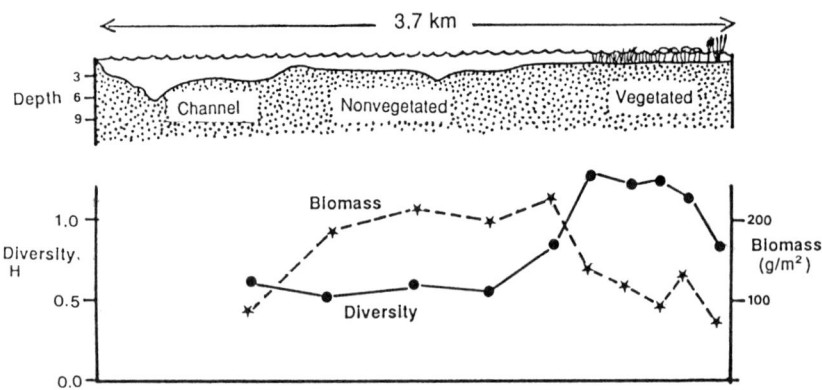

Figure 6.2 Comparison of the biomass and diversity distribution at the sampling stations along transect 3 in Pool 19 of the Upper Mississippi river (from Anderson and Day, 1986)

Conversely, some ecotones, characterized by spatial homogeneity and/or peculiar and severe environmental conditions are less rich in species than the adjacent aquatic environment. However, they harbour a specialized fauna composed of species which have developed morphological and biological adaptations. This situation may be observed in the groundwater ecotones, in the hypogean interstitial biotopes occurring along the margins of running water, in the madicolous (hygropetric) habitats, and also in temporary waters that could be considered as temporal aquatic–terrestrial ecotones.

It seems that the surface water–groundwater ecotone is not a richer zone but corresponds to a zone of intermediate richness between the aquatic surface environments and the underground ones (Gibert *et al.*, 1990).

The damp sandy beaches stretching along streams represent a peculiar ecotonal zone that was studied by Angelier (1953), Danielopol (1976), Danielopol and Hartmann (1986) and Ruttner-Kolisko (1961) in Europe. This subsurface environment is inhabited by the interstitial fauna, or psammon, in which the Hydrachnellidae (water mites) is often the best represented group. The psammic community is made up of two faunal elements (Angelier, 1959a). The psammophilous species are of epigean origin and live in the interstices that are formed in the mixture of coarse sand, gravel and pebbles. Many small larvae of Diptera (Chironomidae), Ephemeroptera (Habrophlebiidae), Plecoptera (Leuctridae), Coleoptera (Elmidae) are psammophilous. The psammobiont species, which only occur in the psammon, are much more specialized and are adapted to an interstitial life; they are often blind and move slowly between sand grains. Compared with their close epigean relatives, interstitial Ostracoda show morphological reductions, e.g. in eye structure, in body pigments, in carapace size, in the number of the setae (Danielopol and Hartmann, 1986).

In Corsica, Angelier (1959b, 1959c) investigated Hydrachnellidae communities in both epibenthic and interstitial stream habitats. We refer to the

Table 6.2 Species richness values for epibenthic and interstitial habitats in Corsican rivers (data from Angelier, 1959b, 1959c)

Sampling sites	Epibenthic habitats	Interstitial habitats	Species occurring in both habitats
Site no 5	9	4	0
Site no 12	14	9	2
Site no 13	13	5	2
Site no 24	16	10	1
Site no 38	18	7	3
Total richness	28	24	
Mean richness	14	7	

samples collected by Angelier in the lower reaches of five mountain streams (these samples are richer in species and individuals than the others collected in the middle and upper reaches). In this study the epibenthic samples and the endogean ones were taken at the same time. Results are presented in Table 6.2. The total species richness is nearly the same in the epibenthic and in the ecotonal interstitial environments, but at the scale of a sample the richness is lower in the ecotone. This means that the two communities are quite different in their faunal composition.

So, according to these examples, the epigean environment is obviously richer in species than the subsurface ecotonal one. This can be explained by the following features of this type of ecotone: obscurity, variations in temperature, water flow and other factors of a lower amplitude than in neighbouring surface habitats, and impoverished trophic resources; moreover, the epigean environment is more diverse in its habitat types than the hypogean one.

Many of the invertebrates living in the madicolous habitats are peculiarly modified for remaining in the thin water film produced at the edge of swift stony streams. Vaillant (1955) collected about 300 invertebrate species in 74 madicolous sites located in southern France, in Corsica and in North Africa; 83 species occurred only in this type of ecotonal habitat.

In lentic ecotones of the littoral lacustrine zone, sandy beach and stony shores are the lowest in species abundance; here, the edge effect is not observed, probably owing to the scarcity of detritic organic matter (Pieczyńska, 1990). Conversely, an increase in invertebrate abundance can be observed on lake shores with muddy substrates and rich vegetation. Semi-aquatic forms which are typical of these shore zones are very abundant: Oligocheta, Coleopterans and mainly Dipterans occur in high density and richness (Pieczyńska, 1990).

From these examples it appears that two describers of the ecotonal community may be considered. First, the qualitative faunistic characteristics (what taxa are present only in the ecotonal area?) and second, the taxonomic richness (how many taxa are present?). These are the main biotic parameters of an ecotonal habitat. As an illustration we consider a regulated river with a system of canals (a head race upstream of a power station, and a diversion section downstream) and a by-passed section (the 'old' course), such as the Rhine river in the Rhinau-Kappel reach (France). In this situation we observed (Bournaud *et al.*, 1991) either faunistic convergences or divergences between the bank ecotonal habitats and the channel of the river. In the concrete banks of the canals, with only one to four different microhabitats, the macroinvertebrate fauna is quite different from that in the channel; this is a faunistic divergence (Figure 6.3). Under these conditions, there is a high specialization in the bank community. On the contrary, in the by-passed section, in the old course of the river, the two communities are very similar: there is a faunistic convergence between them.

Moreover we observe an unusual correlation between taxonomic richness and habitat heterogeneity, measured by means of the number of microhabitats. Now it is important to observe that, as far as the taxonomic richness is concerned, the banks and the channel of the canals have nearly the same richness, while it is the opposite in the by-passed section where the banks are twice as rich as the channel. So we observe that in ecotonal habitats with extreme and severe spatially homogeneous conditions (as concrete banks of canals) the ecotonal fauna is anomalous in the nature of the species recorded and the low taxonomic richness. On the contrary, in spatially heterogeneous ecotonal habitats, with more natural microhabitats, the fauna is not specific to these habitats, but has a high taxonomic richness.

So, the taxonomic richness of an ecotonal habitat is linked not only to its spatial heterogeneity but also to its ecological singularity (extreme conditions of life).

In conclusion, taxonomic richness and specificity in the ecotonal communities vary in opposite ways. They are linked to the originality of the habitat; a good describer of this originality may be the spatial homogeneity of the habitat. A similar result has been obtained in the slikke area, the ecotonal habitat between salt and freshwater in an estuary, where the life conditions are particularly extreme and homogeneous; in this area only four macroinvertebrate species make more than 99% of the total faunistic density (Peres, 1961). In our opinion, the biodiversity of an ecotone depends on the number of connections and gradients between adjacent ecosystems. The more numerous and spatially and temporally diverse these connections are the more complex and richer in trophic resource the ecotonal system will be, which will be inhabited, in such circumstances, by a rich and varied fauna. This explains why the richest communities are those of the river ecotones, particularly those of the floodplain.

Invertebrate diversity

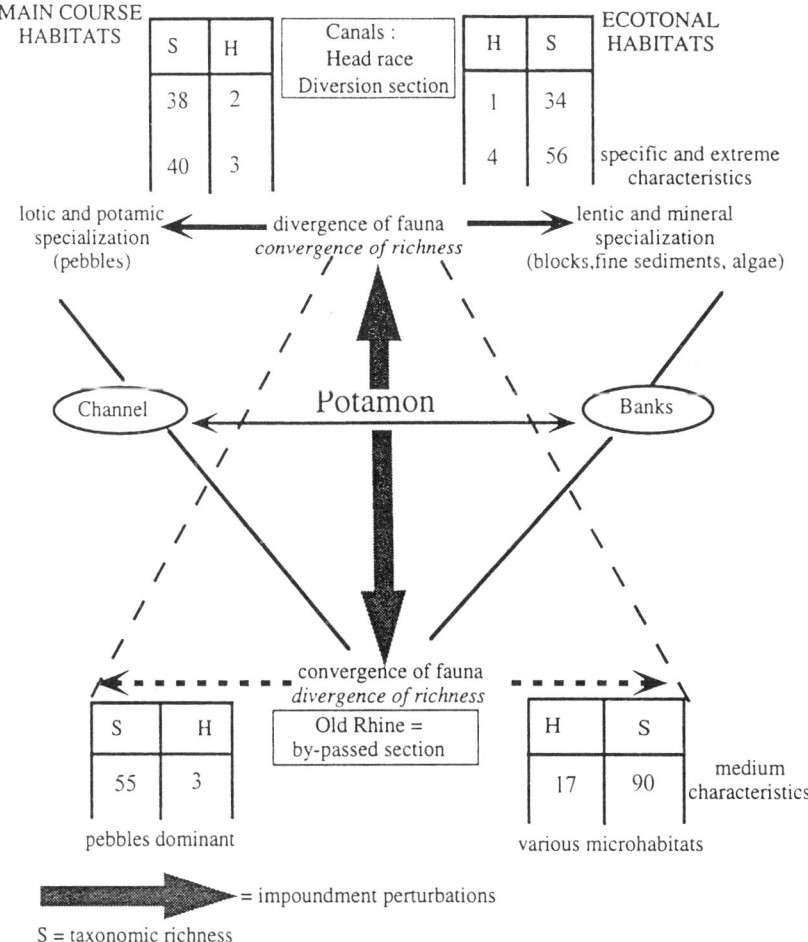

Figure 6.3 Characteristics of the aquatic banks, as ecotonal habitats, of a large regulated river, the Rhine river at Rhinau-Kappel. Comparison with the channel

CONCLUSIONS

Ecotones are very important to maintain biodiversity of adjacent open water ecosystems; they are also involved in providing trophic resources.

Floodplains and riparian zones can be very important in regulating the nutrient economy of streams (Meyer *et al.*, 1988) which are sources of carbon for lakes and river systems. The riparian ecotones largely determine the quantity, quality and retention of allochthonous organic material received by streams. Chironomids, Tipulids, burrowing mayflies and caddisfly larvae that live along the edges of the rivers, are involved in organic matter movement between ecotones and open water. Most of all, riparian and floodplain habitats provide a lot of invertebrates to the stream or to the main channel. High faunistic densities and richness of the ecotonal systems of large watercourses enable them to play an essential role as reservoirs for the potamon.

Some examples illustrate the role of floodplain and riparian aquatic habitats in maintaining the biodiversity and productivity of river and fluvial systems. So, macroinvertebrates and zooplankton, which are abundant in the backwaters of the floodplain, can drift into the main channel of large river systems. Shaeffer and Nickum (1986) observed that backwater areas in the upper Mississippi river which are very productive, have beneficial effects on the downstream main channel habitats. Indeed, many of the backwater invertebrate taxa have been found to drift out of these areas into the main channel, providing food for larvae, juvenile and adult fishes. Similarly, floodplains associated with the Murray river (Australia) seasonally provide a substantial proportion of the downstream river zooplankton (Shiel, 1986).

Moreover, after a disturbance, the river channel may be colonized by invertebrates from the banks or from the floodplain. In the Nile and Niger river systems, the zooplankton is washed away and eliminated from the main channel during the flood and later on recolonises the river from the floodplain (Dumont, 1986a, 1986b). Recent studies in Mediterranean streams (Badri *et al.*, 1987; Ortega *et al.*, 1991) have emphasized the importance of marginal ponds on floodplains after catastrophic flooding as temporary shelters for invertebrates displaced from the main channel during the flood period. During the spate, invertebrate populations are washed away from the river bed and deposited lower down in the floodplain where the velocity of the flow is lower. In this ephemeral ecotonal environment, invertebrates, which passively entered the catastrophic drift through bottom physical disturbance, can find shelters in the riparian vegetation and under rocks and cobbles scattered in the floodplain. The ordinary drift of benthic organisms dispersed in the floodplain has been recognized as the main factor in the recolonization of the channel bottom that had been denuded by the flood.

The high taxonomic richness observed in the aquatic banks of large rivers is partly the consequence of the role as refuge for the macroinvertebrate populations that is played by the microhabitats of the bank against the hydraulic stresses during floods and spates (Gaschignard, 1984).

Therefore, a simplification of the heterogeneity of the ecotone (through modification of the hydrological regime, or river regulation) will involve a decline in faunistic richness. Restructuring stream courses will reduce the possibility of recolonization by riparian fauna, and this will have detrimental effects on the biological integrity of the river system.

REFERENCES

Anderson, N.H. and Cummins, K.W. (1979). Influences of the diet on the life histories of aquatic insects. *Journal of Fishery Research Board of Canada*, **36**, 335–42

Anderson, R.V. and Day, D.M. (1986). Predictive quality of macroinvertebrate–habitat associations in lower navigation pools in the Mississippi River. *Hydrobiologia*, **136**, 101–12

Angelier, E. (1953). Recherches écologiques et biogéographiques sur la faune des sables submergés. *Archives de Zoologie Expérimentale et Générale*, **90**, 37–162

Angelier, E. (1959a). Les eaux douces de Corse et leur peuplement. In Angelier, E. (ed.) *Hydrobiologie de la Corse*, pp. 1–56. Hermann, Paris

Angelier, E. (1959b). Ecologie et biogéographie des Acariens (Hydrachnellae et Porohalacaridae) des eaux superficielles. In Angelier, E. (ed.) *Hydrobiologie de la Corse*, pp. 139–60. Hermann, Paris

Angelier, E. (1959c). Acariens psammiques (Hydrachnellae et Porohalacaridae). In Angelier, E. (ed.) *Hydrobiologie de la Corse*, pp. 161–95. Hermann, Paris

Badri, A., Giudicelli, J. and Prévot, G. (1987). Effets d'une crue sur la communauté d'invertébrés benthiques d'une rivière méditerranéenne, Le Rdat (Maroc). *Acta Oecologica, Oecologia Generalis*, **8**(4), 481–500

Berly, A. (1989). Distribution spatio-temporelle des peuplements macrobenthiques prélevés par dragages dans une station du Haut Rhône. Thèse en sciences, Université de Lyon I

Biggs, B.J. and Malthus, T.J. (1982). Macroinvertebrates associated with various aquatic macrophytes in the backwaters and lakes of Upper Clutha Valley, New Zealand. *New Zealand Marine and Freshwater Research*, **16**, 81–8

Boumezzough, A. (1988). Contribution à la connaissance des invertébrés ripicoles épigés et endogés en zone méditerranéenne. Etude des peuplements ripicoles de deux réseaux hydrographiques du Haut Atlas marocain. Thèse de Doctorat ès Sciences, Université d'Aix-Marseille III

Bournaud, M. and Cogérino, L. (1986). Les microhabitats aquatiques des rives d'un grand cours d'eau: approche faunistique. *Annales de Limnologie.*, **22**(3), 285–94

Bournaud, M., Tachet, M., Doledec, H. and Cellot, B. (1991). Étude de l'état actuel des écosystèmes rhénans dans le secteur de Rhinau-Kappel et suivi de leur reconstitution de 1987 à 1990 au moyen des descripteurs macroinvertébrés. Rapport final Convention C.N.R.S.-Sandoz, URA 1451, Ecologie des Eaux douces et grands Fleuves, Université de Lyon I

Byers, G.W. (1978). Tipulidae. In Merritt, R.W. and Cummins, K.W. (eds.) *An Introduction to the Aquatic Insects of North America*, pp. 285–310. Kendall and Hunt Publ., Dubuque, Iowa

Carothers, S.W., Johnson, R.R. and Aitchison, S.W. (1974). Population structure and social organisation of southwestern riparian birds. *American Zoologist*, **14**, 97–108

Castella, E., Richardot-Coulet, M., Roux, C. and Richoux, P. (1991). Aquatic macroinvertebrate assemblages of two contrasting floodplains: the Rhône and Ain rivers. *Regulated Rivers*, **6**(4), 289–300

Cellot, B. and Bournaud, M. (1986). Modifications faunistiques engendrées par une faible variation de débit dans une grande rivière. *Hydrobiologia*, **135**, 223–32

Chapman, D.W. and Demory, R.L. (1963). Seasonal changes in the food ingested by aquatic insects, larvae and nymphs in two Oregon streams. *Ecology*, **44**, 140–6

Cogérino, L. (1989). Les rives aquatiques de grands cours d'eau: caractérisation mésologique et faunistique. Thèse en Sciences, Université de Lyon I

Cummins, K.W. (1973). Trophic relations of aquatic insects. *Annual Review of Entomology*, **18**, 183–206

Cummins, K.W., Minshall, G.W., Sedell, J.R., Cushing, C.E. and Petersen, R.C. (1984). Stream ecosystem theory. *Verhandlungen der Internationalen Vereinigung für Theoretische und Angewandte Limnologie*, **22**, 1818–27

Danielopol, D.L. (1976). The distribution of the fauna in the interstitial habitats of riverine sediments of the Danube and the Piesting (Austria). *International Journal of Speleology*, **8**, 23–51

Danielopol, D.L. and Hartmann, G. (1986). Stygobiont Ostracoda from inland subterranean waters. In Botosaneanu, L. (ed.) *Stygofauna Mundi*, pp. 265–78. E.J. Brill/ Dr W. Backhuys, H.S. Leiden

Doyen, J.T. and Ulrich, G. (1978). Aquatic Coleoptera. In Merritt, R.W. and Cummins, K.W. (eds.) *An Introduction to the Aquatic Insects of North America*, pp. 203–31. Kendall and Hunt, Dubuque, IA

Dumont, H.J. (1986a). Zooplankton of the Niger system. In Davies, B.R. and Walker, K.F. (eds.) *The Ecology of River Systems*, pp. 49–59. Dr W. Junk, Dordrecht

Dumont, H.J. (1986b). Zooplankton of the Nile system. In Davies, B.R and Walker, K.F. (eds.) *The Ecology of River Systems*, pp. 75–87. Dr W. Junk, Dordrecht

Dussart, B. (1966). *Limnologie. L'étude des Eaux Continentales*. Gauthier-Villars, Paris

Erman, N.A. (1983). The use of riparian systems by aquatic insects. In Warner, R.E. and Hendrix, K.M. (eds.) *California Riparian Systems*, pp. 177–82. University of California Press, Berkeley

Gaschignard, O. (1984). Impact d'une crue sur les macroinvertébrés benthiques d'un bras du Rhône. *Verhandlungen der Internationalen Vereinigung für Theoretische und Angewandte Limnologie*, **22**, 1997–2001

Gerken, B., Dörfer, K., Buschmann, M., Kamps-Schwob, S., Berthelmann, J. and Gertenbach, D. (1991). Composition and distribution of Carabid communities along rivers and ponds in the region of Upper Weser (NW/NDS/FRG) with respect to protection and management of a floodplain ecosystem. *Regulated Rivers*, **6**(4), 313–20

Gibert, J., Dole-Olivier, M.J., Marmonier, P. and Vervier, P. (1990). Subsurface water–groundwater ecotones. In Naiman, R.J. and Décamps, H. (eds.) *The Ecology and Management of Aquatic–Terrestrial Ecotones*, pp. 199–225. MAB Book Series 4. UNESCO/Parthenon, Paris/Carnforth

Giudicelli, J. (1964). L'oviposition chez les Blépharocérides (Diptera). *Revue Française d'Entomologie*, **31**, 116–9

Goulding, M. (1980). *The Fishes and the Forest: Explorations in Amazonia Natural History*. University of California Press, Los Angeles

Guill, G.N. (1985). Differences in the distribution of benthic macroinvertebrates between the banks of the Savannah River. *Proceedings of Annual Meeting of North American Benthological Society*, Corvalis, Oregon, **33**, 1–7

Hawkins, C.P. and Sedell, J.R. (1981). Longitudinal and seasonal changes in functional organization of macroinvertebrate communities in four Oregon streams. *Ecology*, **62**(2), 387–97

Holland, D.G. (1972). *A Key of the Larvae, Pupae and Adults of the British Species of Elminthidae.* Freshwater Biological Association, Scientific Publication No. 26

Hubault, E. (1927). Contribution à l'étude des invertébrés torrenticoles. *Bulletin Biologique de France et Belgique*, Suppl. **9**, 1–390

Jenkins, R.A., Wade, K.R. and Pugh, E. (1984). Macroinvertebrate–habitat relationships in the River Teifi catchment and the significance to conservation. *Freshwater Biology*, **14**(1), 23–42

MacHaffey, D.G. (1972). Photoperiod and temperature influences on diapause in eggs of the floodwater mosquito, *Aedes vexans*. Meigen. *Journal of Medical Entomology*, **9**, 564–71

Meyer, J.L., McDowell, W.H., Bott, T.L., Elwood, J.W., Ishikazi, C., Melack, J.M., Peckarsky, B.L., Peterson, B.J. and Rublee, P.A. (1988). Elemental dynamics in streams. *Journal of the North American Benthological Society*, **7**(4), 410–32

Minshall, G.W., Cummins, K.W., Peterson, R.C., Cushing, C.E., Bruns, D.A., Sedell, J.R. and Vannote, R.L. (1985). Developments in stream ecosystem theory. *Canadian Journal of Fisheries and Aquatic Sciences*, **42**, 1045–55

Moore, K.M.S. (1987). Ecology of aquatic habitats associated with stream margins. Thesis DSc, Oregon State University

Mordukhai-Boltovskoi, D. (1979). The river Volga and its life. In Illies, J. (ed.) *Monographiae Biologicae*, pp. 235–94. Dr W. Junk, The Hague

Naiman, R.J., Décamps, H., Pastor, J. and Johnston, C.A. (1988). The importance of boundaries to fluvial ecosystems. *Journal of the North American Benthological Society*, **7**, 289–306

Odum, E.P. (1971). *Fundamentals of Ecology*. W.B. Saunders, Philadelphia, London

Ortega, M., Suarez, L., Vidal-Abarca, R., Gomez, R. and Ramirez-Diaz, L. (1991). Aspects of postflood recolonization of macroinvertebrates in a 'Rambla' of South-East Spain ('Rambla del Moro': Segura River Basin). *Verhandlungen der Internationalen Vereinigung für Theoretische und Angewandte Limnologie*, **24**, 1994–2001

Peres, J.M. (1961). *Océanographie Biologique et Biologie Marine. I. La Vie Benthique.* Presses Universitaires de France, Paris

Peterson, D.G. and Wolfe, L.S. (1958). The biology and control of black-flies (Diptera: Simuliidae) in Canada. *International Congress of Entomology*, **10**(3), 551–64

Pieczyńska, E. (1990). Lentic aquatic–terrestrial ecotones: their structure, functions and importance. In Naiman, R.J. and Décamps, H. (eds.) *The Ecology and Management of Aquatic–Terrestrial Ecotones*, pp. 103–40. MAB Book Series 4. UNESCO/Parthenon, Paris/Carnforth

Playoust, C. (1988). Etude d'un réseau hydrographique méditerranéen de basse altitude: l'Arc (Bouches-du-Rhône). Hydrochimie, communautés benthiques et ripicoles, impact des perturbations. Thèse en Sciences, Université d'Aix-Marseille III

Playoust, C., Musso, J.J. and Prévot, G. (1989). Étude comparée des communautés benthiques et ripicoles endogées d'un réseau méditerranéen perturbé: l'Arc (Bouches-du-Rhône, France). *Revue des Sciences de l'Eau*, **2**, 587–605

Prévot, G., Musso, J.J., Legier, P. and Playoust, C. (1993). Analysis of damp endogenous riparian communities of a mediterranean hydrographic network: the Arc (Bouches-du-Rhône, France). *Hydrobiologia*, **263**, 45–53

Regulated Rivers (1991). Special issue on surface–water invertebrates of European alluvial floodplains. **6**, 239–350

Rivosecchi, L. (1978). Simuliidae Diptera Nematocera. In *Fauna d'Italia*, 13. Ed. Calderini, Bologna

Ruttner-Kolisko, A. (1961). Biotop und Biozönose des Sandufers einiger osterrischer Flüsse. *Verhandlungen der Internationalen Vereinigung für Theoretische und Angewandte Limnologie* **14**, 362–8

Shaeffer, W.A. and Nickum, J.G. (1986). Relative abundance of macroinvertebrates found in habitats associated with backwater area confluences in Pool 13 of the Upper Mississippi River. In Smart, M.M., Lubinski, K.S. and Schnick, R.A. (eds.) *Ecological Perspectives of the Upper Mississippi River*, pp. 113–20. Dr W. Junk, Dordrecht

Shiel, R.J. (1986). Zooplankton of the Murray–Darling system. In Davies, B.R and Walker, K.F. (eds.) *The Ecology of River Systems*, pp. 661–77. Dr W. Junk, Dordrecht

Spangler, P.J. (1986). *Insecta: Coleoptera*. In Botosaneanu, L. (ed.) *Stygofauna Mundi*, pp. 622–31. E.-J. Brill, Dr W. Backhuys H.S. Leiden

Tachet, H., Bournaud, M., Berly, A. and Cellot, B. (1991). Le macrobenthos d'un fleuve et son milieu: interactions spatio-temporelles à l'échelle d'une station. *Bulletin d'Ecologie* **22**(1), 187–94

Vaillant, F. (1955). Recherches sur la faune madicole de France, de Corse et d'Afrique du Nord. Thèse de Doctorat ès Sciences, Faculté des Sciences de l'Université de Paris

Verrier, M.L. (1956). *Biologie des Ephémères*. Collection Armand Colin, Paris

Welcomme, R.L. (1979). *Fisheries Ecology of Floodplain Rivers*. Longman, London

Welcomme, R.L. (1986). The Niger River system. In Davies, B.R and Walker, K.F. (eds.) *The Ecology of River Systems*, pp. 9–23. Dr W. Junk, Dordrecht

Wiggins, G.B. (1973). A contribution to the biology of Caddisflies (Trichoptera) in temporary pools. *Life Sciences Contributions Royal Ontario Museum*, **88**, 1–28

CHAPTER 7

AMPHIBIAN DIVERSITY AND LAND–WATER ECOTONES

Pierre Joly and Alain Morand

INTRODUCTION

A high diversity of species is a main feature of ecotonal habitats. Such a high diversity relies on the presence at the ecotone of species specific to boundary habitats, which coexist with species from the two adjacent patches. Because their complex life cycle includes an aquatic phase, which usually takes place in water–land ecotonal habitats, many amphibian species participate in this diversity. In most amphibiotic species, egg deposition and larval development take place in shallow water at the boundary of aquatic ecosystems. These habitats may be as diverse as lakes, fish ponds, springs, streams or pools in marshes. Except in small sites, or in very specialized ones (such as caves), the presence of amphibians is restricted to the edges of the sites where they form a part of the specific biotic components of the ecotone.

Usually, the ecotone is the place where the larvae develop. However, in some species, such as water frogs, the whole cycle takes place at the ecotone, despite a crucial change of the life style of the individuals after they have experienced metamorphosis. In other species, such as newts, both adults and larvae occupy the ecotone, but only seasonally.

After metamorphosis, the landward migration of the juveniles contributes to the export of organic matter from the aquatic towards the terrestrial ecosystems. Because of the great abundance of tadpoles in the littoral zone of many ponds, this export may represent an important part in the matter and energy fluxes across the ecotone, especially in such abundant species as land frogs and toads.

The stability of ecotonal habitats mainly depends on hydrologic processes, the temporal pattern of which often appears unpredictable, therefore the spatial–temporal variability constitutes a source of selective pressure acting on the success of amphibian reproduction. The responses to this selective pressure are diverse and affect both larval and adult stages.

These considerations rest on a basic knowledge of amphibian biology. However, the diversity of amphibian species has never been analysed in an ecotonal perspective. The aim of the present chapter is to survey the present knowledge on the following questions:

(1) Can we define the terrestrial–aquatic ecotone on the basis of amphibian habitat use?

(2) What are the ecological implications of such an ecotone?

(3) What are the relationships between species diversity and spatial–temporal variability at the ecotone? Which adaptive traits may be identified as responses to the constraints of ecotonal habitats?

(4) What is the share of the amphibians in the productivity and matter fluxes across the ecotone?

(5) What are the impacts of human activities on ecotonal habitats used by the amphibians and what are the possibilities of management of amphibian diversity?

TYPOLOGY OF THE ECOTONES USED BY AMPHIBIANS

Water–land ecotones may be regarded as a functional concept which may apply at different scales. At the scale of the landscape, the patches may be the functional units of an alluvial floodplain. In this context, the composition of the amphibian communities depends on the degree of spatial–temporal variability experienced by each site within a functional unit (Joly and Morand, 1994; Morand and Joly, 1995). In the Rhône valley, the most unstable sites are occupied by *Bombina variegata*. Several species, such as *Bufo calamita*, *Hyla arborea* or *Pelodytes punctatus*, are characteristic of an intermediate level of instability. Other species, such as *Bufo bufo*, *Rana temporaria* or *Rana lessonae*, prefer sites where the water level is more predictable. Some species, such as *Salamandra salamandra* or *Alytes obstetricans*, are missing in the alluvial valley, probably because the adults do not survive submersion during the floods. A last group of ubiquitous species (*Rana dalmatina*, *R. ridibunda* and *Triturus helveticus*) breed in a great diversity of habitats. The amphibian species richness reaches its highest values in sites experiencing an intermediate level of variability. In the Rhône valley, such sites are mesotrophic temporary ponds the water of which mainly comes from the aquifer. These sites are occasionally flooded.

At the scale of the habitat, two kinds of ecotones may be distinguished according to the longitudinal or lateral structure of the river and of its dependent habitats within the floodplain. In longitudinal structure, the ecotones of the headwaters of the streams show specific physical features. In these habitats, the water level varies greatly throughout the year including periods of drying out. The upstream location of these habitats implies a high probability of oligotrophy. Several amphibian species, such as *Salamandra salamandra* (Thiesmeier, 1991) or *Euproctus asper* (Clergue-Gazeau and Martinez-Rica, 1978), breed or lay their clutches in these ecotones.

Most other amphibiotic species breed in habitats such as old channels or oxbow lakes which may be qualified as riparian on the lateral structure of the stream (Alford, 1986; Griffiths and Mylotte, 1986; Augert and Joly, 1994). The land–water ecotonal gradient covers a greater length because of gentle slopes at the end of oxbow lakes or of lateral channels of a river than are found along the lateral banks. These riparian habitats show a great diversity of spatial structure and of water discharge in relation to the connection of the water body with the active channel(s) of the river. Only a few species reproduce directly along the banks of the active channels. Most species inhabit satellite ponds. These ponds may be as diverse as rock pools in the beds of mountain streams, depressions in pebble beds or in levees of braided sectors, or oxbow lakes more or less connected with the active channel.

In large ponds or lakes, amphibians inhabit only the littoral zone where the macrophytes make up dense patches. In shallow ponds, this macrophytic zone may occupy the whole area of the pond. In this particular case, the whole pond shows the ecological structure of a riparian ecotone, but the pelagic component of the ecotone is missing. The continuation of this chapter will focus more precisely on the amphibian communities inhabiting the littoral zones of still waters.

REASONS FOR THE PRESENCE OF AMPHIBIANS IN ECOTONAL HABITATS

Only in such species as the water frogs, which have an amphibious style of life, is the ecotone inhabited throughout the active period. In most other species, the ecotone is used only seasonally for clutch laying and larval growth. In order to explain why the presence of amphibian larvae is restricted to ecotonal habitats, this section explores the relationships between ecological characteristics of the littoral zone and specific requirements of larvae for optimizing their survival.

Temperature

During spring and the beginning of summer, which correspond to the period of larval development, water temperature reaches higher values in riparian zones because of the shallow depth. The presence of dense vegetation in such zones provides supports making it possible for many species to lay their clutch just below the water surface where the temperature reaches its highest values. Under these high thermal conditions, the length of both embryonic and larval periods proves to be shorter than in colder water. A reduction of the risk of the site drying out or of the risk of predation represent the benefits derived from a shorter larval period (Wilbur and Collins, 1973).

Larval diet and ecotones

The anuran larvae exhibit little variation in diet among species. Most of them feed on attached, mainly epiphytic, microalgae (Savage, 1952; Kenny, 1969). The tadpoles can also feed on planktonic algae by means of their oro-branchial filter (Wassersug, 1972; Polls Pelaz and Pourriot, 1988). However, the efficiency of such a mechanism varies among species (Viertel, 1990, 1992).

The high density of macrophytes makes the availability of support for epiphytic algae highest in the littoral zone of lentic habitats. In such micro-habitats, the large surface available for colonization on submersed macrophytes causes the high contribution of the littoral zone to the total primary productivity of many freshwater ecosystems. However, algal production shows high seasonal variation in the shore zone, with its highest levels during the spring (Adams and Prentki, 1982; Gons, 1982; Wetzel, 1983). During this season, planktonic algae and algae loosely attached to the substrate can also become abundant in the littoral zone (Felföldy, 1958; Kann, 1982).

The causes of the high density of algae in these ecotonal habitats are the same as those explaining the abundance of macrophytes: (i) availability of nutrients imported from the land (litterfall, dissolved nutrients), (ii) favourable conditions for the retention and transformation of these nutrients in ecotonal habitats (Peterjohn and Correll, 1984), and (iii) availability of light due to the shallowness of the water. In addition, the productivity of the littoral zone is enhanced by periodic physical disruptions (water-level fluctuations, wave action, grazing)(pulse stability: Odum, 1971).

Littoral ecotones are also inhabited by newts and several salamanders (such as *Ambystomatidae*), but only seasonally during the breeding period. As in anurans, the development of the larvae takes place in the littoral zone. Though the adults feed on a great diversity of aquatic prey, such as cladocerans, copepods, chironomid larvae, and isopods, the most important prey biomass is exogenous and terrestrial, such as earthworms, caterpillars, and other insects (Joly, 1987; Joly and Giacoma, 1992). The high occurrence of such terrestrial prey in their diet shows the relationship of newts during their aquatic phases with the ecotone.

The diet of urodele larvae is restricted to microcrustaceans and small chironomids (Kuzmin, 1991; Braz and Joly, 1994). As for anuran tadpoles, the aquatic period of amphibiotic urodeles coincides with the period of highest productivity of the ecotone, probably because prey density reaches its highest levels during the period of highest vegetation productivity. Densities of microcrustaceans feeding on epiphytic microflora and detritus are indeed positively correlated with epiphytic algal densities. On the other hand, the accumulation of pelagic plankton along lake shores is well known in the limnological literature as a 'shore effect'. Indeed, plankton mainly accumulate during algal blooms combined with strong wave action.

The most abundant density of potential prey for urodeles is observed in small lakeside pools. Near-shore shallows and small lakeside water bodies partly separated from the main water body have special physicochemical characteristics (smaller volume of water, higher concentrations of organic and inorganic substances and more frequent oxygen debts than in deeper parts of the lake, and higher decomposition rates which enhance productivity), (Pieczyńska, 1972, 1990). However, Dvorak (1978) found that the quantity of macroinvertebrates was greater in the littoral section of large fish ponds than in the rather isolated biotopes.

Ecotones and the availability of refuges against predation

The presence of fish influences the composition of amphibian communities (Kats et al., 1988). Predation by fishes is one of the major causes of tadpole mortality (Morin, 1983, 1986; Van Buskirk, 1988). The high density of macrophytes in the littoral zone provides many shelters allowing the tadpoles to avoid fish predation.

As with tadpoles, the predation by fish on urodeles can explain the restriction of the distribution of the urodeles to the littoral zone (Le Parco et al., 1981). The same factor explains the presence of urodeles in the headwaters of streams where the instability of the water level prevents the presence of large predatory fish. The ability of amphibians to move across the terrestrial habitats surrounding a stream makes the colonization of these headwaters possible for them, whereas waterfalls stop the upward movement of fish (Thiesmeier, 1991).

In conclusion, temperature, food and defense against predation appear as major forces driving the reproduction of amphibians to take place in the littoral zone.

ECOLOGICAL CONSTRAINTS ON THE USE OF ECOTONAL HABITATS AS BREEDING SITES

The exploitation of ecotonal habitats obliges the amphibians to conform to many ecological constraints, which are specific to both headwater and littoral zones. The restriction of the surface of suitable sites, their patchy distribution and the degree of disturbance by hydrological events are the main factors the amphibian populations have to deal with.

Fragmentation and connectivity

In the alluvial floodplains, small surface and a high fragmentation are two main features of the ecotonal patches used by the amphibians for breeding. In the Rhône valley, the largest aquatic area occupied by breeding amphibians reaches 2 ha and many sites do not exceed a few square meters. Such a small surface is indeed a limiting factor for population dynamics.

When densities of anuran larvae are high, both exploitation and interference competition interact with environmental stochasticity in increasing extinction risk. Interference competition mediated by the colonization of the gut by protothecan parasitic algae has been described as reducing the survival of the smallest tadpoles (Richards, 1962; Beebee, 1991). Competition results in subtle micro-habitat partitioning among the larvae of different species (Alford, 1986). As in anuran tadpoles, micro-habitat preference may be detected among larvae of different urodele species. In Europe, *Triturus cristatus* larvae appear to be the most nectonic and *T. alpestris* larvae the most benthic. But such differences are to be considered as trends at the scale of large water bodies because the great majority of larvae are found in the littoral zone (Braz and Joly, 1994).

For most species, such as toads or tree frogs, in which the maintenance habitat of the adult stage is terrestrial, the difference in availability between favourable terrestrial and aquatic habitats points out the smallness of favourable aquatic sites as the main limiting factor (Wilbur, 1980). In amphibious species, such as water frogs, the surface of the suitable aquatic sites always constitutes the main regulating factor in population dynamics, acting both on adult and larval stages.

Amphibian breeding sites are patchily distributed. Even within the network of the channels that are permanently connected, the sites that provide favourable features for amphibian reproduction appear as isolated patches because deep or running waters are not used by moving animals.

The definition of a functional degree of fragmentation depends on the ability to measure the permeability of non-breeding habitats for migrating animals. In many species, adult individuals are able to cover distances as great as several kilometres between breeding sites (13 km in the tree frog *Hyla arborea* (Fog, 1993), 15 km in *Rana 'esculenta'* (Tunner, 1992)). However, we lack data for a clear definition of the factors of connectedness that influence the intersite migrations for each species.

Disturbance through hydrological processes

According to the geomorphology of the alluvial valley, the sites the shape of which may be suitable for amphibian breeding, experience different levels of hydraulic disturbance. Because of the morphology of the valley (slope, width), the intensity of the influence of floods decreases from the braided sectors to the meandered or anastomosed sectors (Petts, 1990; Bravard and Gilvear, 1993).

In sites experiencing the action of strong water current during floods, both organic matter and fine particles of the substrate are swept along, leading to the maintenance of a juvenile stage in the ecological succession. In such highly unstable sites, it is likely that the density of predators and competitors of amphibian larvae remains low, making colonization by species that have good colonizing ability possible.

The joint action of erosion and deposition processes causes the channels to modify their course. According to these processes, and at the scale of decades, suitable sites for amphibian breeding may disappear and new sites may appear in places that are not easily predictable. Thus, in sectors where the stream dissipates a large amount of kinetic energy, the amphibians have to face both spatial and temporal unpredictability (Morand and Joly, 1995).

In sectors where floods are less drastic, several factors join in conferring a higher stability. The clogging of the substrate with both allochthonous and autochthonous alluvium leads to an increase in the ability of the site to maintain a stable water level. The level of the aquifer, which is a source of water in the breeding sites, varies less than in the vicinity of the active channels. In such sites, the high specific richness of the resident communities is expected to promote the success of species that show competitive ability and traits protecting them against predation (Townsend, 1989).

PRODUCTION AND FLUXES

The deposition of eggs in the littoral zone may be regarded as a contribution by adult amphibians to the importation of organic matter from land across the ecotone. In contrast, landward migration of post-metamorphic animals is a component of organic matter exportation from water to land. In this connection, the work of Seale (1980) on a pond near St Louis, Missouri, constitutes a first study establishing the balance of these fluxes.

In this ecosystem in spring, the total tadpole nitrogen twice exceeded the residual pond nitrogen. Peaks in tadpole growth rate corresponded with conditions with a high level of particulate organic nitrogen. In the presence of tadpoles, the particulate organic nitrogen was reduced relative to both dissolved inorganic nitrogen and dissolved organic nitrogen. Such a reduction of the concentration of particulate organic nitrogen was caused by its ingestion by the tadpoles and accompanied by the release of ammonia as the end product of tadpole catabolism. On the other hand, the tadpoles fed on blue green algae, which ensure the fixation of atmospheric nitrogen in the pond.

In the pond she investigated, Seale estimated the amount of nitrogen exported through post-metamorphic landward migration as 60 g N for *Rana pipiens*, 50 g N for *R. catesbiana*, 1 g N for both *Hyla versicolor* and *Acris crepitans*, and 10 g N for *Ambystoma* sp. In most species, the amounts exported exceeded the amounts imported in the shape of eggs (9.4 g N in *R. pipiens*, 0.4 g N in *R. catesbiana*, and 0.9 g N in *Ambystoma*). Only in *Bufo americanus* did the nitrogen amount imported exceed the amount exported, probably because of high predation on the tadpoles of this species.

Tadpoles contribute to the reduction of the natural eutrophication of ecotonal habitats by three mechanisms (Seale *et al.*, 1975). The first rests on the reduction of the input of nitrogen from the atmosphere by reducing the biomass of nitrogen-fixing blue-green algae such as *Anabaena*. The second is the reduction of primary production by feeding on algae. The third is the export of nitrogen assimilated from the pond ecosystem *via* the outward migration of post-metamorphic animals across the ecotone.

ADAPTATION OF AMPHIBIANS TO THE ECOLOGICAL CONSTRAINTS OF LITTORAL ECOTONES

Adaptation to the seasonality of feeding resources

The production of microalgae in the littoral zone of most continental waters shows high seasonal variation (Wetzel, 1983, 1984). The algal production remains very low during autumn and winter, but reaches high levels at the end of spring and the beginning of summer. The transient morphological adaptation of the mouth and digestive tract of tadpoles to algal feeding involves the cost of morphological, physiological, and behavioural changes through metamorphosis, then allowing each individual to exploit food other than algae. If the gain obtained by each individual in feeding in the littoral zone on algae during their highest production compensates the cost of metamorphosis, then this feeding adaptation becomes a key factor explaining the evolution of complex life cycles in amphibians (Wassersug, 1974, 1975; Wilbur, 1980). The ephemeral character of this feeding resource of anuran tadpoles makes paedogenesis maladaptive and probably has prevented the elimination of the adult stage during the course of evolution in all the known groups of anurans (Slade and Wassersug, 1975).

Because variation in prey availability is less than variation in algal density, the predatory behaviour of urodeles provides higher flexibility with respect to food availability than does the algal-specialized diet of the anuran tadpoles. If we extend the reasoning of Slade and Wassersug (1975), who expected that paedogenesis never occurred in anurans because of the seasonality of the feeding resources, we can expect paedogenesis to occur in amphibiotic urodeles. Paedomorphosis has in fact been observed in many species belonging to both Ambystomatidae and Salamandridae (Semlitsch, 1987; Breuil, 1992). However, paedomorphosis remains optional in most species that inhabit more or less unstable habitats, probably because of the temporariness of water in most aquatic sites and presence of fish, which dominate the urodeles in exploiting the pelagic habitat (Semlitsch and Gibbons, 1990).

Adaptation to the unpredictability and transience of suitable habitats

Hydrological dynamics influence tadpole survival which shows high variation according to the water discharge in the spawning site. Quantitative variations of several life history traits contribute to the adjustment of reproduction to the unpredictability and the transience of suitable conditions for development of larvae.

Flexibility of the reproductive period. In many species, the temporal flexibility of the breeding period appears as the major adaptation to the unpredictability of water level (Joly and Morand, 1994). In Europe, such a temporal flexibility of the breeding period is observed in such species as *Bufo calamita*, *Pelodytes punctatus*, or *Bombina variegata* (Diaz Paniagua, 1992). The egg laying behaviour of several species, such as *Ambystoma opacum*, may also be regarded as an adaptation to these factors. In these species, the eggs are laid in a den in the pond bank above the water surface. Hatching may be delayed until the eggs become soaked with water. This behaviour ensures that reproduction fits the date of water filling. The anticipation of the duration of water retention relies on the height of the den in the bank (Jackson *et al.*, 1989).

Clutch partitioning among several sites may also be regarded as another adaptive trait used to overcome the unpredictability of water retention (Silverin and Andren, 1992; Joly and Morand, 1994).

Length of the larval period. The advantage of fast-developing larvae in temporary water has been the subject of many papers (Travis, 1983; Werner, 1986; Travis *et al.*, 1987; Newman, 1988). On first examination, the evolution of development period has been considered as a trade-off between the selective pressure of the desiccation risk and the advantage of attaining a large body size at metamorphosis and expressed as a higher survival of large-sized individuals (Wilbur and Collins, 1973; Travis, 1984; Semlitsch *et al.*, 1988). However, intra- and interspecific competition may also influence the duration of the larval period and the size at metamorphosis (Warner *et al.*, 1993). The shortening of development length may be mediated by genetic factors and/or by maternal effect through a higher investment in egg size (Berven, 1987).

The variation in egg size within a single clutch has been assumed to be an adaptive trait in unpredictable environments (Crump, 1981; Cooper and Kaplan, 1982; Kaplan and Cooper, 1984). According to this assumption, variation in egg size should induce a variation in the development time within a single clutch which may be considered as a response to the unpredictability of the length of wetting ('diversified bet-hedging': Philippi and Seger, 1989).

Despite numerous studies, the adaptation of larval development to transience and unpredictability of wetting remains a problem that is not completely understood. The norms of reaction of the length of the larval period according to environmental factors is a parameter that raises great interest (Alford and Harris, 1988; Blouin, 1992).

Other life-history traits. The variation of a large set of life-history traits may have an adaptive influence on fitness when the animal is faced with environmental stochasticity. However, our present understanding of this topic lacks a clear overview of the relationships between environmental variability and demographic strategies. With regard to the theory of *r/K* selection (Pianka, 1970), the amphibian species of the Rhône floodplain show a great diversity in the ratio of clutch size/relative egg size (egg size/female size) (Morand and Joly, 1995). But the relationship between the position on the *r/K* gradient and the variability of the habitats where a species reproduces is tenuous. The clearest evidence is the absence of species showing parental care (*Salamandra salamandra* and *Alytes obstetricans*) in sites located in the frequently inundated floodplain. In the other species, several combinations of demographic traits may represent responses to environmental variability. Phylogenetic origin appears to heavily constrain these combinations.

Adaptation to foraging in ecotones

Except for some species in which the tadpoles are predacious, tadpoles of most species feed on algae, fine particulate matter, detritus or macrophytes. One singular feature of the nutrition of tadpoles lies in the high speed of transit through the intestinal tract. The mean digestion time at 16–18°C was 6.25 h in *Rana temporaria* tadpoles and 4.75 h in *Bufo bufo* tadpoles (Savage, 1952). Such a high speed of transit explains why many food organisms pass through the tadpole's gut without being damaged. Such a result suggests that bacterial fermentation does not occur in the intestine. It is likely that the main food of the tadpoles is small organisms the cells of which are damaged by the mechanical action of the jaws and associate structures (beak and labial teeth). As a consequence, the more important food sources are algae, either planktonic or epiphytic, and detritus with their associate microfauna (protozoans) and microflora (fungi and bacteria).

The modes of food ingestion by tadpoles are diverse, particularly under tropical climates (review in Duellman and Trueb, 1986, pp. 159–62). In the holarctic zone two main feeding modes prevail. The first mode consists of scraping off the algal film on the surface of rocks or plants, with the beak and the labial teeth (Savage, 1952; Kenny, 1969; Wassersug, 1980). The second mode consists of filtering water, the movement of which through the buccal cavity is ensured by the pharyngeal pump. Both the vellums, which precede the gills, and the gill filter, contribute to the entrapment of planktonic microalgae and other small particles (Savage, 1952; Kenny, 1969; Wassersug, 1972).

Both modes may be regarded as adaptations to the exploitation of the algal blooms which occur in the ecotonal habitats. The prevalence of one mode or the other depends on habitat variability. In temporary pools, the density of planktonic algae may reach higher values than in permanent sites (Savage,

1952), favouring filtering species such as *Hyla arborea* (Diaz Paniagua, 1985, 1989), *Bufo calamita* or *Bombina variegata* (Viertel, 1987). In such temporary sites, the scarcity of predators and competitors allows these filtering tadpoles to shift from littoral microhabitats to pelagic microhabitats.

In more permanent sites, the density of planktonic algae is often depressed by the abundance of a large community of plankton consumers (microcrustaceans, fish). On the other hand, the littoral zone is occupied by dense vegetation, which provides support for a thick layer of epiphytic algae. Such epiphytic algae constitute the main part of the diet of grazer tadpoles, such as those of *Rana temporaria* or *Bufo bufo* (Savage, 1952; Diaz Paniagua, 1985).

However, this classification of species according to their mode of feeding is somewhat arbitrary. In fact, most species of pond tadpoles show a great flexibility of feeding mode from filtering to scraping, and the differences in feeding mode should be regarded more as trends than as narrow specializations (Waringer-Löschenkohl, 1988; Viertel, 1990, 1992).

In contrast with the anuran larvae, the urodele larvae do not show any specialized mode of feeding. They essentially feed on microcrustaceans (cladocerans, copepods, ostracods) and chironomids, which show high densities in the littoral zone of most still waters (Brophy, 1980; Taylor *et al.*, 1987; Kuzmin, 1991; Fasola, 1993; Braz and Joly, 1994).

Adaptations of the amphibian larvae to escape fish predation

The use of aquatic ecotonal habitats for spawning implies that the larvae are exposed to the risk of predation by fish. A first adaptation against predation lies in microhabitat selection. The vegetation of the littoral zone provides shelters against large fishes for both anuran and urodele larvae (Holomuzki, 1989; Warkentin, 1992).

However, defence against fish predation also involves other mechanisms. The tadpoles of most species show cryptic colours. In many species, the ability to avoid predation is based on the secretion of noxious and/or repellent substances, which make the tadpole unpalatable to fish or other vertebrate predators (Licht, 1968; Brodie and Formanovicz, 1987; Peterson and Blaustein, 1991). However, the efficiency of unpalatability depends on the tasting ability of the predator. Chemicals that act as repellents for fish appear ineffective in insects (Voris and Bacon, 1966; Kruse and Francis, 1977; Brodie *et al.*, 1978).

Behavioural responses to fish presence have been investigated by several authors. The tadpoles detect the presence of fish through chemical, visual and tactile cues (Petranka *et al.*, 1987; Kats *et al.*, 1988; Semlitsch and Reyer, 1992). Locomotor activity and space use are modified in the presence of fish. Swimming activity is reduced (Lawler, 1989; Semlitsch and Reyer, 1992) and the period of activity shifts from diurnal to nocturnal (Taylor, 1983). In the presence of fish, or in the presence of the chemical mediated cues from fish, the tadpoles of *Bufo bufo* or *Rana esculenta* modify their use of space and seek refuges (Semlitsch and Gavasso, 1992; Stauffer and Semlitsch, 1993). Schooling appears a social response to fish predation (Wassersug, 1973; Waldman, 1982; Caldwell, 1989).

In summary, a lot of physiological and behavioural adaptations have evolved in anuran tadpoles, making it possible to avoid predation by fish. Such adaptations may be regarded as contributions to the use of littoral ecotones for larval development.

Demographic adaptation to the fragmentation of ecotones

If we plot on a map the ecotones that constitute suitable sites for the reproduction of amphibians, we are aware of the patchiness of such habitats in the landscape. Despite great variation, the area of each patch usually appears small, under 1 ha. In species that exhibit breeding site fidelity (Twitty, 1959; Heusser, 1969; Joly and Miaud, 1989), the group of animals breeding in a site may be considered as a local population and the probability of extinction of such a population essentially depends on its number (Shaffer and Samson, 1984; Shaffer, 1987). Hence, in small sites, the extinction probability of each local amphibian population remains relatively high.

The risk of extinction due to the vulnerability of small populations when faced with environmental stochasticity may be balanced by immigration from other populations experiencing reproductive success. Many amphibian species appear as good examples for illustrating such metapopulation functioning (Gill, 1978; Sjögren, 1991; Edenhamn, 1993; Miaud *et al.*, 1993).

In amphibians, the ability of individuals to live most of the time in the terrestrial environment (which may be considered an intersite environment) confers special characteristics on metapopulation functioning. Low mortality in the terrestrial environment, high longevity and/or a long delay in maturation ensure the maintenance of a stock of individuals that compensate for the uncertainty of the aquatic breeding sites. Many species of amphibiotic amphibians show such characteristics. The life span of a crested newt or a common toad in the field can reach 17 years and 9 years respectively (Miaud *et al.*, 1993; Hemelaar, 1988) and the length of the juvenile period in *Notophtalmus viridescens* or *Triturus helveticus* reaches 4–5 years (Gill, 1978; Miaud, 1991).

Metapopulation functioning assumes that immigration flow is to be ensured by movement of individuals between different sites. Site fidelity constitutes a behavioural barrier to dispersion. However, despite astonishing homing performance in some individuals, both homing experiments and long-term metapopulation studies show that a regular and non-negligible proportion of adult individuals moves from one site to another. Such changes may occur within the breeding season in cases of site disturbance, or during the interbreeding terrestrial stage (Fog, 1993; Joly, unpublished data). The dispersal range during the juvenile period is poorly known. In a study of the dispersal of juvenile common toad in Great Britain, 19% of individuals were found to breed in a pond other than the pond where they were born, suggesting that the juveniles play an important role in ensuring immigration fluxes (Reading *et al.*, 1991).

IMPACT OF HUMAN ACTIVITIES AND GUIDELINES FOR ECOTONE MANAGEMENT

The huge increase in the influence of human engineering on wild fluvial ecosystems has profound consequences for the configuration and the quality of the ecotones inhabited by amphibians.

Fragmentation and isolation of ecotonal habitats

With regard to configuration, the main impacts of human activities have led to reduction and a greater fragmentation of ecotone surface, thus increasing the isolation of suitable breeding habitats. In the Upper Rhône valley, the first significant impact of human activities on ecotones were the consequences of embankment works that occurred during the seventeenth century. These works caused a decrease in the surface of dissipation for kinetic energy of the river and, consequently, incision processes occurred. One consequence has been the decrease in the whole ecotone surface in sections where these embankments were erected, causing a reduction in the number of active channels and braided sections and the drying up of peripheral sites following the lowering of the water table (Bravard, 1987; Roux *et al.*, 1989). More recently, the modifications of water discharge by the construction of dams for flood regulation or electrical power generation caused a new reduction in the surface of energy dissipation, implying a new reduction in the ecotone surface (Bravard *et al.*, 1986).

In the braided sections, the reduction of the energy discharge in the by-passed courses caused alluvium deposition in the secondary channels, which was accelerated by the growth of dense vegetation that trapped this alluvium. These processes ensured a rapid reduction in the surface of water, and consequently of the surface of ecotonal habitats.

In the Rhône alluvial valley, the number of aquatic sites suitable for amphibian breeding has also been reduced through drainage works executed by farmers and by the agencies for mosquito limitation.

The use of the terrestrial habitats that separate the aquatic sites has also been modified. In many places, the original alluvial forest has been replaced by crops or intensive poplar cultures, both of which reduce the connectivity of the terrestrial environment for the amphibians. In other places the connectivity has been reduced by the construction of roads, highways or railways, which cause high mortality in migrating amphibians. All these human modifications to terrestrial habitats reduce the ability of amphibians to move between suitable ecotonal habitats for breeding, leading to an increase in the isolation of local populations and to an increased threat of extinction.

Quality of ecotonal habitats

Water–land ecotones play an important role through different water filtering and entrapment mechanisms for both suspended and dissolved matter (Gibert et al., 1990). When the site is adjacent to crops, pesticides and heavy metals may accumulate in the vegetation and the animals living in the ecotone (Cosson, 1987). In places where poplar culture is intensive, the accumulation of leaf litter in the ecotone causes a 'dystrophic' effect by the loss of oxygen in the water following the oxidative condensation of polyphenolic substances (Trémolières and Carbiener, 1981, 1982).

Many pollutants are known to act on the survival of amphibians. Heavy metals accumulate in amphibian larvae and bind with a metallothionein protein (Goldfischer et al., 1983; Suzuki and Tanaka, 1983; Hall and Mulhern, 1984; Suzuki et al., 1986). The survival of eggs and larvae may be altered by the presence of detergents (Plötner and Günther, 1987) or by the synergy of aluminium and low pH (Clark and Lazerte, 1985; Andren et al., 1988). However, we lack data on the importance of the filtering and entrapment processes that may take place in ecotonal habitats of lentic waters, thereby influencing amphibian population viability.

In many species, the selection of oviposition sites in ecotonal habitats exposes the eggs to solar radiation. Recent studies reveal the differential ability among amphibian species to repair UV-B radiation damage to DNA (Blaustein et al., 1994). These authors consider the global decline in certain amphibian species to be caused by an increase in UV-B radiation at the surface of the earth. If this assumption proves to be true, solar radiation will become a new constraint on amphibian reproduction in ecotonal habitats.

Guidelines for the restoration of ecotonal habitats suitable for amphibian breeding

Habitat fragmentation can be an advantage for reproduction of amphibians in ecotonal habitats (Mann et al., 1991). However, excessive habitat fragmentation leads to greater risk of extinction of local populations. The main problem appears to be the maintenance of a network of aquatic sites that provide suitable ecotonal habitats for amphibian breeding, that is sufficiently dense to ensure individual fluxes between them. The alluvial forest constitutes the terrestrial habitat of many species, but is also used as corridor allowing individuals to migrate from one site to another. However, open habitats such as heathlands on gravelly soils also constitute terrestrial habitats and corridors, particularly for burrowing species or tree frogs.

The littoral zone of many large water bodies may be used as a breeding site provided that gently sloping banks make the growth of a large belt of submersed or floating macrophytes possible. Satellite sites of large water bodies, such as fishless shallow pools, provide optimum growing conditions for a great diversity of species.

In the alluvial valley the restoration of breeding sites for the amphibians must take into account the hydraulic functioning of the river. With regard to floods, the highest amphibian diversity is observed at an intermediate level of disturbance. In the Rhône valley, greatest richness is measured in mesotrophic temporary sites with water mainly originating from the water table (Morand and Joly, 1995). These sites rarely overflow. Successional processes occur in these sites, leading to their rapid filling by both allochthonous and autochthonous matter. Man-made embankments affect the persistence of such sites in the valley of most holarctic rivers which now mainly depend on human activities, such as the extraction of pebbles and gravel. However, the success of the colonization of such man-made sites depends on their topographical profile (width of the ecotonal habitat, transience of water filling) and their location with respect to the other ecological components of the amphibian cycle and to metapopulation functioning.

CONCLUSIONS

The ecotone concept is particularly relevant in explaining the evolutive ecology of many amphibian species. These species play an important part in the functioning of water–land ecotones, especially in the headwaters and in the littoral zone of many aquatic systems. In most amphibiotic species of anurans the larvae exploit the transient resource of littoral micro-habitats. The morphology of the tadpoles is specialized for exploitation of both epiphytic and planktonic algae, which show highest productivity during the spring. The urodele larvae occupy the same micro-habitat, feeding on microcrustaceans that also show cyclic variation in density, with highest density in spring. The metamorphosis and the landward migration of the postmetamorphic juveniles represent an export of nutrients from water to land, and may contribute significantly to limiting eutrophication.

Different adaptive traits take part in the exploitation of transient ecotonal resources by amphibians. The first is the evolutive stability of the complex life cycle in many species. In the context of the variability of habitats, other traits contribute to the success of the complex life cycle. The temporal flexibility of reproduction, clutch partitioning and life span may be regarded as responses to the uncertainty of sites in which water filling depends on river discharge. The variation in the length of the larval period is related to the uncertainty of water retention in the habitats occupied by each species.

Local populations are small and show high extinction risk because of the small surface of the ecotonal habitats used for reproduction. The ability to move across different terrestrial environments contributes to the maintenance of immigration fluxes, thus compensating for the risk of extinction of local populations.

The amphibians may be regarded as important components of the functional biodiversity of aquatic ecosystems because of the ability their larvae to exploit transient resources in ecotonal habitats and because of their ability to export organic matter toward land. The maintenance of viable populations depends on the design of landscape management schemes for large floodplains and other aquatic systems.

ACKNOWLEDGEMENTS

A first version of the manuscript has been revised by Drs Volker Mahnert and Trevor J.C. Beebee. We are indebted to Dr Janine Gibert for preliminary discussions and to Professor Eric Pattee for his critical reading of the manuscript.

REFERENCES

Adams, M.S. and Prentki, R.T. (1982). Biology, metabolism and function of littoral submersed weedbeds of Lake Wingra, Wisconsin, USA: a summary and review. *Archiv für Hydrobiologie*, Supplement, **62**, 333–409

Alford, R.A. (1986). Habitat use and positional behaviour of anuran larvae in a northern Florida temporary pond. *Copeia*, 408–23

Alford, R.A. and Harris, R.N. (1988). Effects of larval growth on anuran metamorphosis. *American Naturalist*, **131**, 91–106

Andren, C.L., Henrikson, M., Olsson, G. and Nilson, G. (1988). Effects of pH and aluminium on embryonic and early stages of Swedish brown frogs, *Rana arvalis*, *R. temporaria* and *R. dalmatina*. *Holarctic Ecology*, **11**, 127–35

Augert, D. and Joly, P. (1994). Dispersal of *Rana temporaria* tadpoles in large fish-ponds. *Alytes*, **12**, 31–40

Beebee, T.J.C. (1991). Purification of an agent causing growth inhibition in anuran larvae and its identification as a unicellular unpigmented alga. *Canadian Journal of Zoology*, **69**, 2146–53

Berven, K.A. (1987). The heritable basis of variation in larval developmental patterns within populations of the wood frog (*Rana sylvatica*). *Evolution*, **41**, 1088–97

Blaustein, A.R., Hoffman, P.D., Hokit, D.G., Kiesecker, J.M., Walls, S.C. and Hays, J.B. (1994). UV repair and resistance to solar UV-B in amphibian eggs: a link to population declines. *Proceedings of the National Academy of Sciences*, **91**, 1791–5

Blouin, M.S. (1992). Comparing bivariate reaction norms among species: time and size at metamorphosis in three species of *Hyla* (Anura, Hylidae). *Œcologia*, **90**, 288–93

Bravard, J.P. (1987). *Le Rhône, du Léman à Lyon*. La Manufacture, Lyon

Bravard, J.P., Amoros, C. and Pautou, G. (1986). Impacts of civil engineering works on the succession communities in a fluvial system: a methodological and predictive approach applied to a section of the Upper Rhône River. *Öikos*, **47**, 92–111

Bravard, J.P. and Gilvear, D.J. (1993). Structure hydro-géomorphologique des hydrosystèmes. In Amoros, C. and Petts, G.E. (eds.) *Hydrosystèmes Fluviaux*, pp. 83–100. Masson, Paris

Braz, E. and Joly, P. (1994). Micro-habitat use, resource partitioning and ecological succession in a size-structured guild of newt larvae (g. *Triturus, Caudata, Amphibia*). *Archiv für Hydrobiologie*, **131**, 129–39

Breuil, M. (1992). La néoténie dans le genre *Triturus*: mythe et réalités. *Bulletin de la Société Herpétologique de France*, **61**, 11–44

Brodie, E.D. Jr and Formanovicz, D.R. Jr. (1987). Antipredator mechanisms of larval anurans: protection of palatable individuals. *Herpetologica*, **43**, 369–73

Brodie, E.D. Jr, Formanovicz, D.R. Jr and Brodie, E.D. III. (1978). The development of noxiousness of *Bufo americanus* tadpoles to aquatic insect predators. *Herpetologica*, **34**, 302–5

Brophy, T.E. (1980). Food habits of sympatric larval *Ambystoma tigrinum* and *Notophthalmus viridescens*. *Journal of Herpetology*, **14**, 1–6

Caldwell, J.P. (1989). Structure and behaviour of *Hyla geographica* tadpole schools, with comments on classification of group behaviour in tadpoles. *Copeia*, **4**, 938–50

Clark, K.L. and Lazerte, B.D. (1985). A laboratory study of the effects of aluminium and pH on amphibian eggs and tadpoles. *Canadian Journal of Fisheries and Aquatic Sciences*, **42**, 1544–51

Clergue-Gazeau, M. and Martinez-Rica, J.P (1978). Les différents biotopes de l'urodèle pyrénéen *Euproctus asper*. *Bulletin de la Société d'Histoire Naturelle de Toulouse*, **114**, 461–71

Cooper, W.S. and Kaplan, R.H. (1982). Adaptive 'coin-flipping': a decision-theoretic examination of natural selection for random individual variation. *Journal of Theoretical Biology*, **94**, 135–51

Cosson, R.P. (1987). Influences des pratiques agricoles sur la contamination des écosystèmes aquatiques par les métaux lourds: approche écotoxicologique. Thèse de Doctorat d'Etat, Université Paris Sud Orsay

Crump, M.L. (1981). Variation in propagule size as a function of environmental uncertainty for tree-frogs. *American Naturalist*, **117**, 724–37

Diaz Paniagua, C. (1985). Larval diets related to morphological characters of five anuran species in the biological reserve of Donana (Huelva, Spain). *Amphibia–Reptilia*, **6**, 307–22

Diaz Paniagua, C. (1989). Larval diets of two anuran species, *Pelodytes punctatus* and *Bufo bufo* in SW Spain. *Amphibia–Reptilia*, **10**, 71–5

Diaz Paniagua, C. (1992). Variability in timing of larval season in an amphibian community in SW Spain, *Ecography*, **15**, 267–72

Duellman, W.E. and Trueb, L. (1986). *Biology of Amphibians*. McGraw-Hill, New York

Dvorak, J. (1978). Macrofauna of invertebrates in helophyte communities. In Dykyjova, D. and Kvet, J. (eds.) *Pond Littoral Ecosystems: Structure and Functioning*, pp. 389–95. Springer Verlag, Berlin

Edenhamn, P. (1993). Metapopulation dynamics in an amphibian perspective. In Stumpel, A.H.P. and Tester, U. (eds.) *Ecology and Conservation of the European Tree Frog*, pp. 65–70. Institute for Forestry and Nature Research, Wageningen

Fasola, M. (1993). Resource partitioning by three species of newts during their aquatic phase. *Ecography*, **16**, 73–81

Felföldy, L.J.M. (1958). A contribution to the ecology and biological productivity of the diatom mass–vegetation on the stony shores of Lake Balaton. *Annales Instituti Biologici (Tihany) Hungaricae Acadiamiae Scientarum*, **25**, 331–42

Fog, K. (1993). Migration in the tree frog *Hyla arborea*. In Stumpel, A.H.P. and Tester, U. (eds.) *Ecology and Conservation of the European Tree Frog*, pp. 55–64. Institute for Forestry and Nature Research, Wageningen

Gibert, J., Dole-Olivier, M.J., Marmonier, P. and Vervier, P. (1990). Surface water–groundwater ecotones. In Naiman, R.J. and Décamps, H. (eds.) *The Ecology and Management of Aquatic–Terrestrial Ecotones*, pp. 103–40. MAB Book Series 4. UNESCO/Parthenon, Paris/Carnforth

Gill, D.E. (1978). The metapopulation ecology of the red-spotted newt, *Notophthalmus viridescens* (Rafinesque). *Ecological Monographs*, **48**, 145–66

Goldfischer, S., Schiller, B. and Sternwieb, I. (1983). Copper in hepatocyte lysosomes of the toad *Bufo marinus* L. *Nature*, **228**, 172–3

Gons, H.J. (1982). Structural and functional characteristics of epiphyton and epipelon in relation to their distribution in Lake Vechten. *Hydrobiologia*, **95**, 79–114

Griffiths, R.A. and Mylotte, V.J. (1986). Observations on the dispersal of common frog tadpoles *Rana temporaria* from the spawn site. *Bulletin of the British Herpetological Society*, **18**, 21–3

Hall, R.J. and Mulhern, B.M. (1984). Are anuran amphibians heavy metal accumulators? In Seigel, R.A., Hunt, L.E., Knight, J.L., Malaret, L. and Zuschlag, N.L. (eds.) *Vertebrate Ecology and Systematics,* pp. 123–33. Museum of Natural History of the University of Kansas, Lawrence

Hemelaar, A. (1988). Demographic study of *Bufo bufo* L. (Anura, Amphibia) from different climates, by means of skelettochronology. PhD Thesis, University of Nijmegen

Heusser, H. (1969). Die Lebensweise der Erdkröte, *Bufo bufo* (L.). Das Orientierungsproblem. *Revue Suisse de Zoologie*, **76**, 443–518

Holomuzki, J.R. (1989). Predation risk and macroalgal use by the stream-dwelling salamander *Ambystoma texanum*. *Copeia*, **1989**, 22–8

Jackson, M.E., Scott, D.E. and Estes, R.A. (1989). Determinants of nest success in the marbled salamander (*Ambystoma opacum*). *Canadian Journal of Zoology*, **67**, 2277–81

Joly, P. (1987). Activité prédatrice sur le site de reproduction chez le triton alpestre *Triturus alpestris* (Amphibien Urodèle). *Colloque National C.N.R.S. de Biologie des Populations*, Lyon, 4–6 septembre 1986, 635–43

Joly, P. and Giacoma, C. (1992). Limitation of similarity and feeding habits in a guild of three syntopic species of newt. *Ecography*, **15**, 401–11

Joly, P. and Miaud, C. (1989). Fidelity to the breeding site in the alpine newt *Triturus alpestris*. *Behavioural Processes*, **19**, 47–56

Joly, P. and Morand, A. (1994). Theoretical habitat templets, species traits and species richness: amphibians in the Upper-Rhône floodplain. *Freshwater Biology*, **31**, 455–68

Kann, E. (1982). Qualitative Veranderungen der litoralen Algenbiocönose östereichischer Seen (Lunzer Untersee, Traunsee, Attersee) im Laufe der letzen Jahrzehnte. *Archiv für Hydrobiologie*, Suppl., **62**, 440–90

Kaplan, R.H. and Cooper, W.S. (1984). The evolution of developmental plasticity in reproductive characteristics: an adaptation to the 'adaptive coin-flipping' principle. *American Naturalist*, **123**, 393–410

Kats, L.B., Petranka, J.W. and Sih, A. (1988). Antipredator defenses and the persistence of amphibian larvae with fishes. *Ecology*, **69**, 1865–70

Kenny, J.S. (1969). Feeding mechanisms in anuran larvae. *Journal of Zoology, London*, **157**, 225–46

Kruse, K.C. and Francis, M.G. (1977). A predation deterent in larvae of the bullfrog, *Rana catesbiana*. *Transactions American Fisheries Society*, **106**, 248–52

Kuzmin, S.I. (1991). Food resource allocation in larval newt guilds (genus *Triturus*). *Amphibia–Reptilia*, **12**, 293–304

Lawler, S.P. (1989). Behavioural responses to predators and predation risk in four species of larval anurans. *Animal Behaviour*, **38**, 1039–47

Le Parco, Y., Carton, Y. and Thouveny, Y. (1981). Recherches préliminaires sur la dynamique et la génétique des populations de *Triturus alpestris* Laurenti dans le Parc National des Ecrins. *Travaux Scientifiques du Parc Naturel des Ecrins*, **1**, 177–206

Licht, L.E. (1968). Unpalatability and toxicity of toad eggs. *Herpetologica*, **24**, 93–8

Mann, W., Dorn, P. and Brandl, R. (1991). Local distribution of amphibians: the importance of habitat fragmentation. *Global Ecology and Biogeography Letters*, **1**, 36–41

Miaud, C. (1991). Essai de synthèse sur les caractéristiques démographiques des tritons du genre Triturus. *Bulletin de la Société Herpétologique de France*, **59**, 1–18

Miaud, C., Joly, P. and Castanet, J. (1993). Variation of age structures in a subdivided population of *Triturus cristatus*. *Canadian Journal of Zoology*, **71**, 1874–9

Morand, A. and Joly, P. (1995). Habitat variability and space utilization by the amphibian communities of the French Upper-Rhône floodplain. *Hydrobiologia*, **300/301**, 249–57

Morin, P.J. (1983). Predation, competition and the composition of larval anuran guilds. *Ecological Monographs*, **53**, 119–38

Morin, P.J. (1986). Interactions between intraspecific competition and predation in an amphibian predator–prey system. *Ecology*, **67**, 713–20

Newman, R.A. (1988). Genetic variation for larval anuran (*Scaphiopus couchii*) development time in an uncertain environment. *Evolution*, **42**, 763–73

Odum, E.P. (1971). *Fundamentals of Ecology*. 2nd edn. W.B. Saunders, Philadelphia

Peterjohn, W.T. and Correll, D.L. (1984). Nutrient dynamics in an agricultural watershed: observations on the role of a riparian forest. *Ecology*, **65**, 1466–75

Peterson, J.A. and Blaustein, A.R. (1991). Unpalatability in anuran larvae as a defence against natural salamander predators. *Ethology, Ecology and Evolution*, **3**, 63–72

Petranka, J.W., Kats, L.B. and Sih, A. (1987). Predator–prey interactions among fish and larval amphibians: use of chemical cues to detect predatory fish. *Animal Behaviour*, **35**, 420–5

Petts, G.E. (1990). The role of ecotones in aquatic landscape management. In Naiman, R.J. and Décamps, H. (eds.) *The Ecology and Management of Aquatic–Terrestrial Ecotones,* pp. 227–60. MAB Book Series 4, UNESCO/Parthenon, Paris/Carnforth

Philippi, T. and Seger, J. (1989). Hedging one's evolutionary bets, revisited. *Trends in Ecology and Evolution,* **4**, 41–4

Pianka, E.R. (1970). On *r*- and *K*- selection. *American Naturalist,* **104**, 592–7

Pieczyńska, E. (1972). Ecology of the eulittoral zone of lakes. *Ekologia Polska,* **20**, 637–732

Pieczyńska, E. (1990). Lentic aquatic–terrestrial ecotones: their structure, functions and importance. In Naiman, R.J. and Décamps, H. (eds.) *The Ecology and Management of Aquatic–Terrestrial Ecotones,* pp. 103–40. MAB Book Series 4, UNESCO/Parthenon, Paris/Carnforth

Plötner, J. and Günther, R. (1987). Toxicity of an anionic detergent to the spawn and larvae of anurans (Amphibia). *Internationale Revue der Gesamten Hydrobiologie,* **72**, 759–71

Polls Pelaz, M. and Pourriot, R. (1988). Une approche de la fonction de filtration des têtards de grenouille verte (*Rana ridibunda*). *Vie Milieu,* **38**, 293–8

Reading, C.J., Loman, J. and Madsen, T. (1991). Breeding pond fidelity in the common toad, *Bufo bufo. Journal of Zoology, London.,* **225**, 201–11

Richards, C. (1962). The control of tadpole growth by alga-like cells. *Physiological Zoology,* **35**, 285–96

Roux, A.L., Bravard, J.P., Amoros, C. and Pautou, G. (1989). Ecological changes of the French upper Rhône river. In Petts, G.E., Möller, H. and Roux, A.L. (eds.) *Historical Changes of Large Alluvial Rivers in Western Europe,* pp. 323–50. John Wiley, London

Savage, R.M. (1952). Ecological, physiological and anatomical observations on some species of anuran tadpoles. *Proceedings of the Zoological Society of London,* **122**, 467–514

Seale, D.B. (1980). Influence of amphibian larvae on primary production, nutrient flux and competition in a pond ecosystem. *Ecology,* **61**, 1531–50

Seale, D.B., Rodgers, E. and Boraas, M.E. (1975). Effects of suspension feeding frog larvae on limnological variables and community structure. *Verhandlungen Internationale Vereinigung für Theoretische und Angewandte Limnologie,* **19**, 3179–84

Semlitsch, R.D. (1987). Paedomorphosis in *Ambystoma talpoideum*: effect of density, food and pond drying. *Ecology,* **68**, 994–1002

Semlitsch, R.D. and Gavasso, S. (1992). Behavioural responses of *Bufo bufo* and *Bufo calamita* tadpoles to chemical cues of vertebrate and invertebrate predators. *Ethology, Ecology and Evolution,* **4**, 165–73

Semlitsch, R.D. and Gibbons, J.W. (1990). Paedomorphosis in *Ambystoma talpoideum*: maintenance of population variation and alternative life history pathways. *Evolution,* **44**, 1604–13

Semlitsch, R.D. and Reyer, H.V. (1992). Modification of antipredator behaviour in tadpoles by environmental conditioning. *Journal of Animal Ecology,* **61**, 353–60

Semlitsch, R.D., Scott, D.E. and Pechmann, H.K. (1988). Time and size at metamorphosis related to adult fitness in *Ambystoma talpoideum. Ecology,* **69**, 184–92

Shaffer, M.L. (1987). Minimum viable populations: coping with uncertainty. In Soulé, M.E. (ed.) *Viable Populations for Conservation*, pp. 69–86. Cambridge University Press, Cambridge

Shaffer, M.L. and Samson, F.B. (1984). Population size and extinction: a note on determining critical population sizes. *American Naturalist*, **125**, 144–52

Silverin, B. and Andren, C. (1992). The ovarian cycle in the natterjack toad, *Bufo calamita* and its relation to breeding behaviour. *Amphibia–Reptilia*, **13**, 177–92

Sjögren, P. (1991). Extinction and isolation gradients in metapopulations: the case of the pool frog (*Rana lessonae*). *Biological Journal of the Linnean Society*, **42**, 135–47

Slade, N.A. and Wassersug, R.J. (1975). On the evolution of complex life cycles. *Evolution*, **29**, 568–71

Stauffer, H.P. and Semlitsch, R.D. (1993). Effect of visual, chemical and tactile cues of fish on the behavioural response of tadpoles. *Animal Behaviour*, **46**, 355–64

Suzuki, K.T., Iroh, N., Ohta, K. and Sunaga, H. (1986). Amphibian metallothionein. Induction in the frogs: *Rana japonica*, *R. nigromaculata* and *Rhacophorus schlegelli*. *Comparative Biochemistry and Physiology*, **83C**, 253–9

Suzuki, K.T. and Tanaka, Y. (1983). Induction of metallothionein and effect on essential metals in cadmium-loaded frog *Xenopus laevis*. *Comparative Biochemistry and Physiology*, **74C**, 311–17

Taylor, B.E., Estes, R.A., Pechmann, J.H.K. and Semlitsch, R.D. (1987). Trophic relations in a temporary pond: larval salamanders and their microinvertebrate prey. *Canadian Journal of Zoology*, **66**, 2191–8

Taylor, J.T. (1983). Orientation and flight behaviour of a neotenic salamander (*Ambystoma gracile*) in Oregon. *American Midland Naturalist*, **109**, 40–9

Thiesmeier, B. (1991). Biomasse, Produktion und ökologische Einnischung der Larven des Feuersalamanders (*Salamandra salamandra terrestris*) in einem fischfreien Mittelgebirgs-Quellbach. *Acta Biologica Benrodis*, **3**, 87–98

Townsend, C.R. (1989). The patch dynamics concept of stream community ecology. *Journal of the North American Benthological Society*, **8**, 36–50

Travis, J. (1983). Variation in development patterns of larval anurans in temporary ponds. I. Persistent variation within a *Hyla gratiosa* population. *Evolution*, **37**, 496–512

Travis, J. (1984). Anuran size at metamorphosis: experimental test of a model based on intraspecific competition. *Ecology*, **65**, 1155–60

Travis, J., Emerson, S.B. and Blouin, M. (1987). A quantitative-genetic analysis of larval life-history traits in *Hyla crucifer*. *Evolution*, **41**, 145–56

Trémolières, M. and Carbiener, R. (1981). Aspects de l'impact de litières forestières sur des écosystèmes aquatiques. I. L'effet désoxygénant de systèmes polyphénoloxydase-polyphénols. *Acta Oecologica*, **2**, 199–212

Trémolières, M. and Carbiener, R. (1982). Aspects de l'impact de litières forestières sur des écosystèmes aquatiques. II. Evolution à moyen terme de l'activité de désoxygénation de quelques propriétés des phytomélanines formées. *Acta Oecologica*, **3**, 241–57

Tunner, H.G. (1992). Locomotory behaviour in water frogs from Neusiedlersee (Austria, Hungary), 15 km migration of *Rana lessonae* and its hybridogenetic associate *Rana esculenta*. In Korsos, Z. and Kiss, I. (eds.) *Proceedings of the Sixth*

Ordinary General Meeting of the Societa Europaea Herpetologica, pp. 449–52. Budapest (1991)

Twitty, V.C. (1959). Migration and speciation in newts. *Science*, **130**, 1735

Van Buskirk, J. (1988). Interactive effects of dragonfly predation in experimental pond communities. *Ecology*, **69**, 857–67

Viertel, B. (1987). The filter apparatus of *Xenopus laevis*, *Bombina variegata* and *Bufo calamita* (Amphibia, Anura): a comparison of different larval types. *Zoologischer Jarbuch für Anatomie*, **115**, 425–52

Viertel, B. (1990). Suspension feeding of anuran larvae at low concentrations of *Chlorella* algae. *Oecologia* **85**, 167–77

Viertel, B. (1992). Functional response of suspension feeding anuran larvae to different particle sizes at low concentrations (Amphibia). *Hydrobiologia*, **234**, 151–73

Voris, H.K. and Bacon, J.P. Jr (1966). Differential predation on tadpoles. *Copeia*, **1966**, 594–8

Waldman, B. (1982). Sibling association among schooling toad tadpoles: field evidence and implications. *Animal Behaviour*, **30**, 700–13

Waringer-Löschenkohl, A. (1988). An experimental study of micro-habitat selection and microhabitat shifts in European tadpoles. *Amphibia–Reptilia*, **9**, 219–36

Warkentin, K.M. (1992). Microhabitat use and feeding rate variation in green frog tadpoles (*Rana clamitans*). *Copeia*, **1992**, 731–40

Warner, S.C., Travis, J. and Dunson, W.A. (1993). Effect of pH variation on interspecific competition between two species of Hylid tadpoles. *Ecology*, **74**, 183–94

Wassersug, R. (1972). The mechanism of ultraplanktonic entrapment in anuran larvae. *Journal of Morphology*, **137**, 279–88

Wassersug, R. (1973). Aspect of social behaviour in anuran larvae. In Vial, J.L. (ed.) *Evolutionary Biology of the Anurans: Contemporary Research on Major Problems*, pp. 273–97. University of Missouri Press, Columbia, MO

Wassersug, R. (1974). The evolution of anuran life cycles. *Science*, **185**, 377–8

Wassersug, R. (1975). The adaptive significance of the tadpole stage with comments on the maintenance of complex life cycles in anurans. *American Zoologist*, **15**, 415–17

Wassersug, R. (1980). Internal oral features of larvae from eight anuran families: functional, systematic, evolutionary and ecological considerations. *Miscellaneous Publication(s) of the Museum of Natural History*. University of Kansas, **68**, 1–146

Werner, E.E. (1986). Amphibian metamorphosis: growth rate, predation risk and optimal size at transformation. *The American Naturalist*, **128**, 319–41

Wetzel, R.G. (1983). *Limnology*. 2nd edn. Saunders College Publishing, Philadelphia

Wetzel, R.G. (1984). A comparative study of the primary productivity of higher aquatic plants, periphyton and phytoplankton in a large, shallow lake. *Internationale Revue der Gesampten Hydrobiologie*, **49**, 1–61

Wilbur, H.M.,(1980). Complex life cycles. *Annual Review of Ecology and Systematics*, **11**, 67–93

Wilbur, H.M. and Collins, J.P. (1973). Ecological aspects of amphibian metamorphosis. *Science*, **182**, 1305–14

CHAPTER 8

FISH DIVERSITY AND ECOTONAL HABITAT

Maciej Zalewski

INTRODUCTION

The most distinctive effect of increasing human activity around the globe is steady reduction of environmental diversity. The most common examples are conversion of forest to fields and pastures, or desertification of eroded land due to poor land management practices. Those processes are as old as human civilization. However, while in the past they appeared in restricted areas, now they are occurring at a dramatic rate on a global scale affecting aquatic ecosystems situated in depressions in the landscape, and their top biotic component – fish.

In the case of fluvial ecosystems, one of the most important factors responsible for the sharp decline in biodiversity has been channelization combined with wetland degradation. This is due to the reduction of water retention in the catchment, reduction of flow variation and loss of habitats resulting in increased abiotic stress (Ward and Stanford, 1989). In lentic ecosystems such as lakes and reservoirs, the recently increased abiotic stress is mostly due to point source and non-point source pollution leading to eutrophication and a dramatic reduction in fish diversity. Also, evidence has emerged on the role of atmospherically dispersed contaminants, e.g. PCBs whose concentration in fish increases with length of the food chain and decreases with distance from urban and industrial centres (Rasmussen *et al.*, 1990).

In most areas of the world, it will be very difficult to change human activity and its impact on river and lake catchments, in the short term, for the benefit of the aquatic ecosystem and the maintenance of fish, because of the exponential increase of human population.

The holistic school of ecological thinking assumes that systems are more than the sum of their parts and thus possess new emergent properties which cannot be predicted from analyses of their component parts, no matter how precise and complete. In consequence, such systems are hierarchically organized, with new properties at each level of the hierarchy (Solbrig, 1991). The above approach has been expressed in terms of the cybernetic nature of the ecosystem (Patten and Odum, 1984), where the feedback regulatory mechanisms based on biotic interactions maintain homeostatic equilibrium of the ecosystem and its biodiversity. Those processes are often determined by ecotonal structure, its utilization by fish and the effect of ecotone/fish interactions. In an example of a tropical swamp and stream, Winemiller (1990) demonstrated that detritus from aquatic macrophytes and terrestrial origin was the most important pathway of energy flow to fishes, and that between site variation in food web parameters was associated with differences in species richness. Fish may regulate and stabilize the whole system energy flow, e.g. 'top-down effect' (Northcote, 1988), which has significant consequences for homeostatic equilibrium in ecosystems (Fausch et al., 1990). It might be especially significant in global change as the predicted increase in major greenhouse gases (CO_2, CH_4, N_2O, and CFCs) and a doubling in carbon dioxide concentration expected by the year 2050, may raise average global temperature by 2–5°C. That will dramatically change the functioning of main aquatic systems and their biotic components (Zalewski, 1995). However, our predictive ability is still poor (Smith, 1990) and long-term ecological research (Magnuson, 1990) may elucidate these processes. The present chapter considers the complexity of interactions between land–water ecotones and top trophic levels in aquatic food webs (fish) and its influences on various levels of organization of freshwater ecosystems from a holistic perspective. Consequently, it is an attempt to highlight the range of questions, from: how a particular form of ecotonal degradation influences fish populations, to the specific analysis of processes at whole ecosystem interactions initiated at ecotones. Considering the accelerating environmental degradation of river and lake catchment, it is necessary to create a realistic strategy which can be implemented in the short term. The significant role of land–water ecotones in the dynamics of aquatic systems and their top component – fish (Zalewski et al., 1991b; Schiemer and Zalewski, 1992) seems to be one of the keys to creating an efficient method towards conservation of aquatic ecosystems and maintenance of their biodiversity.

INCREASE OF FISH DIVERSITY WITH SHIFT FROM ABIOTIC TO BIOTIC MODE OF ECOSYSTEM REGULATION

The first question to answer in analysing the relations between land–water ecotones and fish diversity seems to be: what are the factors which stimulate the increase in biodiversity of ecosystems? There are two principal factors: the area of the ecosystem (MacArthur and Wilson, 1967; Magurran, 1988) and its latitudinal position (Pianka, 1983). The first is well supported by experimental evidence. The second seems to be much more complicated and ectothermal vertebrates (e.g. fish) are useful organisms for the critical evaluation of its importance. Among all aquatic habitats, rivers, due to their spatial and temporal heterogeneity, are most appropriate ecosystems for the analyses of the relationships between fish biodiversity and environment properties.

Analysing data from different rivers, Zalewski and Naiman (1985) concluded that, contrary to common opinion, species richness in riverine fish communities increased from ecosystems where abiotic factors play the prevailing role (severity of environment), to ecosystems where a biotic mode of community regulation prevails. Abiotic factors are also of primary importance in regulating fish communities. However, in many types of predictable, tropical riverine ecosystems which fluctuate within the range of physiological tolerance of ectothermal fish, the biotic factors are the prevalent regulators of fish communities, enhancing their diversity. The factor of primary importance determining fish biodiversity is longevity of the ecosystem and possibility of colonization after large geo-climatic events, e.g. glaciation (Power et al., 1973).

Abiotic factors are chiefly a function of climate and geologic history. They determine the character and productivity of terrestrial ecosystems in the river catchment, and the annual pattern of temperature, flow, oxygen and pH, e.g. the higher coefficient of variation of annual flows in boreal and semi-arid ecosystems (MacMahon, 1982), and consequent trophy and habitat diversity for fish. Biotic factors (e.g. competition, predation) depend chiefly on community composition, which is a result of the phylogeny specific for the particular geographic area. The course of phylogeny has been a function of the strength of abiotic factors (e.g. catastrophic events), the range of physiological tolerance of fish and their compensatory abilities (e.g. tolerance of low oxygen content under ice, or high temperatures) and behavioural adaptation, e.g. migration (Thorpe, 1988). In consequence, the lowest biodiversity occurs in harsh abiotically regulated environments where there is often strong trophic limitation. In contrast, the highest fish diversity occurs in biotically regulated rivers where the main abiotic factors fluctuate around the fishes' physiological optima.

The increase in fish species richness from abiotic to biotically regulated communities contradicts the ecological handbooks (e.g. Pianka, 1983; Magurran, 1988) which suggest that, in general, biodiversity increases from north to south. The confirmation of abiotic–biotic regulation of fish diversity is lowest in abiotically regulated rivers, e.g. boreal (Moisie, Labrador) and Colorado River in the Nevada Desert (Naiman and Soltz, 1981). In both systems abiotic factors controlled fish communities and temperature was the main axis of freshwater fish diversity distribution (Figure 8.1). The minimum biodiversity occurs at both the lowest and the highest temperatures. However, the hierarchy of factors and physiological compensatory mechanisms differ. In boreal ecosystems, long periods of low temperature during winter, poor energy supply (Cunjak and Power, 1987), low oxygen concentrations (Reynolds, pers. comm.) and catastrophic spring floods are the main selective factors. In desert streams, channel desiccation, variability of oxygen concentration, temperature, osmotic pressure and catastrophic floods play the main roles in regulating fish community dynamics (Naiman and Soltz, 1981).

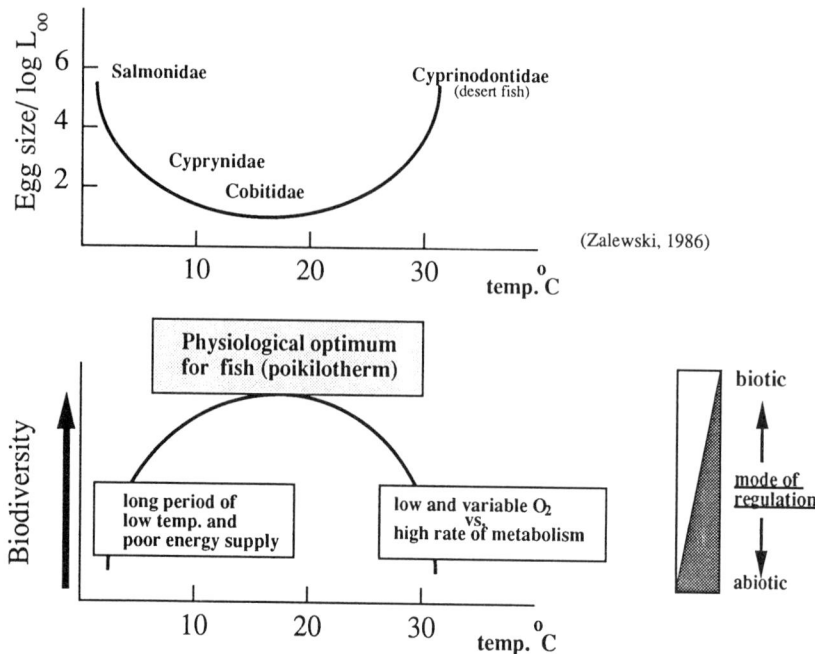

Figure 8.1 The relation between mode of community regulation (abiotic–biotic) and fish biodiversity exemplified by morphophysiological adaptation – relative egg size. L_{∞} is a parameter from the von Bertalanffy equation describing growth of fish, calculated using the Ford–Walford method

Despite the various abiotic factors (temperature *vs.* oxygen) the main compensatory physiological mechanism in both abiotically controlled environments is similar: accumulation in tissues and eggs of a large amount of highly efficient energetic substances – lipids. This is reflected in the tendency to produce larger eggs, (Figure 8.1), which is probably due to compensation for trophic limitation in boreal rivers or for sharp declines in oxygen and increased energy expenditure on osmoregulation by glycolysis, a less oxygen efficient metabolic process in a desert stream. Hence, large reserves of energetic substances are necessary.

However, fish species with parental protection during early ontogenesis (e.g. Cichlidae), do not show this tendency. Also, in the oldest tropical ecosystems, many fish possess morphological adaptations for breathing atmospheric air (Weber and Kramer, 1983). Such physiological mechanisms compensate for the effects of abiotic factors based on long-term coevolutionary processes, which might be considered in the context of the primary axis determining fish diversity in freshwater ecosystems.

However, many field observations in systems with similar climate, hydrological regime, and hierarchy of regulatory factors reveal a broad range of species richness. These are attributable to aquatic habitat character, and in rivers, they increase downstream due to declining abiotic stresses (e.g. hydraulics, temperature) and increasing habitat heterogeneity (Horowitz, 1978). Both of which are due mostly to increasing space in land–water ecotones – by transmission of the riparian zone into the floodplain. The highest biodiversity occurs in the biggest and most complex floodplain river (Lowe-McConnell, 1987). Fish diversity declines sharply when rivers are channelized (Portt et al., 1986) or impounded (Schiemer and Spindler, 1989). So, the tendency for fish diversity to increase downstream in natural river ecosystems is not only the result of the reduction in environment harshness (temperature and hydraulic extremes) but also of an increase in riverine habitat complexity by riparian/floodplain interactions.

Another important factor determining fish diversity is age of the ecosystem. In relatively young riverine ecosystems, colonized postglacially (Power et al., 1973) diversity is much lower than in much older tropical ecosystems. Also beta and gamma diversity (Cody and Diamond, 1975) between ecosystems is much lower there: species community composition in Polish upland, lowland and coastal rivers is much lower than in analogous types of streams in Venezuela (Zalewski, in preparation).

THE CHANGE OF BIODIVERSITY AS AN EFFECT OF THE ECOSYSTEM EUTROPHICATION

On a geological time scale, lakes are temporary environments or impoundments of rivers, while fish assemblages originated in rivers (Kitchell et al., 1977; Fernando and Holcik, 1981) and consequently mostly utilize littoral ecotone zone resources.

One of the most important processes of degradation of lakes and reservoirs, which dramatically reduces fish diversity, is eutrophication. It results from increasing human activity in river and lake catchments, which in turn amplifies point and non-point source pollution. Increase in ecosystem trophy is reflected in the diversity and productivity of fish communities. The dependence of diversity and productivity of the lake fish community on lake trophic status through nutrient load or transparency is parabolic. In general the main effect of the succession of lake ecosystems from oligo- to eu/hypotrophy on fish communities is replacement of species of high quality fishery yield, such as salmonids and coregonids, by centrarchids, percids and finally by eutrophy-tolerant but low value dwarfed cyprinids (Colby et al., 1972).

In oligotrophic alpine and arctic lakes, fish diversity and productivity are low due to low nutrient input and limited energy supplied during the short summer period. In such systems energy from primary producers is very efficiently transferred to the top of the pyramid – fish – so conversion of phytoplankton to fish production is about 1%. Due to a limited autochthonal food supply, the role of terrestrial food from the land–water interface zones might be high. The development of the shoreline may stimulate lake trophy as some positive correlations between fish productivity and shoreline development have been found (Backiel et al., 1982). In hypereutrophic lakes, primary productivity several orders of magnitude higher than in oligotrophic systems is transferred mostly to the sediments. Anoxic conditions in the hypolimnion, lack of spawning substrate for lithophil and sometimes for phytophil fish, occurrence of toxic algal blooms (Kotak et al., 1993) combined with poor control by predators, cause unbalanced low diversities dominated by dwarfed fish communities. In consequence fish productivity declines, with an efficiency of phytoplankton production transfer to the fish of as little as 0.002% (Downing et al., 1990).

The highest diversity and optimal fish productivity are achieved in fish communities at intermediate stages of eutrophication. As world-wide agricultural activities in the catchment increase the process is dramatically spread and accelerated. In consequence, many systems are reaching the last degraded phase of succession. Because it seems unrealistic to stop or sharply reduce human activities in lake and river catchments, the most promising way forward is to create efficient buffering zones between land and water (Naiman et al., 1989). The littoral ecotone zone has been an important trap for nutrients, provided as non-source pollution from the catchment (Jørgensen, 1990). Lewis et al. (1984) demonstrated that almost all particulate matter provided from the terrestrial system accumulates in the littoral zone. This form of nutrient transfer provides 20–50% of the nitrogen and 30–90% of the phosphorus entering the lake. Another function of the littoral ecotone zone is as a spawning and rearing area for many fish species, and a habitat for some predators, e.g. pike (Grimm, 1989), which, in modern management of lakes under eutrophication, can be used for regulation of planktivorous fish density and provides a refuge for cladocerans which may control green algal density efficiently.

The importance of the littoral zone declines inversely to lake size. It can also be expected that efficient manipulation of the biotic structure of the lake ecosystem for reversing symptoms of eutrophication will depend on two factors: first, it will be inversely related to the size of the lake, and second, it will decline above a certain degree of eutrophication. On the basis of existing literature data (Gulati et al., 1990) it can be expected that biomanipulation appears to be most efficient up to 100 mg total P l^{-1}. However, some success has been observed in the range from 100 to 500 mg P l^{-1}, but at the highest levels of P concentrations it is not often satisfactory, probably due to strong interference of physical and hydrochemical processes (e.g. oxygen deficit). Also, the terrestrial component of the land–water ecotone may play an important role in regulating symptoms of eutrophication. High trees along the shoreline stabilize the epilimnion's internal nutrient load which is reduced during the summer: the N:P ratio might be disturbed and consequently blooms of Cyanobacteria appear which are toxic to vertebrates.

The most distinctive process, which occurs in fish communities of the lakes and streams under increasing eutrophication, is decline of fish community diversity as an effect of the elimination of specialist species and increasing dominance of generalists such as roach, *Rutilus rutilus* (L.). Analysing the success of this most widely distributed species in European waters, Schiemer and Wieser (1992) concluded that increasing uniformity of the aquatic habitat facilitates spread of this successful generalist. The experimental data demonstrated that roach possess functional adaptations such as: flexibility in food choice and microhabitat use, non-specialized swimming performance and better adaptation of digestive ability to temperature, high feeding efficiency of larvae and juveniles for zooplankton in dim light and darkness. Many questions remain to be answered, such as, does the roach only fill the empty niche or does it actively eliminate specialized species which are in the process of decline with progressive eutrophication?

Natural systems are always fluctuating over certain ranges of their biotic characteristics. In lakes under eutrophication such fluctuations exceed the tolerance limits of some animals, and so reduce the biodiversity of the system. The various forms of biomanipulation in which ecotone management may play a key role are techniques to maintain those parameters (e.g. oxygen level, H_2S concentration) at close to optimal levels and restorative measures undertaken in the catchment for preservation of the aquatic environment.

SHORELINE ECOTONES – FISH AS REGULATORS OF COMMUNITY DYNAMICS AND PATTERNS OF ENERGY FLOW

Reservoirs are traps for nutrient organic matter and mineral sediments carried down by the river. Thus, they tend toward fast eutrophication. These man-made ecosystems are becoming more and more important as sources of water: by the year 2000 almost 70% of the world's total stream flow will be controlled by dams (Petts, 1984).

The effect of dams on diversity of fish populations has two aspects. First, by creation of a discontinuity in the riverine ecosystem, a dam disturbs the natural pattern of physical and biotic processes along the river continuum (Ward and Stanford, 1982). This generates a dramatic shift in fluvial fish communities by lotic/lentic habitat conversion and by disturbance of fish migration. The process of elimination and replacement of the reophilous fish in impounded rivers occurs in both tropical and temperate systems to similar extents despite higher diversity in the former. This replacement process is similar in the Danube (Schiemer and Waidbacher, 1992) and Parana rivers (Agostinho *et al.*, 1994) because almost the entire upper sections of both are impounded. On the upper Parana one dam remains to be completed and may

Figure 8.2 Cascading effects stimulated by the different complexity of shoreline/riparian ecotones at the lowland reservoir. *Left panel* – low water level, terrestrial vegetation not covered, fish reproductive success low, high biomass of zooplankton, low biomass of phytoplankton, good growth and winter survival of juveniles. *Right panel* – high water level, terrestrial vegetation flooded, fish reproductive success high, low biomass of zooplankton, high biomass of phytoplankton, low growth and winter survival of juveniles. Adapted from Zalewski *et al.* (1990)

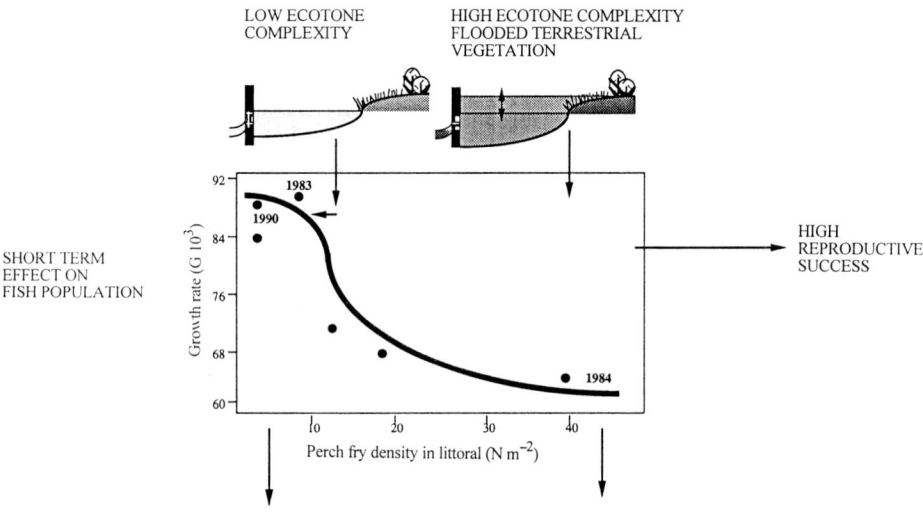

(Continued)

Figure 8.2 (*continued*)

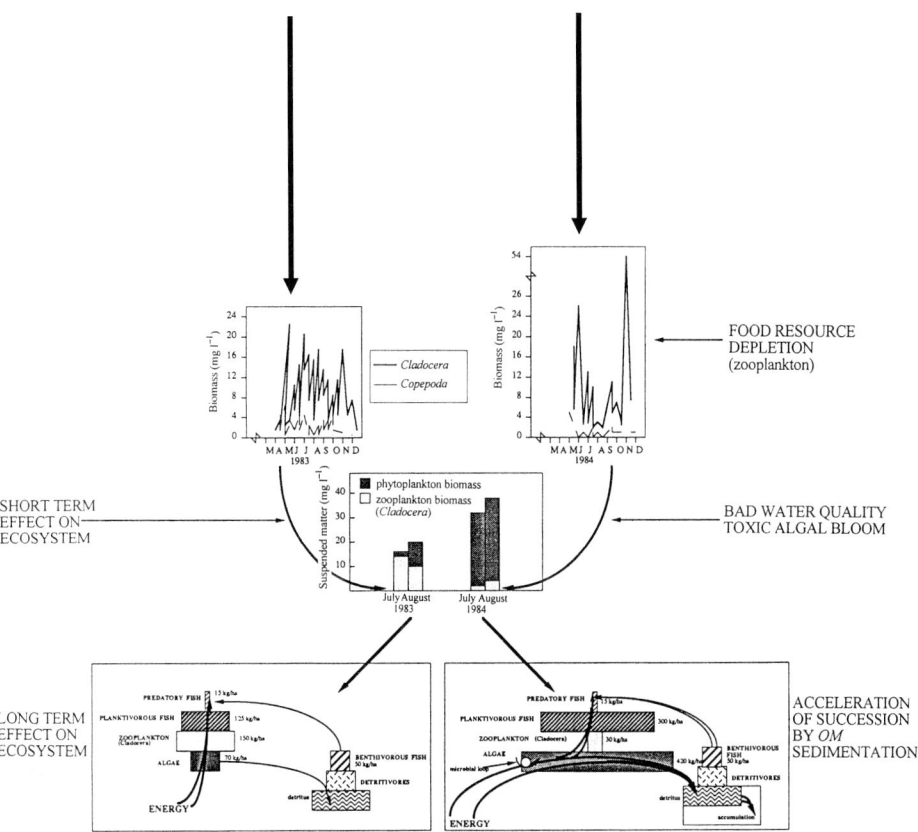

completely eliminate fluvial fish from this distinctive ecosystem (Agostinho and Zalewski, 1995).

The transition of most of the lotic to the lentic environment also leads to eutrophication which, as in the case of lakes, will degrade water quality relatively quickly and so eliminate further fish species. The second result of impoundments is the decline or loss of migratory fish (e.g. Backiel, 1985). In some cases stocking has been undertaken to avoid the extinction of migratory species, but there may then be genetic disturbances in such natural populations (Waples, 1991).

One of the most promising ways to maintain reservoir and lake ecosystems in a state of dynamic equilibrium, and not in an advanced state of succession, and to maintain fish diversity, seems to be reduction of nutrient inputs combined with manipulation of the food web. The biomanipulation technique, or restoring water quality by manipulating biotic components, increases the density of large efficient filtrators such as *Daphnia* sp., thereby reducing many fold the concentration of phytoplankton (e.g. Gulati *et al.*, 1990). In oligotrophic systems (good water quality, low production) the largest portion of energy is diverted from primary producers (algae) to subsequent trophic levels (planktivorous and predatory fish). In eutrophic systems, where large *Daphnia* are eliminated by juvenile fish or pelagic planktivores, high primary production (phytoplankton) is not consumed by filtrators (Figure 8.2). This is because large *Daphnia* are eliminated by dense populations of juvenile percid and cyprinid fishes (Zalewski *et al.*, 1990a, 1990b). The dead algal nutrients are recirculated by the 'microbial loop' and a continuing high nutrient supply maintains concentrations of algae, resulting in deterioration of water quality. Benthic detritivores may be stimulated in the system as long as the bottom oxygen supply of the reservoir is sufficient. In such a system a large part of the energy is accumulated in bottom sediments leading to a decrease in aquatic habitat quality. In shallow lakes resuspension of bottom sediments may additionally amplify symptoms of eutrophication.

Figure 8.2 demonstrates the possibilities of restoring a reservoir ecosystem for improvement of water quality, using food web dynamics, thus regulating access for fish to reproduce in the land–water ecotone which is covered by terrestrial vegetation. In parallel the predatory pike-perch, which reduce pelagic zooplanktivorous fish populations efficiently, should be enhanced. In temperate reservoirs, pelagic predators such as pike-perch or walleye are the most important biomanipulation tools because the second important predator of temperate zone lakes, the northern pike *Esox lucius*, rarely exists in reservoirs as water-level fluctuations have destroyed its littoral habitat. From a dynamic point of view, such restoration can be described in terms of changes of the energy flow pattern from detrital food chain characteristics of a eutrophic ecosystem to a stage of mesotrophic, high water quality and fish yield ecosystems.

Restoring aquatic habitat quality should provide time to create strategies for the conservation of the whole catchment area. These strategies should

RIPARIAN ECOTONES OF INTERMEDIATE COMPLEXITY AS A FACTOR MAINTAINING FISH COMMUNITY DIVERSITY AND PRODUCTION

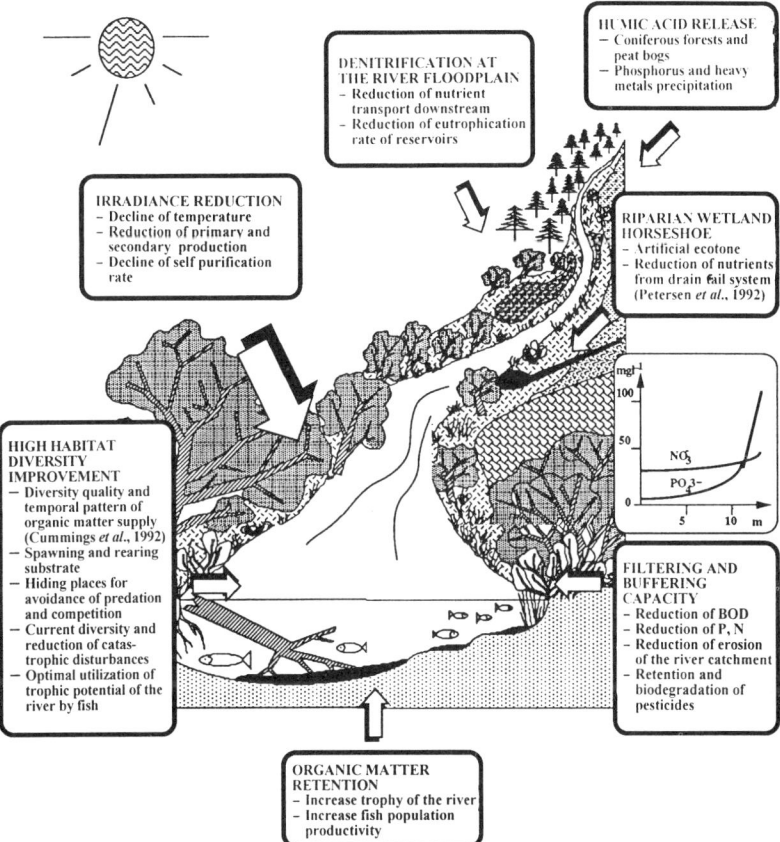

Figure 8.3 The function of restored riparian floodplain ecosystem in agricultural catchment toward maintenance of fish biodiversity. Adapted from Zalewski (1995)

include sewage treatment plants and the reduction of aerial pollution by the creation of riparian ecotone zones.

THE ROLE OF THE LAND–WATER ECOTONES AT THE RIVER IN AGRICULTURAL CATCHMENTS

Considering the role of riparian/floodplain ecotones from the point of view of the maintenance of fish biodiversity, it seems necessary to analyse what should be its optimal composition. The fundamental function is to increase the diversity of stream habitat, reduce non-source pollution, accelerate self purification and improve trophic conditions for fish (Zalewski et al., 1991b, 1994). In Figure 8.3 the hypothetical highly patchy land–water interface ecotone zone is presented. The first group of components create factors which indirectly influence fishes, mostly by improvement of water quality of rivers and reservoirs. Among those, the filtration and retention of the riparian zone is most important. Knauer and Mander (1989), demonstrated that even a 7 m wide bush strip along the bank may reduce up to 60% of phosphorus in rivers. The retention of pesticides in this zone accelerate their decomposition thus reducing input into aquatic systems and accumulation in food webs. Independently, riparian vegetation reduces bank erosion, which has a very negative effect on stream biota eliminating many benthivorous invertebrates and fish (Berkman and Rabeni, 1987). Tree roots, fallen trunks and branches increase retention of organic matter and maintain large amounts of invertebrates which are food for fish (Ward and Stanford, 1982). However, tall trees along the stream banks reduce light and hence primary productivity so influencing trophic conditions for fish (Zalewski et al., 1994). In areas where coniferous forest and peat bogs occur along stream banks the released humic acids in wet periods increase phosphorus and heavy metal precipitation reducing eutrophication of impounded sections. In intensively cultivated farmland the ends of irrigation tubes should lead into small artificial wetlands which retain and reduce nutrients (Petersen et al., 1992).

Fallen trees, beaver dams and small retention facilities may, to a great extent, reduce nutrient transport downstream (Naiman et al., 1987) and increase patchiness of the river valley, however, they may obstruct migration of fish upstream.

The most important direct effect of land–water ecotone complexity is diversified habitats which create conditions for the coexistence of a large number of fish species and their life stages, reduce competitive interactions, pressure of predators and catastrophic disturbances, and provide feeding and spawning/rearing grounds (Schiemer and Zalewski, 1992). The oxbow lakes in the river valley may play a role as genetic reservoirs for populations in harsh or human impacted environments (Bouvet and Pattee, 1991). A further analysis of riparian habitat role is given in Pieczyńska and Zalewski (Chapter 3).

STOCKING AS A METHOD OF COMPENSATION FOR LOSS OF ECOTONAL HABITATS

Fish stocking is a method used widely for improvement of fisheries. During the last fifty years this has been intensified to increase catches but this has led to degradation of freshwater ecotonal habitats. These are the main spawning and rearing areas and require monitoring of the long-term effects on genetic diversity of fish populations. According to recent investigations (Phillip and Whitt, 1991), the stock transfers from one ecosystem to another may lead to a decrease of the tolerance to abiotic factors due to hybridization. Besides, hatchery populations are often characterized by inbreeding depression and have been exposed to conditions very different from those in the natural environment. Mixing hatchery stocks with wild fish may lead to outbreeding depression (Wasser and Price, 1989) and changed selection regimes. Both reduce ability to survive variations in abiotic factors (Waples, 1991). Efficiency of stocking is very variable even within the same ecosystem. Zalewski *et al.* (1985) demonstrated that survival of introduced brown trout declines downstream as an effect of predation increase.

In a lentic environment, with fish of a known tolerance to a range of abiotic factors, stocking efficiency might be described by a curvilinear relation (logistic function) exemplified by the results of Müller (1990) from Lake Hallwil. At low levels stocking may be ineffective because of activity of specific predators but, after saturation, stocking efficiency may increase proportionally to the number of fish stocked up to the carrying capacity of the environment.

One of the important limitations on stocking efficiency is the rearing conditions. Shustov and Shchurov (1989) showed that salmon parr reared in hatchery ponds had much lower stamina than parr from streams. This implies that, to minimize domestication selection during rearing, the hatchery environment should mimic the natural environment. The question remains as to whether stocking is still necessary. In the case of endangered species and stocks, such as Arizona trout (*Salmo apache*, Miller), it may be the only solution remaining (Rinne, 1988).

It is necessary to stress that genetic investigation provides evidence that the best method of fish species conservation is not re-stocking but preventing further deterioration of habitat, and aiding its restoration and management (Soulé and Mills, 1992).

CONCLUSIONS

How is fish diversity affected by degradation of land–water ecotones? In habitats where complexity of the interface zone of the aquatic environment was high, degradation of the ecotone zone was reflected negatively in fish density and diversity (Schiemer and Spindler, 1989; Schiemer and Zalewski, 1992). It is due mostly to reduction of habitat dimension and unstable flow conditions (Schlosser and Karr, 1981). There is some experimental evidence that reduction of the complexity of aquatic–terrestrial interfaces drastically reduces establishment of large specimens (Portt *et al.*, 1986), so that trophic structure of the fish community is modified (Schlosser, 1982) and, even more important, the reproductive potential of the population might be sharply reduced leading to greater variability and/or smaller numbers of specimens in a population. This process is especially important in large seriously polluted rivers where the available space for the population is restricted to the tributaries, and any

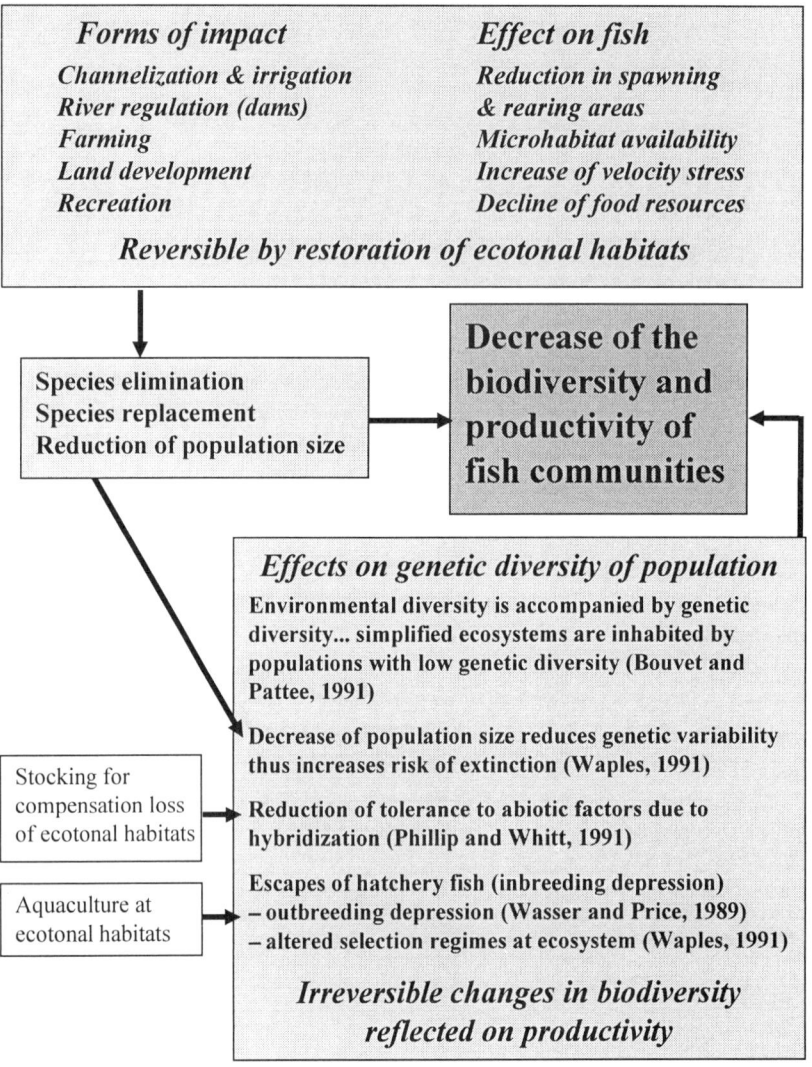

Figure 8.4 The reversible and irreversible effects on fish communities due to degradation of land–water ecotones

reduction of ecotone complexity may drastically reduce population size. Decline of population size is accompanied by reduction of genetic variability, which in turn increases vulnerability of the smaller population to environmental changes (Figure 8.4). The problem becomes self-amplifying as harshness of environmental conditions increases due to reducing riparian ecotone complexity, and population extinction becomes more probable (Coelho and Zalewski, 1995).

Phillip (1994) suggests that species loss is not the only component of biodiversity loss, because there is increasing evidence that species are not single randomly-breeding panmictic units but are rather mosaics of genetically divergent units, i.e. populations or stocks. Therefore species have inherent levels of both intrastock and interstock genetic variation. Thus the most important goal is to protect local populations which typically exhibit a relatively high degree of genetic divergence. According to Ryman et al. (1994) intraspecific genetic variability may be threatened in three ways: (1) local populations may become extinct through such processes as excess harvesting, habitat losses, disease introductions, and displacement by introduction of other populations; (2) losses of genes and gene complexes that distinguish populations may occur through hybridization resulting from intentional stocking, sea ranching, escapes from aquaculture; and (3) loss of selection through reduced population size.

The above assumptions are connected with the concept 'minimum viable population size' (MVP) which, according to Ewens et al. (1987), has two aspects. First, 'genetic' – based on the rate at which genetic variation in a population is lost (fitness decreased) through random genetic drift. Second, 'demographic' which concerns the probability of extinction of the population through random demographic processes. The interconnection of those aspects appears as inbreeding reduces fecundity and increases mortality.

It seems necessary to analyse existing data describing such processes at different scales and levels of integration including physical, chemical, biological and human effects on aquatic systems. The analysis of the cumulative impact (Bedford and Preston, 1989) should consider the discrepancy between what ecologists prefer to observe and analyse (usually single factors vs. single populations), and reality, where many factors act cumulatively in antagonistic and synergistic ways influencing the whole community complex. Consequently, the efficient conservation approach should combine the complexity of interactions which appear in the land–water ecotones in relation to the various forms of cumulative impact and at various levels of integration, assuming that understanding of such complex processes will be helpful in establishing a realistic strategy for conservation of fish diversity in inland waters. For example, this assumption implies that any fish diversity conservation and management practices should focus on certain components of the ecosystem. In abiotically regulated systems, exemplified by the boreal rivers, the effort should be focused on reduction of environmental variability and metabolic stress by improvement of the channel diversity and enhancement of trophic conditions. Both can be done in the most efficient way by creation and enhancement of the diversified interface zone between land and water. In environments with more favourable abiotic conditions, in the sense of a lower range of variabilities and higher predictability, most are biotic interactions such as predation or competition which activate feedback control mechanisms such as resource depletion (effect on lower trophic levels). In such cases the biotic component of the ecosystem should be a subject of manipulation, for maintenance of biodiversity and improvement of fish yield (Zalweski et al., 1994). Again this is the most flexible component of an ecosystem from the point of view of management of land–water ecotones which are feeding, spawning and rearing habitats for many fish species. The dependence of key fish species on the structure and dynamics of land–water ecotones requires the creation or reduction of the given type of ecotone habitats. Manipulation of this interface zone by a series of feedback mechanisms can regulate fish community composition dynamics which is reflected on whole ecosystem processes (e.g. Zalewski et al., 1990a, 1990b).

Integrated strategy toward conservation of fish biodiversity in inland waters should identify the best possibilities and most critical points. Figure 8.4 presents synthesized processes discussed in previous sections. In this respect the most efficient action seems to be the restoration of degraded riparian ecotonal habitats which may reverse negative trends in fish communities. However, irreversible changes appear in populations and communities if genetic structure is modified due to processes such as: species elimination, replacement, reduction of population size or stocking and escapes of domesticated specimens from aquaculture (Waples, 1991).

Subsistence fisheries are an example of a discrepancy between conservation of fish diversity and local economy. Their existence shows the need for realistic strategies. For example, Nile perch were introduced to Lake Victoria. This efficient predator has been a reason for the dramatic reduction in the species richness of endemic cichlid species from over 200 to 100 species. However, the problem is that the introduction (Barel, 1986) increased the catch value from $30 million to $230 million. So, to provide realistic strategies for biodiversity conservation to the regions of the world where the trade-off is maintenance of biodiversity *versus* protein gain, will be difficult. One important solution might be based on a more profound understanding of the relation between fish and ecotones, since these are environmental elements of relatively small area, so easier to reclaim, but extremely active in the sense of ecological processes. In many cases they are at the start of cascades of reactions reflected in the whole ecosystems dynamics, and they influence directly and indirectly the diversity of the top components in aquatic systems – fish.

Consequently there is an urgent need to develop general models of the links between fish occurring in the main aquatic habitats and their aquatic/ terrestrial interface structures, for better understanding of the effects of the degradation of ecotonal habitats on fish diversity. These models should consider various food web alterations, reproduction constraints, population dynamics modifications, and finally genetic effects on populations due to modification of ecotonal structure. The most important advantage of the management and restoration of the ecotone structure as a tool for fish diversity conservation is that it could be done by public involvement, thus might be relatively inexpensive and efficient in the long term. In consequence ecology and environmental conservation will accelerate the transition from a restrictive to a creative phase which, at recent rates of global environment degradation, is urgently need.

ACKNOWLEDGEMENTS

The author wishes to thank John Thorpe for valuable comments and linguistic revision of the manuscript.

REFERENCES

Agostinho, A.A., Julio, H.F. and Petrere, M. (1994). Itaipu Reservoir (Brasil): Impacts of the impoundment on the fish fauna and fisheries. In Cowx, I. (ed.) *Rehabilitation of Freshwater Fisheries,* pp. 171–84. Blackwell Scientific Publications, Oxford

Agostinho, A.A. and Zalewski, M. (1995). The dependence of fish community structure and dynamics on floodplain and riparian ecotone zone in Parana river, Brasil. *Hydrobiologia,* **303**, 141–8

Backiel, T. (1985). Fall of migratory fish populations and changes in commercial fisheries in impounded rivers in Poland. In Alabaster, J.S. (ed.) *Habitat Modifications and Freshwater Fisheries,* pp. 28–41. Butterworths, London

Backiel, T., Thorpe, J.E. and Kitchell, J.F. (1982). The functioning of freshwater ecosystems. In Le Cren, E.D. and Lowe-McConnell, R.H. (eds.) *Fish,* pp. 307–41. Cambridge University Press, Cambridge

Barel, C.D.N. (1986). The decline of Lake Victoria's cichlid species flock. Summary of the flock's current status. *Zoologisch Laboratorium.* University of Leidein

Bedford, B.L. and Preston, E.M. (1989). Developing the scientific basis for assessing cumulative effects of wetland loss and degradation on landscape functions: Status, Perspectives and Prospects. *Environmental Management,* **12**, 751–71

Berkman, H. and Rabeni, C.F. (1987). Effect of siltation on stream fish communities. *Environmental Biology of Fishes,* **18**, 285–94

Bouvet, Y. and Pattee, E. (1991). Ecotones and genetic diversity of fish in the river Rhône. In Zalewski, M., Thorpe, J.E. and Gaudin, P. (eds.) *Fish and Land/Inland Water Ecotones,* pp. 25–7. UNESCO MAB, University of Lodz, Stirling, Lyon

Cody, M. and Diamond, I.M. (eds.) (1975). *Ecology and Evolution of Communities.* Belknap Press, Cambridge, MA

Coelho, M.M. and Zalewski, M. (1995). Evolutionary adaptations by fish to ecotonal complexity in spatially variable landscapes – a perspective. In Schiemer, F., Zalewski, M. and Thorpe, J.E (eds.) The importance of aquatic terrestrial ecotones for freshwater fish. *Hydrobiologia,* **303**, 223–8

Colby, P.J., Spangler, G.R., Hurley, D.A. and McCombie, A.M. (1972). Effects of eutrophication on salmonid communities in oligotrophic lakes. *Journal of the Fisheries Research Board of Canada,* **29**, 975–83

Cunjak, R.A. and Power, G. (1987). The feeding and energetics of stream resident trout in winter. *Journal of Fish Biology*, **3**, 493–511

Downing, J.A., Plante, C. and Lalonde, S. (1990). Fish production correlated with primary productivity, not the morphoedaphic index. *Canadian Journal of Fisheries and Aquatic Sciences*, **47**, 1929–36

Ewens, W.J., Brockwell, P.J., Gani, J.M. and Resnick, S.I. (1987). Minimum viable population size in the presence of catastrophes. In Soule, M. (ed.) *Viable Populations for Conservation,* pp. 59–68. Cambridge University Press, Cambridge

Fausch, K.D., Lyons, J., Karr, J.R. and Angemaier, P.L. (1990). Fish communities as indicator of environmental degradation. *American Fisheries Society*, **8**, 123–44

Fernando, C.H. and Holcik, J. (1981). Fish in reservoirs. *Internationale Revue der Gesamten Hydrobiologie*, **76**, 149–67

Grimm, M.P. (1989). Northern pike (*Esox lucius* L.) and aquatic vegetation, tools in the management of fisheries and water quality in shallow waters. *Hydrobiological Bulletin*, **23**, 61–7

Gulati, R.D., Lammens, E.H.R.R., Meijer, M.-L. and van Donk, E. (eds.) (1990). *Biomanipulation Tool for Water Management.* Proceedings of an International Conference in Amsterdam, The Netherlands, 8–11 August. *Hydrobiologia*, 628 pp.

Horowitz, R.J. (1978). Temporal variability patterns and distribution patterns of stream fishes. *Ecological Monographs*, **48**, 307–21

Jørgensen, S.E. (1990). Erosion and filtration. In Jørgensen, S.E. and Löffler, H. (eds.) *Guidelines of Lake Management.* Vol. 3. *Lake Shore Management,* pp. 13–19. International Lake Environment Committee Foundation and United Nations Environment Programme, Otsu, Japan

Kitchell, J.F., Carpenter, S.R., Bayley, S.E., Ewell, K.C., Howarth, R.W., Nixon, S.W. and Schindler, D.W. (1991). Aquatic ecosystem experiments in the context of global climate change: working group report. In Mooney, H.A., Medina, E., Schindler, D.W., Schulze, E.D. and Walker, B.H. (eds.) *Ecosystem Experiments,* pp. 229–35. SCOPE Published by John Wiley, Chichester

Kitchell, J.F., Johnson, M.G., Minns, C.V.K., Loftus, K.H., Greig, L. and Olver, C.H. (1977). Percid habitat: the river analogy. *Journal of the Fisheries Research Board of Canada*, **34**, 1936–40

Knauer, N. and Mander, U. (1989). Untersuchungen uber die Filterwirkung verschiedener Saumbiotope an Gewassern in Schleswig-Holstein. 1. Mitteilung: Filterung von Stickstoff und Phosphor. *Zietschrift für Kulturtechnik und Landentwicklung*, **30**, 365–76

Kotak, B.G., Kenefick, S.L., Fritz, D.L., Rousseaux, C.G., Prepas, E.E. and Hrudey, S.E. (1993). Occurrence and toxicological evaluation of cyanobacterial toxins in Alberta lakes and farm dugouts. *Water Research*, 495–506

Lewis, W.M., Saunders, J.F., Crumpacker, D.W. and Brendecke, C.M. (eds.) (1984). *Eutrophication and Land Use.* Springer Verlag, New York, Berlin, Heidelberg

Lowe-McConnell, R.H. (1987). *Ecological Studies in Tropical Fish Communities.* 382 pp. Cambridge University Press, Cambridge

MacArthur, R.H. and Wilson, E.O. (eds.). (1967). *The Theory of Island Biogeography.* Princeton University Press, Princeton, NJ

MacMahon, T.A. (1982). Hydrological characteristics of selected rivers of the world. *Technical Documents in Hydrology*, UNESCO, Paris

Magnuson, J.J. (1990). Long-term ecological research and the invisible present. *BioScience* **40**(7), 495–501

Magurran, A.E. (1988). *Ecological Diversity and its Measurement*, 179 pp. Croom Helm, London, Sydney

Müller, R. (1990). Management practices for lake fisheries in Switzerland. In Van Densen, W.L.T., Stainmetz, B. and Hughes, R.H. (eds.) *Management of Freshwater Fisheries*, pp. 477–92. EIFAC Symp., Göteborg, Sweden; PUDOC, Wageningen

Naiman, R.J., Décamps, H. and Fournier, F. (eds.) (1989). The Role of Land–Inland Water Ecotones in Landscape Management and Restoration: a Proposal for Collaborative Research. 93 pp. *MAB Digest 4*, UNESCO, Paris

Naiman, R.J., Malillo, J.M., Lock, M.A., Ford, T.E. and Reice, S.R. (1987). Longitudinal patterns of ecosystem processes and community structure in a subarctic river continuum. *Ecology*, **68**, 1139–56

Naiman, R.J. and Soltz, D.L. (eds.) (1981). *Fishes in North American Deserts*. 552 pp. Wiley Interscience, New York

Northcote, T.G. (1988). Fish in the structure and function of freshwater ecosystem: a 'top-down' view. *Canadian Journal of Fisheries and Aquatic Sciences*, **45**, 361–79

Patten, B.C. and Odum, E.P. (1984). The cybernetic nature of ecosystem. *American Naturalist*, **118**, 886–95

Petersen, R.C., Petersen, L.B.M. and Lacoursiere, J. (1992). A building-block model for stream restoration. In Boon, P.J., Calow, J. and Petts, G.E. (eds.) *River Conservation and Management*, pp. 293–310. John Wiley, Chichester, New York

Petts, G.E. (ed.) (1984). *Impounded Rivers: Perspectives for Ecological Management*. John Wiley, Chichester

Phillip, D.P. (1994). Protection of aquatic biodiversity. In Voigtlander, C.W. (ed.) *Proceedings of the World Fisheries Congress*. pp. 186–9. Oxford and IBH Publishing Co Ltd., New Dehli

Phillip, D.P. and Whitt, P. (1991). Survival and growth of northern pike in Florida and reciprocal F1 hybrid largemouth bass in Central Illinois. *Transactions of the American Fisheries Society*, **120**, 58–64

Pianka, E.R. (ed.) (1983). *Evolutionary Ecology*. 3rd edn. Harper and Row, New York

Pieczyńska, E. and Zalewski, M. (1996). Habitat complexity in land–inland water ecotones. In Lachavanne, J.-B. and Juge, R. (eds.) *Biodiversity in Land–Inland Water Ecotones*, pp. 61–79. UNESCO/Parthenon, Paris/Carnforth

Portt, C.B., Balon, E.K. and Noakes, D.L.G. (1986). Biomass and production of fishes in natural and channelized streams. *Canadian Journal of Fisheries and Aquatic Sciences*, **43**, 1926–34

Power, G., Pope, G.F. and Coad, B.W. (1973). Postglacial colonisation of the Matamek River by fishes. *Journal of the Fisheries Research Board of Canada*, **30**, 1586–9

Rasmussen, J.B., Rowan, D.J., Lean, D.R.S. and Carey, J.H. (1990). Food chain structure in Ontario Lake determines PCB levels in lake trout (*Salvelinus namaycuch*) and other pelagic fish. *Canadian Journal of Fisheries and Aquatic Sciences*, **47**, 2030–8

Rinne, J.N. (1988). Native southwestern (USA) trouts: status, taxonomy, ecology and conservation. *Pol. Arch. Hydrobiol.*, **35**, 305–20

Ryman, N., Utter, F. and Laikre, L. (1994). Protection of aquatic biodiversity: a review of the problem. In Voigtlander, C.W. (ed.) *Proceedings of the World Fisheries Congress*, pp. 92–115. Oxford and IBH Publishing. New Dehli

Schiemer, F. and Spindler, T. (1989). Endangered fish species of the Danube River in Austria. *Regulated Rivers: Research and Management*, **4**, 397–407

Schiemer, F. and Waidbacher, H. (1992). Strategies for conservation of a Danubian Fish Fauna. In Boon, P.J., Calow, P. and Petts, G.E. (eds.) *River Conservation and Management*, pp. 363–82. John Wiley, London

Schiemer, F. and Wieser, W. (1992). Epilogue: food and feeding, ecomorphology, energy assimilation and conversion in cyprinids. *Environmental Biology of Fishes*, **33**, 223–7

Schiemer, F. and Zalewski, M. (1992). The importance of riparian ecotones for diversity and productivity of riverine fish communities. *Netherlands Journal of Zoology*, **42**, 323–35

Schlosser, I.J. (1982). Trophic structure, reproductive success and growth rate of fishes in natural and modified headwater stream. *Canadian Journal of Fisheries and Aquatic Sciences*, **39**, 968–78

Schlosser, I.J. and Karr, J.R. (1981). Water quality in agricultural watersheds: impact of riparian vegetation during baseflow. *Water Resources Bulletin*, **17**, 233–40

Shustov, Y.A. and Shchurov, I.L. (1989). Experimental study of effect of young salmon, *Salmo salar* L. stamina on their feeding rates in a river. *Journal of Fish Biology*, **34**, 959–61

Smith, J.B. (1990). From global to regional climate change: relative knowns and unknowns about global warming. *Fisheries*, **15**, 2–6

Solbrig, O.T. (1991). Biodiversity: Scientific Issues and Collaborative Research Proposals, 79 pp. *MAB Digest 9*, UNESCO, Paris

Soulé, M.E. and Mills, L.S. (1992). Conservation genetics and conservation biology: a troubled marriage. In Sandlund, O.T., Hindar, K. and Brown, A.H.D. (eds.) *Conservation of Biodiversity for Sustainable Development*, pp. 55–69. Scandinavian University Press, Oslo

Thorpe, J.E. (1988). Salmon migration. *Science Progress, Oxford*, **72**, 345–70

Waples, R.S. (1991). Genetic interactions between hatchery and wild salmonids: lessons from the Pacific Northwest. *Canadian Journal of Fisheries and Aquatic Sciences*, **48**, 124–33

Ward, J.V. and Stanford, J.A. (1982). The serial discontinuity concept of lotic ecosystems. In Fontaine, T.D. and Bartell, S.M. (eds.) *Dynamics of Lotic Ecosystems*. Ann Arbor Science Publishers, Ann Arbor, MI

Ward, J.V. and Stanford, J.A. (1989). Riverine ecosystems: the influence of man on catchment dynamics and fish ecology. In Dodge, D.P. (ed.) *Proceedings of the International Large River Symposium. Canadian Special Publication of Fisheries and Aquatic Sciences*, **106**, pp. 56–64.

Wasser, N.M. and Price, M.V. (1989). Optimal outcrossing in *Ipomopsis aggregata*: seed set and offspring fitness. *Evolution*, **43**, 1097–109

Weber, J.H. and Kramer, D.L. (1983). Effects of hypoxia and surface access on growth, mortality and behaviour of juvenile guppies, *Poecilia reticulata*. *Canadian Journal of Fisheries and Aquatic Sciences*, **40**, 1538–88

Wetzel, R.G. (1990). Land–water interfaces: metabolic and limnological regulators. Edgardo Baldi Memorial Lecture, Proceedings of SIL Congress, Munich (1989). *Verhandlungen Internationale Vereinigung für Theoretische und Angewandte Limnologie*, **24**, 6–24

Winemiller, K.O. (1990). Spatial and temporal variation in tropical fish trophic networks. *Ecological Monographs*, **60**, 331–67

Zalewski, M. (1986). Regulacia zespotów ryb w rzekach pnez kontinuum czynnikó abiotycznych i biotycznych. 86 pp. *Acta Universitatis Lodzensis*, Uniwersytet Lodzki, Lodz

Zalewski, M. (1995). Freshwater habitat management and restoration in the face of global change. Key note lecture theme. 1: Condition of the world's aquatic habitats. In Voigtlander, C.W. (ed.) *Proceedings of the World Fisheries Congress*, pp. 354–78. Oxford and IBH Publishing, New Delhi

Zalewski, M., Brewinska-Zaras, B. and Frankiewicz, P. (1990a). Fry communities as a biomanipulation tool in a temperate lowland reservoir. In Lind, O.T. and Sladecek, V. (eds.) Proceedings of the International Conference on reservoir limnology and water quality and its prediction; '*Key factors of reservoir limnology, eutrophication, water quality and its prediction*', pp. 763–74. *Archiv für Hydrobiologie Beih. Ergebn. Limnologie,* Springer Verlag, Stuttgart

Zalewski, M., Brewinska-Zaras, B., Frankiewicz, P. and Kalinowski, S. (1990b). The potential for biomanipulation using fry communities in a lowland reservoir: concordance between water quality and optimal recruitment. *Hydrobiologia*, **200/201**, 549–56

Zalewski, M., Frankiewicz, P. and Brewinska, B. (1985). The factors limiting growth and survival of brown trout *Salmo trutta m. fario* (L.) introduced to different types of streams. *Journal of Fish Biology*, **27** (Suppl. A): 59–73

Zalewski, M. and Naiman, R.J. (1985). The regulation of riverine fish communities by a continuum of abiotic–biotic factors. In Alabaster, J.S. (ed.) *Habitat Modification and Freshwater Fisheries*, pp. 3–9. Butterworths Scientific, London

Zalewski, M., Puchalski, W., Frankiewicz, P. and Nowak, M. (1991a). The relation between primary productivity and fish biomass distribution in an upland river system. *Verhandlungen Internationale Vereinigung für Theoretische und Angewandte Limnologie*, **24**, 2493–6

Zalewski, M, Puchalski, W., Frankiewicz, P. and Bis, B. (1994). Riparian ecotones and fish communities in rivers – intermediate complexity hypothesis. In Cowx, I.G. (ed.) *Rehabilitation of Freshwater Fisheries*, pp. 152–60. Blackwell Scientific Publications, Oxford

Zalewski, M., Thorpe, J.E. and Gaudin, P. (eds.) (1991b). *Fish and Land–Inland Water Ecotones*, 102 pp. UNESCO–MAB. University of Lodz, Stirling, Lyon

CHAPTER 9

BIRD DIVERSITY IN ECOTONAL HABITATS

Kazimierz A. Dobrowolski

Many bird species are native to a particular habitat type, which makes it possible to distinguish particular morpho-ecological types (Dobrowolski, 1969, 1973; Jakubiec, 1978). However, most bird species exhibit high adaptative ability in relation to both breeding and feeding and such ecological plasticity results in overlapping communities in ecotonal habitats. Thus, higher bird diversity is typical of ecotones.

Waterfowl breed and/or feed in various water habitats or wetlands. They consist of species that may differ in their biology and taxonomic position. Kalbe (1978) includes various species belonging to divers (Gaviiformes), grebes (Podicipediformes), Ciconiiformes, Pelecaniformes, Anseriformes, Gruiformes, Charadriiformes, Lariiformes, and also some species of Passeriformes. That list should also contain certain birds of prey, and – taking into account sea habitats – penguins, albatrosses and petrels (Figure 9.1). In Europe, about 43–45% of bird species are dependent on freshwater habitats. Many either nest or feed in areas that can be described as ecotonal habitats (Figure 9.2).

The focus of this report is ecotones of lowland freshwaters, mostly those that occur at lake or big river shores. The area occupied by such ecotones can be substantial. It seems that for birds the delta of a big river can be regarded as an extensive ecotonal zone. For instance, that of the Volga River occupies 13,000 km^2, and is 170 km long. (Lugovoy, 1963) (Figure 9.3). There were 257 bird species of 23 orders recorded, and among them 106 breeding, 129 migratory, nomadic or wintering and 22 sporadic visitors. The most abundant, of course, were Passeriformes (97 species), Charadriiformes (37 species), Anseriformes (28 species), Accipitriformes and Falconiformes (20 species), Lariiformes (16 species) and Ciconiiformes (11 species). They live in different habitats: 131 species belong to waterfowl (among them 73 breed), 54 are forest or bush species (of which 17 breed), 52 are grassland species (27 breed), and in addition, 20 species (15 breed) that cannot be associated with any of the habitats distinguished.

As with deltas of big rivers, extensive swamps on rivers should also be included in ecotonal freshwater–land habitats. The swamps in the valley of the Biebrza River in Poland are an example (Figure 9.4). They cover an area of 1,270 km^2, of which 1,000 km^2 is bogs (Dyrcz *et al.*, 1984). The area is 90 km long and 3 to 25 km wide, out of a total length of 164 km.

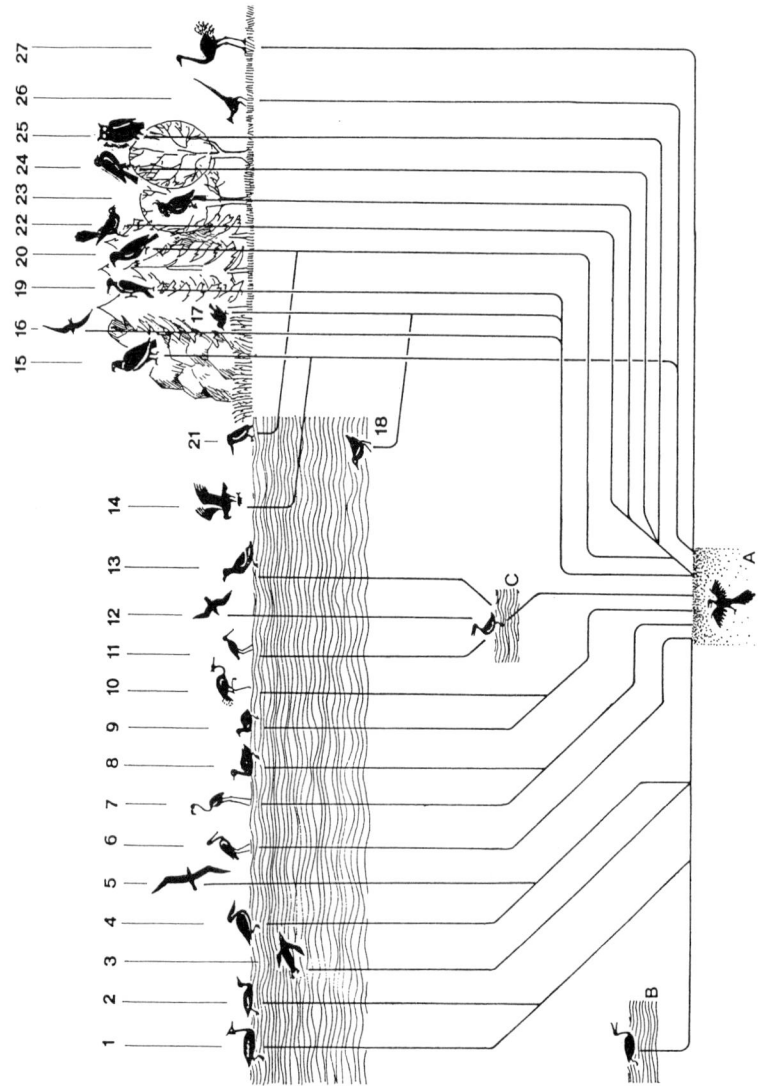

Figure 9.1 A simplified scheme of adaptative radiation of birds, with particular reference to water habitats (Dobrowolski, 1971)
(1) Podicipediformes; (2) Gaviiformes; (3) Sphenisciformes; (4) Pelecaniformes; (5) Procellariiformes; (6) Ciconiiformes; (7) Phoenicopteriformes; (8) Anseriformes; (9) Rallidae; (10) Gruidae (Gruiformes); (11) Charadrii; (12) Lari; (13) Alcae (Charadriiformes); (14) Pandion haliaetos (as an example of quite numerous birds of prey associated with waters); (15) Accipitriformes and Falconiformes; (16) Apodiformes, Caprimulgiformes; (17) Passeriformes; (18) *Cinclus cinclus* (as an example of large passerine group of birds associated with waters); (19) Piciformes; (20) Coraciiformes; (21) Alcedinidae; (22) Cuculiformes; (23) Psittaciformes; (24) Columbiformes; (25) Strigiformes; (26) Galliformes; (27) Struthioniformes; (Extinct birds: (A) Archaeopteryx; (B) Hesperornis + Baptornis + Hargeria; (C) Ichtyornis + Apatornis)

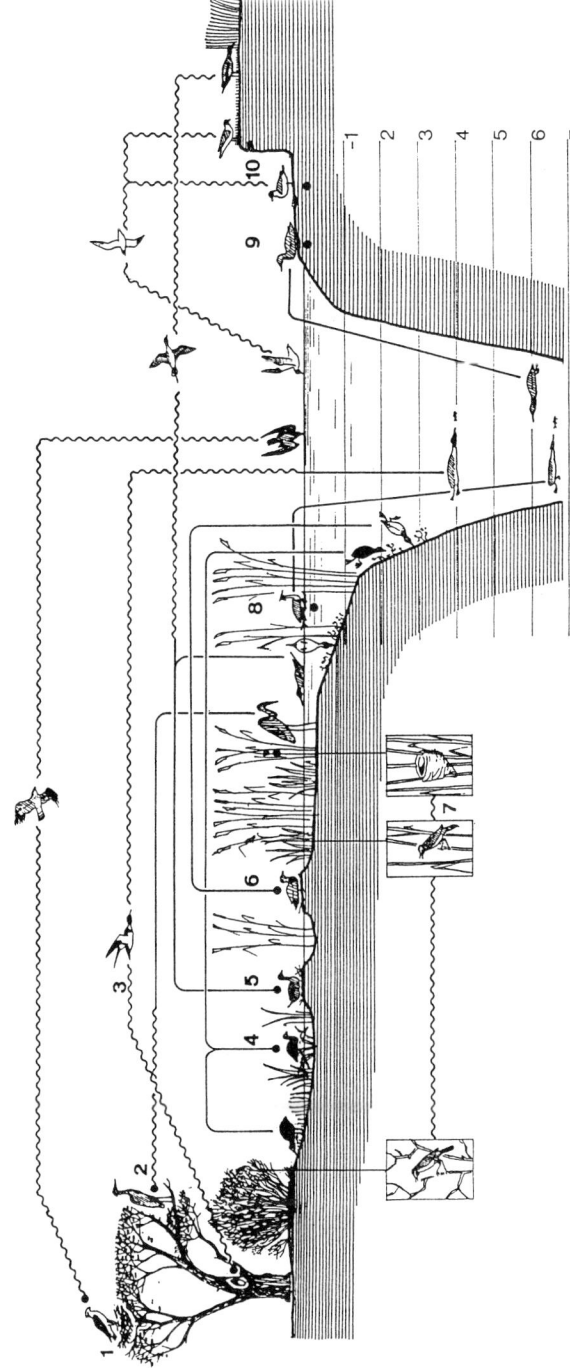

Figure 9.2 Breeding and feeding places of waterfowl as exemplified by lake habitats (after Dunajewski, 1938) Undulating line denotes flight, straight line stands for walking, swimming or wading. Breeding places marked with dots. (1) *Pandion haliaetus*; (2) *Ardea cinerea*; (3) *Bucephala clangula* or *Mergus merganser*; (4) *Fulica atra*; (5) Anatinae; (6) Nyrocinae; (7) *Acrocephalus arundinaceus* or *Acrocephalus scirpaceus*; (8) *Podiceps cristatus*; (9) *Gavia arctica*; (10) *Larus ridibundus*

Biodiversity in land–inland water ecotones

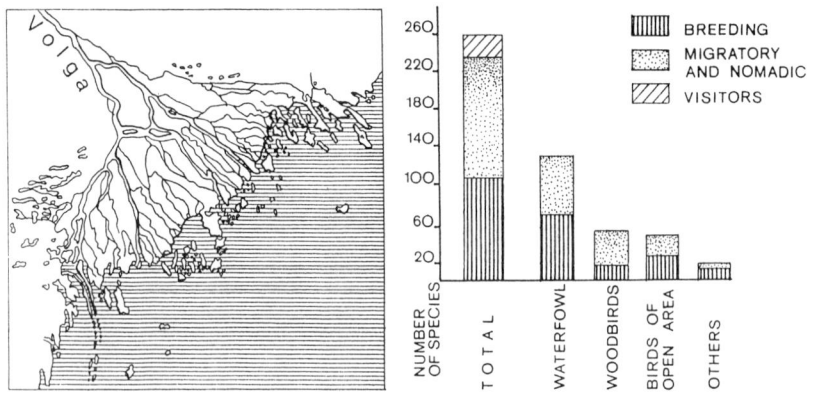

Figure 9.3 The delta of the Volga River and the numbers of bird species present

Smaller streams as well as man-made channels join the river and make an intricate system of swamps, bogs, meadows, inundated forests and estuaries. The entire area can be, supposedly, regarded as an ecotone. There were 235 bird species (that is 58.75% of Polish avifauna) of which 176 breed or probably breed (which makes 76.86% of breeding bird species in Poland). Of the bird species 104 can be classified as waterfowl (among them 66 breed). Forest and bush birds are represented by 86 species (79 breed), grassland birds by 26 species (20 breed), and there are also 19 species (17 breed) not typical of any of the above-mentioned habitats. The number of bird species in Poland is taken from Tomiadojc (1990).

The above examples show that birds make especially intense use of such ecotonal areas. The freshwater–land ecotonal areas provide particularly diversified habitats and this results in exceptionally diverse species composition of birds.

Three different types of ecotones, for lakes and rivers, can be distinguished in relation to land habitats: grasslands (meadows, pastures, fields), large forest complexes, or narrow strips of forest (mostly alder carrs) along river valleys or alder woods adjoining lake shores (Figure 9.5).

For both lakes and rivers it is easy to fix the ecotone border for birds when proceeding from water to land. In the case of lakes it is bounded by the littoral zone characterized by emerging plants. For rivers the ecotonal zone is the bank. The ecotone border, proceeding from land toward water, is more difficult to establish. Only in the case of narrow strips of forests one can postulate the border line between the forest and an open area as the ecotone border. In the cases of extensive forest complexes, or open (grassland) areas, the ecotone border on the 'land side' is unclear. It should be pointed out that for birds islets in lakes or rivers may also provide similar ecotonal zones.

Bird diversity

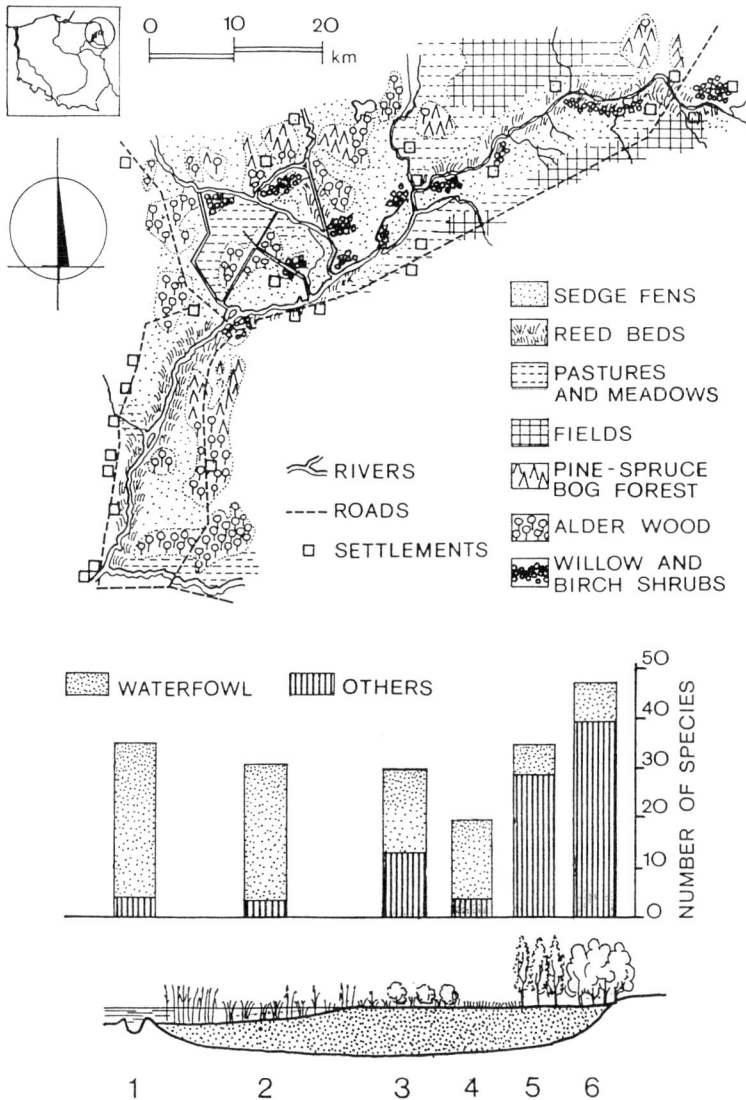

Figure 9.4 The valley of the Biebrza River and the numbers of bird species in different habitats: (1) river banks and oxbow lakes; (2) open flooded alluvial areas near the river; (3) bush overgrown fens; (4) open fens outside flooding zone; (5) birch forest; (6) swampy alder woods

RIVER ECOTONES

Most of the valuable information on ecotonal zones along rivers and riverine forests of Europe, as well as diversity of their bird communities, is available in Imboden (1987), being the result of the fifteenth Conference of the European

Biodiversity in land–inland water ecotones

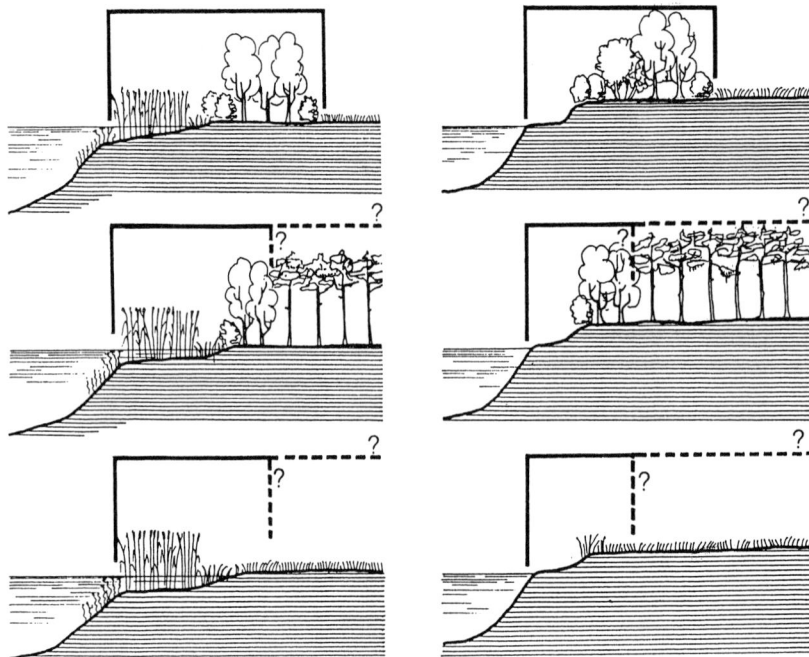

Figure 9.5 The types of lake and river ecotones

Section of ICBP held in Rapperswil (Switzerland) in 1985. That publication provides phytosociological characteristics of the main types of riverine forests in Central Europe (Kuhn, 1987) as well as in Czechoslovakia (Hudec, 1987), Italy (Pandolfi, 1987) and Poland (Wesolowski, 1987). They all draw attention to the impact of forest structure and species composition on the richness of the bird communities involved. According to Kuhn (1987), natural hardwood forests are especially rich in animal life, and natural riparian forests belong to the most interesting habitats of the natural environment in Europe.

The main plant associations are composed of willows, alder forests, and finally hardwood forests consisting primarily of common ash and white poplar, with some oaks. In Poland there are willow–poplar forests (Salici–Populetum) in the valleys of big rivers, ash–alder forests (Ficario–Ulmetum campestris) and alder carrs (Circaeo–Alnetum) growing in long, flat and swampy valleys of slow running rivers. The association of Ficario Ulmetum campestris occupies a transitory place between Salici–Populetum and Tilio–Carpinetum or Galio–Carpinetum on one hand, and between Circaeo–Alnetum and Tilio–Carpinetum on the other.

Reichholf (1987) is of the opinion that, among the basic habitats in Europe, riverine forests have the greatest diversity of bird species. About 62% of all land birds of Europe breed in or regularly visit riverine forests, and he also

Bird diversity

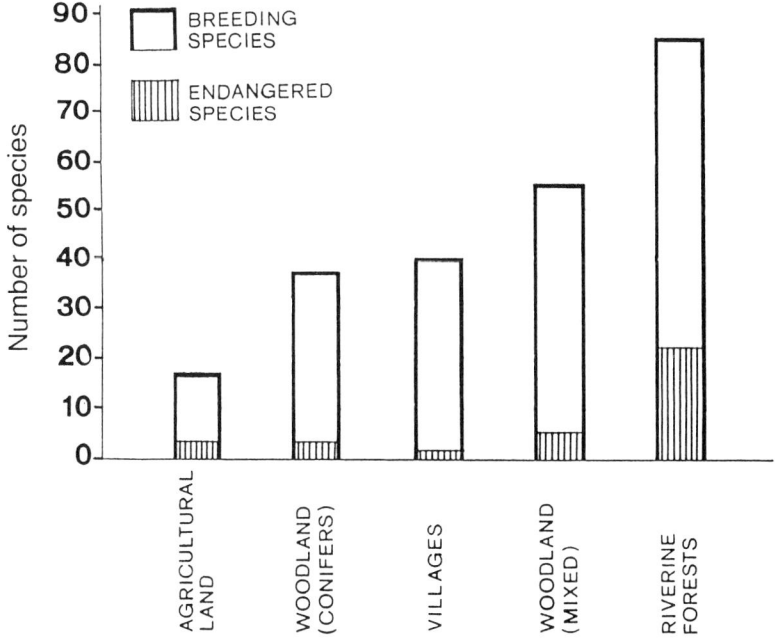

Figure 9.6 Distribution of bird species in different habitat types in the lower Inn valley (Bavaria). (After Reichholf, 1987)

points out the importance of the seasonal function of riverine forests as resting places for migratory and wintering birds. That author also shows an interesting diagram to illustrate numbers of breeding bird species in various habitats which clearly shows the advantage provided by riverine forests (Figure 9.6). Similar conclusions can be drawn from the remaining papers (Hudec, 1987; Randik, 1987; Pandolfi, 1987; Wesolowski, 1987). It should be pointed out that riverine forests can support both forest birds and waterfowl or even species associated with other habitats.

Differences in bird species richness depend also on the type of ecotone. The highest diversity occurs in the ecotone water–shrubs or water–forests. The ecotone water–open area is less rich, as shown by a set of studies conducted in Poland on bird fauna of the Narew River (Lewartowski and Piotrowska, 1987), the Warta River (Czarnecki, 1975) and the Biebrza River (Dyrcz *et al.*, 1984).

Three areas of different habitat diversity were studied at the Narew River. The first was 201 ha in area, and consisted of a river expansion system, extensive reed beds given over to *Phragmites communis*, canary grass and manna grass associations, a sedge association interspersed with willow shrub clumps, an open tract of sedge association (Caricetum elatae), and – on the valley slopes – turf sedge associations, meadows, pastures and arable

Biodiversity in land–inland water ecotones

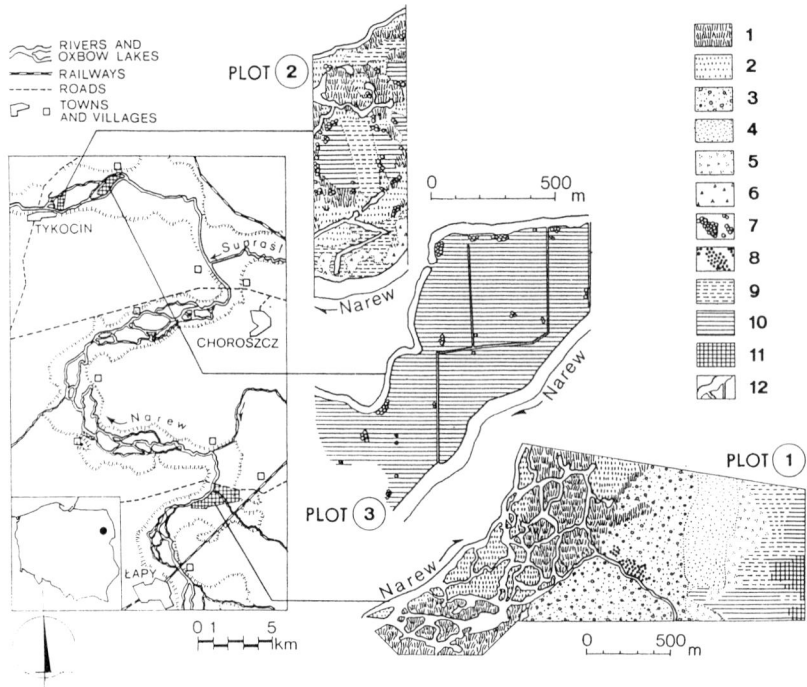

Figure 9.7 The valley of the Narew River and maps of the study areas.
(1) reed beds (*Phragmites communis*); (2) canary grass and manna grass association; (3) sedge association with willow scrub and plums; (4) open tract of sedge association (Caricetum elatae); (5) turf sedge communities; (6) sandy plains; (7) trees or willow scrub clumps; (8) osier beds; (9) pastures; (10) meadows; (11) arable farmland; (12) rivers, oxbow lakes, ditches

farmland. A small area of osier beds was also noted (Figure 9.7). There were 50 species of breeding birds recorded (among them 31 waterfowl), and the density of the breeding population was 32 pairs per 10 ha (Figure 9.8). There are evident differences in abundance and density between the habitats in the study area (Figure 9.9).

The second area (31 ha) is a partly drained, swampy valley of the Narew River with oxbow lakes, sandbanks, areas of reed beds, canary grass, meadows and pastures. There are also sparsely distributed trees and willow scrub clumps. Thirty-six breeding bird species (29 waterfowl) with a density of 44 pairs per 10 ha were found here.

The third area (43 ha) is the reclaimed valley of the Narew River with uniform meadows and few willow scrubs, trees and old river beds. Thirteen breeding bird species occurred there (7 waterfowl) with a density of 11 pairs per 10 ha.

The relatively low number and density of species of waterfowl in meadows, pastures, arable farmland and drained areas should be noted. A similar phenomenon was found in the valley of the Biebrza River (Dyrcz *et al.*, 1985). For each of four partly drained regions (swamp Biele, the Jegrznia

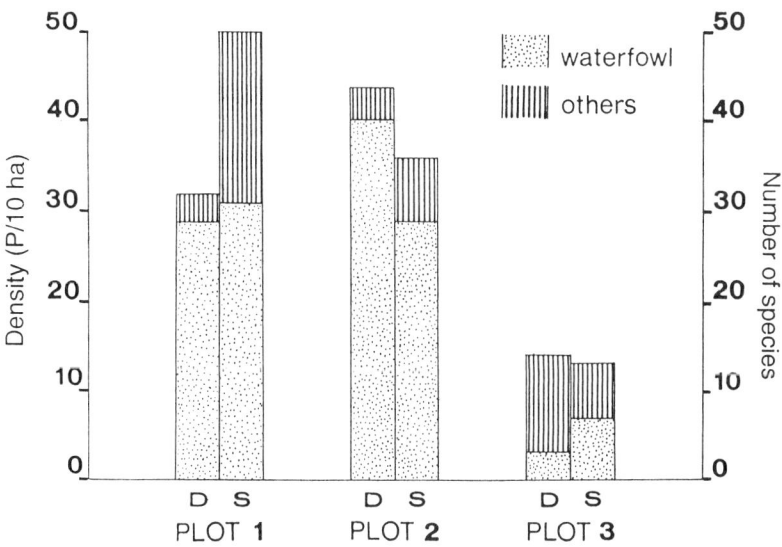

Figure 9.8 Density (D: number of pairs per 10 ha) and number of bird species (S) in the study areas at the Narew River

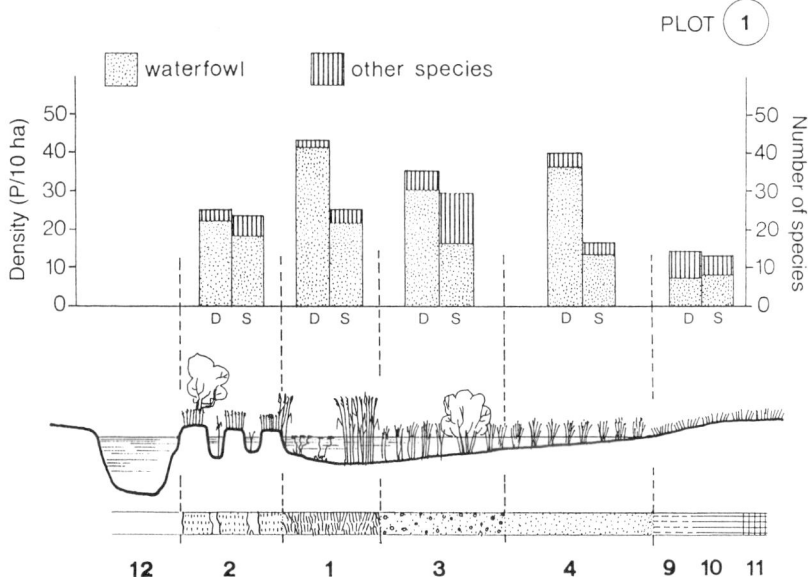

Figure 9.9 Density (number of pairs per 10 ha) and number of bird species in different habitats of the first study area. A schematic cross-section of the Narew valley at the area: (9, 10, 11) valley slope (meadows, pastures, arable farmland); (12) river. (for explanation of other symbols see Figure 9.7)

Biodiversity in land–inland water ecotones

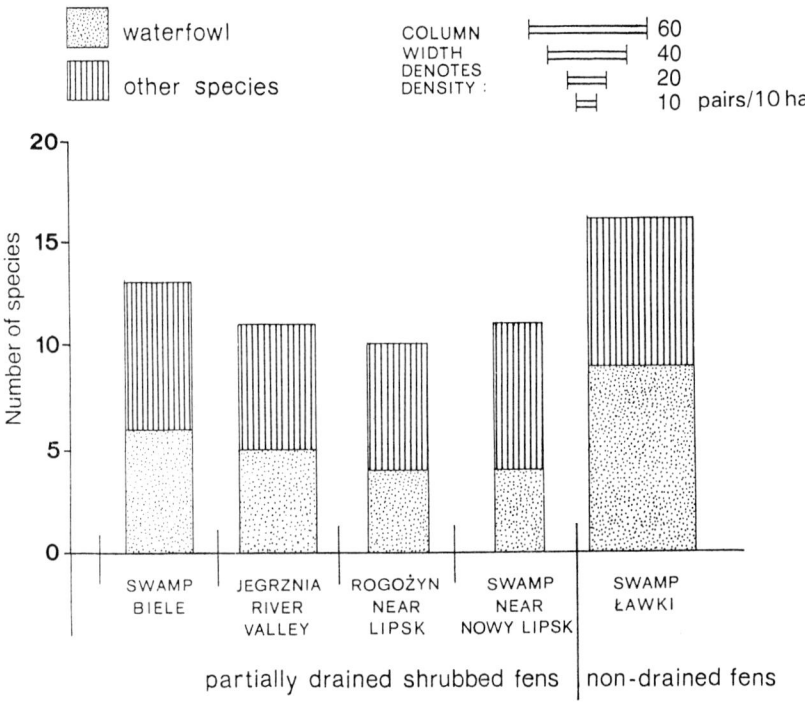

Figure 9.10 Density (in pairs per 10 ha) and number of bird species in partly drained and undrained bogs in the Biebrza valley

River valley, Rogozyn near Lipsk, and Nowy Lipsk) there were: 13 bird species (6 waterfowl) with a density of 35 pairs per 10 ha, 11 species (5 waterfowl) with a density of 39 pairs per 10 ha, 10 species (4 waterfowl) with a density of 29 pairs per 10 ha, and 11 species (4 waterfowl) with a density of 25.5 pairs per 10 ha, respectively. For comparison, the undisturbed swamp at Lawki supports 15 bird species (9 waterfowl) and their density is 55 pairs per 10 ha (Figure 9.10). In both of the above cases (i.e. the Narew and the Biebrza valleys) drainage of the ecotonal zones (that is habitat homogenization) resulted in a decrease of bird species number, particularly of waterfowl, and in a decrease of their breeding population density.

Bird studies of the Biebrza valley (Dyrcz *et al.*, 1984) revealed a dependence of bird species richness on the type of water–land ecotone. In the study area 103 bird species breed (54 waterfowl). Dyrcz *et al.*, (*loc. cit.*) distinguished types of habitats with different bird species numbers. Swampy alder woods appeared to be the richest: 48 species (8 waterfowl). The second richest were river banks and oxbow lakes – 36 species (31 waterfowl), as well as birch forest – 35 species (only 6 waterfowl). Open flooded alluvial areas on the river supported 31 species (27 waterfowl), and bush overgrown fens supported 30 species (17 waterfowl). Among the poorest are open fens outside the flooding zone where 20 species live, of which 16 are waterfowl.

The studies by Czarnecki (1975) conducted at the Warta River showed that in an ecotone with willows there were 34 breeding bird species, of which 15 represented waterfowl. The author noted the diverse use of that ecotone by various bird species. He concluded that the studied ecotone type is the encounter ground for 56 bird species; 46 of them regularly feed there, 5 species form overnight flocks, and 6 are ephemeral visitors.

Ecotonal zones, both river–land and lake–land, provide especially attractive resting grounds and overnight shelters, particularly in autumn and winter, for many bird species. They are used, for instance, by rooks (*Corvus frugilegus*), various herons, swallows (*Hirundo rustica*) and starlings (*Sturnus vulgaris*). At Swidwie Lake near Szczecin (a site considered by the Ramsar Convention as a reserve) starling flocks gathering there overnight reach a gigantic size, exceeding 3 million individuals (Dr. Giergielewicz, personal communication).

Riverine forests are used by many bird species as ecological corridors for migration. Thus, in the migration period, bird species richness substantially increases in that ecotone. At Wilga settlement, located at the Vistula banks, south of Warsaw, autumnal ringing (from mid-August till mid-October) of birds caught in nets has been conducted since 1983 (Keller *et al.*, 1986; Keller *et al.*, 1988; Keller and Matyjasik, 1989) in close proximity to the Vistula River, i.e. within the inundation terrace covered moderately with willow scrub, and also within an alder carr depression with predominant willows and poplars. Such tree stands are typical of the mid-course of the Vistula River. During a 5 year catching programme there were in total almost 11,000 birds caught. The numbers of species caught in different years are presented in Table 9.1.

Some of the species caught normally reside in that type of habitat (Table 9.2). There were, however, species ecologically strange for that habitat type, among them typical inhabitants of coniferous forests. Only about 10 species could be classified as associated with water habitats.

Occurrence of birds on the 850 ha area of inundation terrace located within the administrative border of Warsaw was studied in 1987 and 1988 (Gorzelski, 1989). A central zone (47 ha in downtown) and peripheral zones (803 ha) stretching to the south and to the north of the central zone were distinguished and had distictive habitats including meadows with willow scrub and sparsely distributed trees (mainly poplars), clumps of willow scrub of height up to 2.5 m, areas with continuous cover of willow scrub reaching more than 2.5 m in height, small gardens and urban areas. There were 73 breeding species of birds and 2 species of visitors recorded. Among them 14 species could be associated with water habitats. In the peripheral zone 71 bird species, and in the central zone (characterized by absence of meadows, and presence of small gardens and clumps of willow scrub up to 2.5 m tall) only 21 species were nesting. The same regularity appears when similar habitats of the peripheral and the central zones are

Table 9.1 Pluriannual evolution of birds species richness at the Vistula banks south of Warsaw. (Total caught: 11,000 birds)

Year	Number of species
1983	46
1984	45
1985	54
1987	65
1988	68

Table 9.2 Example of usage intensity of an alder carr and willow scrub by 17 most numerous bird species. (F and R stand for forest and reed)

Species	Numbers caught	Ecountered in other habitats	Encountered in the studied habitat
Phylloscopus collybita	3059		+
Prunella modularis	1610		+
Erithacus rubecula	1220	F	
Parus major	531		+
Parus coeruleus	491		+
Sylvia atricapilla	476	F	
Acrocephalus palustris	461	R	
Phylloscopus trochilus	299		+
Regulus regulus	297	F	
Sylvia borin	283	F	
Troglodytes troglodytes	245	F	
Muscicapa striata	211	F	
Turdus philomelos	165		+
Actrocephalus scirpaceus	133	R	
Sylvia curruca	105		+
Sylvia communis	102		+
Hippolais icterina	100		+

compared. In areas with continuous cover of willow scrub of over 2.5 m in height 34 species reach a breeding density of 26 pairs per 10 ha in the peripheral zone, and the figures for the central zone were 10 species with a density of 29 pairs per 10 ha. In the alder carrs of the peripheral zone 61 species occurred (35 pairs per 10 ha), and in the central zone – 21 species reaching a density equal to 74 pairs per 10 ha (Figure 9.11).

In urban areas of the peripheral zone 37 species reaching a density of 41 pairs per 10 ha occurred, and in the central zone 17 species with a population density equal to 67 pairs per 10 ha were found. So high densities are due to the presence of synanthropic species, which are lacking in the peripheral zone. The highest number of species (61) were found in alder carrs of the peripheral zone. The second highest number, 43 species, occurred in meadows with shrubs. Thirty-seven species were recorded in urban areas with a high

Bird diversity

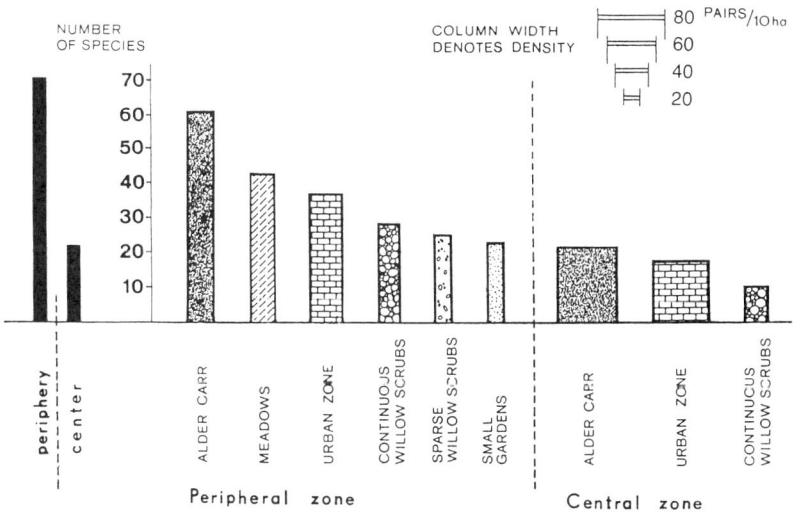

Figure 9.11 Number of bird species and their population density in various habitats of the inundation terrace of the Vistula River in Warsaw

proportion of natural vegetation. Thirty-four species occurred in areas continuously covered by willow scrubs, 25 in clumps of willow scrub, 23 in small gardens, 21 species in alder carrs of the central zone, 17 within urban areas, and 10 species were found in the areas with continuous cover of willow scrub located in the central zone. Species that prefer the central, urbanized, zone were distinguished (reaching higher densities than in the peripheral zone), as well as 23 species that did not occur in the central zone. Apart from clear differences in biological diversity related to habitat type, the differences due to the degree of urbanization of the river banks were also pointed out. One can suppose that, for some species that penetrate into the cities and are the subject of advancing synanthropization (or synurbization), the peripheral zone of ecotonal habitats makes an appropriate base for that process.

LAKE ECOTONES

In both lake and river ecotones the diversity and species richness of bird communities depend on the types of neighbouring habitats. An example is provided by the studies at Luknajno Lake in 1982–1985 (Bukaciński and Jabłoński, 1992a, 1992b) which is a shallow, eutrophic water basin of 630 ha in area with rich avifauna. One hundred and forty-one bird species were found there, among them 119 were breeding. It is a nature reserve in terms of the Ramsar Convention. A wide strip of reed adjoins the diverse land habitats and invites comparisons of bird richness in different ecotones. The number of breeding bird species was estimated for the following habitats: reed beds – 25

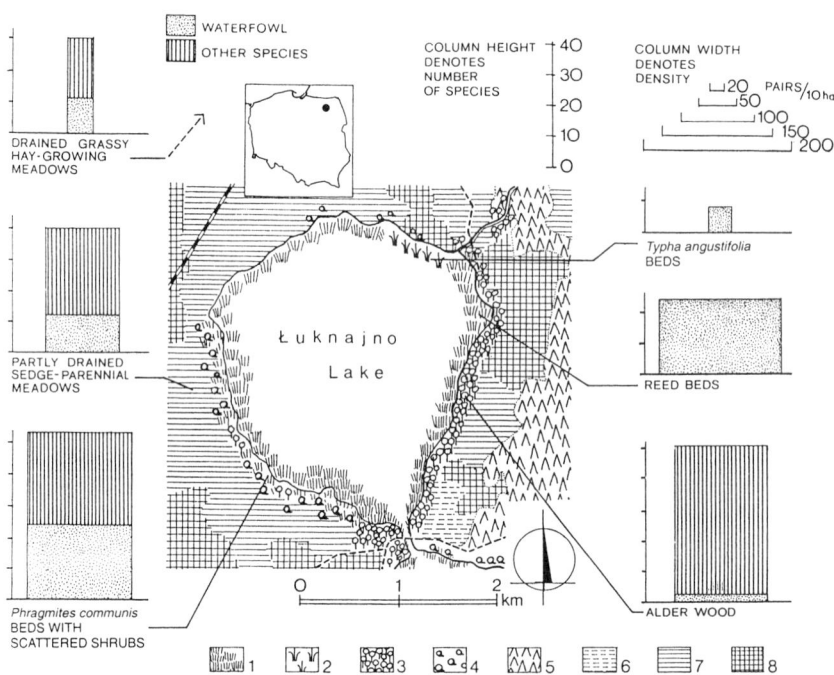

Figure 9.12 Number of bird species and their population density (in pairs per 10 ha) in various habitats of Luknajno Lake. (1) reed beds (*Phragmites communis*); (2) *Typha angustifolia* beds; (3) alder wood; (4) willow shrubs; (5) pine forest; (6) pastures; (7) meadows; (8) arable farmland

species (all waterfowl) with densities varying from 57 to 170 pairs per 10 ha; a mosaic of *Phragmites communis* beds with scattered shrubs and trees – 54 species (24 waterfowl) with densities ranging from 55 to 154 pairs per 10 ha. In *Typha angustifolia* beds there were 8 species (all waterfowl) with a density equal to 36 pairs per 10 ha, and in partly drained sedge–perennial meadows – 41 species (13 waterfowl) reaching a density of 104 pairs per 10 ha. Drained grassy hay-growing meadows supported 33 bird species (12 waterfowl) and density equal to 34 pairs per 10 ha. In alder woods there were 52 species (3 waterfowl) with a density of 116–127 pairs per 10 ha (Figure 9.12). The highest biological diversity, as in riverine forests, occurred in alder woods, and was lower in more homogeneous habitats like *Typha angustifolia* beds, reed beds or hay-growing meadows.

An attempt to estimate differences between bird communities in various ecotones at 4 lakes of the Masurian Lakeland was made between 1983 and 1984 (Filcek, 1985). The studies were located at Gardyńskie Lake (ecotone lake shore–alder wood), Majcz Wielki Lake (ecotone lake shore–coniferous forest), Talty Lake (ecotone lake shore–meadow), and Inulec Lake (ecotone lake shore–arable farmland and a village). The highest number of breeding bird species (16) occurs in the ecotone lake shore–coniferous forest, followed by 11

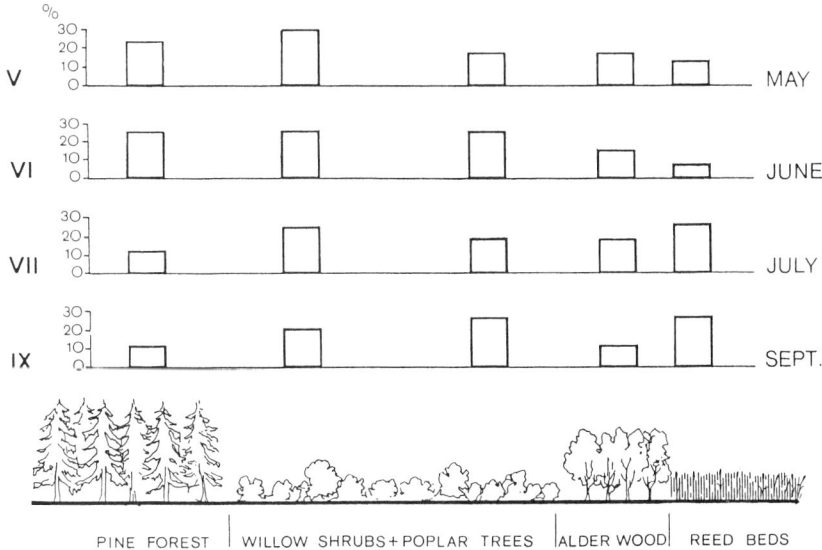

Figure 9.13 Intensity of bird penetration (as a percentage of the entire population) into various habitats not associated with reed at Sniardwy Lake in various months

for the ecotone lake shore–alder wood, 10 for that of lake shore–arable farmland, and 7 for lake shore–meadow. The bird communities are mutually dissimilar because of the different land habitats involved. An attempt to delimit the range of bird penetration into different zones of each ecotone type was also made: 4 forest species of birds penetrated into reed beds in search of food, but bird species of reed beds were not seen in forest. A narrow belt of alder wood and reed beds was only rarely visited by forest birds in the area of the coniferous forest ecotone. However, catches of birds in nets conducted at the shores of Sniardwy Lake (Dmowski and Kozakiewicz, 1990) have shown that the littoral zone neighbouring a coniferous forest was regularly visited by forest birds, especially by chiffchaff (*Phylloscopus collybita*), blue tit (*Parus coeruleus*), robin (*Erithacus rubecula*), willow warbler (*Phylloscopus trochilus*), and also by goldcrest (*Regulus regulus*), lesser whitethroat (*Sylvia curruca*) and wren (*Troglodytes troglodytes*). These visits were infrequent in the breeding period (as in the previously mentioned studies) and rather regular in the preceding and following periods (Figure 9.13). On the other hand, the birds inhabiting the littoral zone (i.e. reed beds) hardly penetrate into riverine forests and bushes.

The ecotones of both a river and a lake are characterized by a high diversity of avifauna which depends on the ecotone type and the phenological period used by birds in different seasons. As various ecotone types generally form a mosaic along the shores, one should expect this area to have a particularly high diversity of bird communities.

REFERENCES

Bukacinski, D. and Jablonski, P. (1992a). Avifauna legowa jeziora Luknajno i terenow przyleglych (Breeding avifauna of Luknajno Lake and its vicinity). *Notatki Ornitologiczne/Ornithological Notices*, **33**, 5–45

Bukacinski, D. and Jablonski, P. (1992b) Sezonowa zmiennosc zespolu ptakow wodno-blotnych na jeziorze Luknajno w latach 1982–1984 (Seasonal changes in water and marsh bird community on the Luknajno Lake). *Notatki Ornitologiczne / Ornithological Notices*, **33**, 185–226

Czarnecki, Z. (1975). Studia nad znakowanymi populacjami ptakow gniezdzacych sie w wiklinach nadrzecznych (Studies concerning banded bird populations nesting in riverside willows). *Acta Ornithologica*, **15**, 1–79

Dmowski, K. and Kozakiewicz, M. (1990). Influence of a shrub corridor on movements of passerine birds to a lake littoral zone. *Landscape Ecology*, **4**, 99–108

Dobrowolski, K.A. (1969). Structure of the occurrence of waterfowl types and morpho-ecological forms. *Ekologia Polska, Ser. A*, **17**, 29–72

Dobrowolski, K.A. (1971). *Jak plywaja zwierzeta* (The Way Animals Swim), 218 pp. PZWS, Warszawa

Dobrowolski, K.A. (1973). Ptaki wodne i ich rola w ekosystemie jeziornym (Waterfowl and their role in lake ecosystem). *Wiadomosci Ekologiczne / Ecological News*, **19**, 353–71

Dunajewski, A. (1938) *Fauna Slodkowodna Polski. Ptaki (Aves)* (Freshwater fauna of Poland. Birds (Aves)). 426 pp. Kasa im. Mianowskiego, Warszawa

Dyrcz, A., Okulewicz, J., Witkowski, J., Jesionowski, J., Nawrocki, P. and Winiecki, A. (1984). Ptaki torfowisk niskich kotliny biebrzanskiej. opracowanie faunistyczne (Birds of fens in Biebrza Marshes. Faunistic approach). *Acta Ornithologica*, **20**, 1–108

Dyrcz, A., Okulewicz, J. and Witkowski, J. (1985). Changes in bird communities as the effect of peatland management. *Polish Ecological Studies*, **11**, 79–85

Filcek, K. (1985). Porownanie zespolow ptakow na wybranych typach pobrzezy jezior (A comparison of bird communities in chosen lake shores). MSc thesis, Warsaw University, typescript, 49 pp

Gorzelski, W. (1989). Awifauna legowa tarasu zalewowego Wisly w Warszawie w latach 1987–1988 oraz czynniki ja ksztaltujace (Breeding avifauna of the inundation terrace in Warsaw in 1987–1988 and the effective factors). MSc thesis, Warsaw University, typescript 57 pp

Hudec, K. (1987). The Morava–Dyie riverine forest system, Czechoslovakia. In Imboden, E. (ed.). *Riverine Forests in Europe, Status and Conservation*, pp. 38–42. Report of the 15th Conference of the European Continental Section, Rapperswil, Switzerland, 20–25th February (1985). ICBP

Imboden, E. (ed.) (1987). *Riverine Forests in Europe, Status and Conservation*. 64 pp. Report of the 15th Conference of the European Continental Section, Rapperswil, Switzerland, 20–25th February (1985). ICBP

Jakubiec, Z. (1978). Zroznicowanie morfologiczno-ekologiczne ptakow wodno-blotnych (Morpho-ecological differentiation of aquatic-swamp birds). *Wiadomosci Ekologiczne / Ecological News*, **24**, 99–107

Kalbe, L. (1978). *Okologie der Wasservogel* (The Waterfowl Ecology). 116 pp. Die Neue Brehm-Bucherei. Ziemsen Verlag, Wittenberg Lutherstadt

Keller, M., Kruszewicz, A., Kraska, R., Konofalski, M., Zawadzki, J. and Kamola, M. (1986). Sprawozdanie z akcji obraczkowania ptakow wroblowatych w centralnej Polsce podczas ich wedrowki jesiennej w latach 1983–1985 (Report on passerine ringing in central Poland during autumn migration in years 1983–1985). *Notatki Ornitologiczne / Ornithological Notices*, **29**, 182–7

Keller, M., Kraska, R. and Matyjasiak, P. (1988). Sprawozdanie z akcji obraczkowania ptakow wroblowatych w centralnej Polsce podczas ich wedrowki jesiennej w 1987 roku (Report on passerine ringing in central Poland during autumn migration in 1987). *Notatki Ornitologiczne / Ornithological Notices*, **29**, 88–91

Keller, M. and Matyjasiak, P. (1989). Sprawozdanie z akcji obraczkowania ptakow wroblowatych w centralnej Polsce podczas ich wedrowki jesiennej w roku 1988 (A report on passerine bird ringing in central Poland during the 1988 autumn migration). *Notatki Ornitologiczne / Ornithological Notices*, **30**, 116–20

Kuhn, N. (1987). Distribution, general ecology and characteristics of European riparian forests. In Imboden, E. (ed.). *Riverine Forests in Europe, Status and Conservation,* pp. 7–15. Report of the 15th Conference of the European Continental Section, Rapperswil, Switzerland, 20–25th February (1985). ICBP

Lewartowski, Z. and Piotrowska, M. (1987). Zgrupowania ptakow legowych w dolinie Narwi (Breeding birds in the valley of the Narew River, Poland). *Acta Ornithologica*, **23**, 215–72

Lugovoy, A.E. (1963). Pticy delty reki Wolgi (Birds of the Volga delta). Fauna i ekologia ptic delty Wolgi i pobierezij Kaspia (The fauna and ecology of birds of the Volga delta and the Caspian sea shores). *Astrakhan, Transactions of the Astrakhan Wildlife State Reserve*, **8**, 9–185

Pandolfi, M. (1987). An overview of Italian riverine forests with special reference to Populetalia albae associations. In Imboden, E. (ed.). *Riverine Forests in Europe, Status and Conservation,* pp. 29–37. Report of the 15th Conference of the European Continental Section, Rapperswil, Switzerland, 20–25th February (1985). ICBP

Randik, A. (1987). The bird fauna on the riverine forests along the Danube in Czechoslovakia. In Imboden, E. (ed.). *Riverine Forests in Europe, Status and Conservation,* pp. 43–47. Report of the 15th Conference of the European Continental Section, Rapperswil, Switzerland, 20–25th February (1985). ICBP

Reichholf, J.H. (1987). Composition of bird fauna in riverine forests. In Imboden, E. (ed.). *Riverine Forests in Europe, Status and Conservation,* pp. 16–21. Report of the 15th Conference of the European Continental Section, Rapperswil, Switzerland, 20–25th February (1985). ICBP

Tomiaojc, L. (1990). *Ptaki Polski, Rozmieszczenie i liczebnosc* (The birds of Poland, their distribution and abundance), 462 pp. PWN, Warszawa

Wesolowski, T. (1987). Riverine forests in Poland and the German Democratic Republic – their status and avifauna. In Imboden, E. (ed.). *Riverine Forests in Europe, Status and Conservation,* pp. 48–54. Report of the 15th Conference of the European Continental Section, Rapperswil, Switzerland, 20–25th February (1985). ICBP

CHAPTER 10

MAMMAL DIVERSITY IN INLAND WATER ECOTONE HABITATS

Rüdiger Schröpfer

INTRODUCTION

The term ecotone is defined by Di Castri *et al.* (1988) as a 'zone of transition between adjacent ecological systems, having a set of characteristics, uniquely defined by space and time scales and by the strength of the interactions between adjacent ecological systems'. In many cases the term ecotone is equally used for field or forest edges, hedgerows, boundaries or transition zones (see Hansen *et al.*, 1992). If understood in this sense, the cultivated landscape would be the actual ecotone landscape. However, it is questionable whether the term 'ecotone' may be used as such a collective term. We think that this extension of the term is not correct when looked at from the point of mammalian fauna. We shall demonstrate this in the group of semiaquatic mammals. Also we would like to make it clear that a discussion about species diversity in an ecotone should not be confused with a reflection on the edge effect.

THE BANK AS A TYPICAL ECOTONE

Some of the few natural ecotones are banks, especially of running waters, where a riparian forest often covers the banks like a gallery. It is here where the two media, water and land, meet. The bank as aquatic–terrestrial habitat stretches linearly for miles and miles, that is to say it is extensive in length, however minor in breadth, an aquatic–terrestrial corridor, and a typical ecotone form.

THE DIVERSITY OF PHENOTYPE OF SEMIAQUATIC MAMMALS

The only vertebrates typical for the ecotone 'bank' are the semiaquatic mammals: they are the real occupants of this ecotone (Schröpfer, 1985).

To describe the biodiversity of a fauna group it is, among other things, of interest to show its diversity of form, of size, and of behaviour strategy, as these attributes point at the life-form diversity of this group.

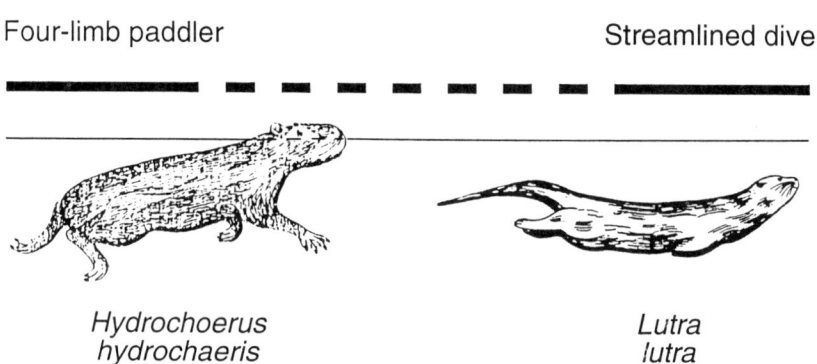

Figure 10.1 Between the two extremes, the capybara and the otter, lies the spectrum of the form diversity of semiaquatic mammals

DIVERSITY OF FORM

Among semiaquatic mammals all transition stages occur, starting from long-legged land walkers using four limbs when swimming with little swimming adaptability up to the graceful fast swimmers preferring the water element for their locomotion (Figure 10.1). Depending on their phylogenetic origin the tailless species are fitted with a powerful armstroke for water treading. The other species with laminar form, possessing a tail, move forward with a complete winding movement of the body. These different morphological adaptations provide no exact clue as to whether the species concerned lives mainly in the water or on land, rather do they point at the completely different foraging strategies (see below).

Referring to ecotones in general

Amongst the ecotone inhabitants one finds form adaptations pertaining to either one or the other of the adjacent ecosystems. The two ecosystems linked together in the ecotone in a community manner favour fauna elements which complement each other in their morphological functioning.

DIVERSITY OF SIZE

At first sight the semiaquatic mammals show a considerable diversity in body size: the smallest is the water shrew, the largest is the hippopotamus (Figure 10.2). When comparing within closely related groups, one finds that the semiaquatic living species are always the biggest (Table 10.1). This is probably due to energetic reasons, because these animals are able to live in the

Mammal diversity

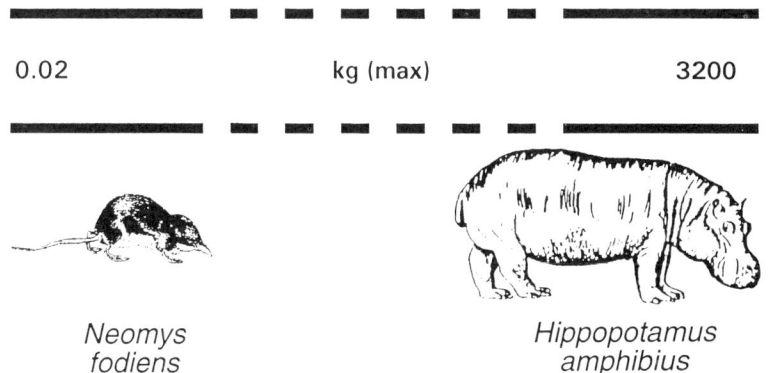

Figure 10.2 Between the two extremes, the water shrew and the hippopotamus, lies the spectrum of size diversity of semiaquatic mammals

water medium which has a much stronger cooling effect. Naturally, buoyancy will be important, too.

Species living in ecotones vary in body size, just like other species in other communities; however, they show physiological adaptations which are more pronounced when tending towards the more extreme of two adjacent ecosystems, in this case towards water. That means that for a syntopy within an ecotone the different species should stem from as many different taxa as possible in order to make use of the different niches. One and the same taxon can obviously produce very few species which can meet the necessary requirements of the respective ecotone. This reduces the species richness.

DIVERSITY OF STRATEGIES

Specific indications are provided when comparing the behaviour strategies; such a comparison proves the functional significance of a species within their habitat. Possible niche occupations also become evident. An example of this is the semiaquatic mammalian species living in Europe. An indication for the

Table 10.1 The biggest species in the taxa mentioned live semiaquatically

kg (max)	Biggest species	Taxon
0.02	*Neomys fodiens*	Soricidae
0.09	*Limnogale mergulus*	Oryzorictinae
0.5	*Desmana moschata*	Talpidae
2.4	*Ondatra zibethicus*	Arvicolidae
32.0	*Pteronura brasiliensis*	Mustelidae
35.0	*Castor fiber*	Sciuromorpha
50.0	*Hydrochoerus hydrochaeris*	Caviomorpha
3200.0	*Hippopotamus amphibius*	Artiodactyla

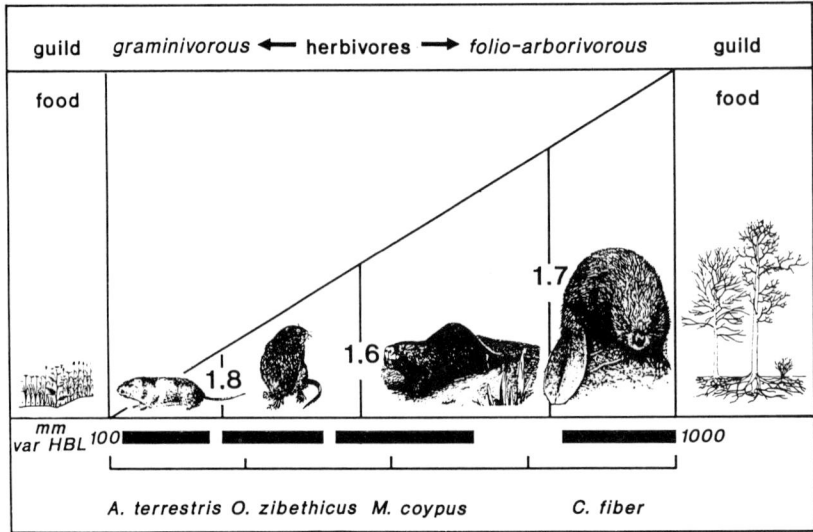

Figure 10.3 The niche occupation of the guild of herbivorous semiaquatic mammals on the banks of European running waters. Var: variation of the species; HBL: head–body length

occupation of a niche is the body size, in this case the head–body length, has been chosen to present a better overview. In Figure 10.3 the number between the species specifies the size difference of the two species.

According to the theriocoenosis model (Schröpfer, 1990), the herbivores form a guild in which probably all niches were occupied after the two species *Ondatra zibethicus* and *Myocastor coypus* had been imported to Europe and set free. Only the water voles *Arvicola terrestris/sapidus* and the beaver *Castor fiber* are autochthonous species. Through their foraging strategies the mammals of this guild are always producing new succession stages on the banks (McGinley and Whitham, 1985; Pagel, 1989).

The carnivorous semiaquatic mammals form two guilds according to their food preference and foraging strategies. The macrocarnivorous species feed on the aquatic–terrestrial macrofauna of the banks and the bottom of the waterbody (Figure 10.4a), often digging out prey in specific areas (Ruthardt, 1990). The megacarnivorous species prefer the aquatic–terrestrial megafauna (Figure 10.4b) which they hunt. When taking the size relationship as a simple indication for the niche occupation (compare Hutchinson, 1959), it becomes obvious that the niches for carnivorous mammals on the banks of European waters have always been and still are completely occupied, with no gaps whatsoever. Two recent examples exist in Europe showing the niche exclusion of similar sized semiaquatic mammals, belonging to the same guild: the competitive exclusion of the European mink (*Mustela lutreola*) by the American mink (*M. vison*) (Schröpfer and Paliocha, 1989), and the forcing

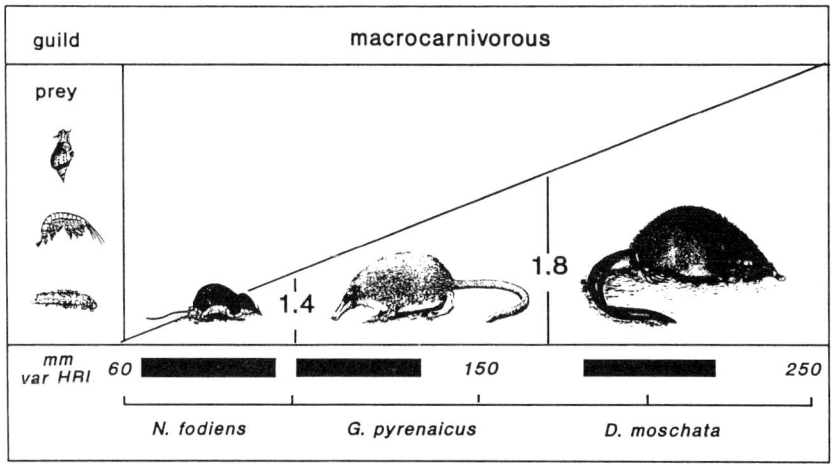

Figure 10.4a The niche occupation of the guild of the macrocarnivorous semiaquatic mammals on the banks of European running waters (assuming the situation as it existed in Central Europe during the Pliocine and Pleistocene (see Juckwer, 1990))

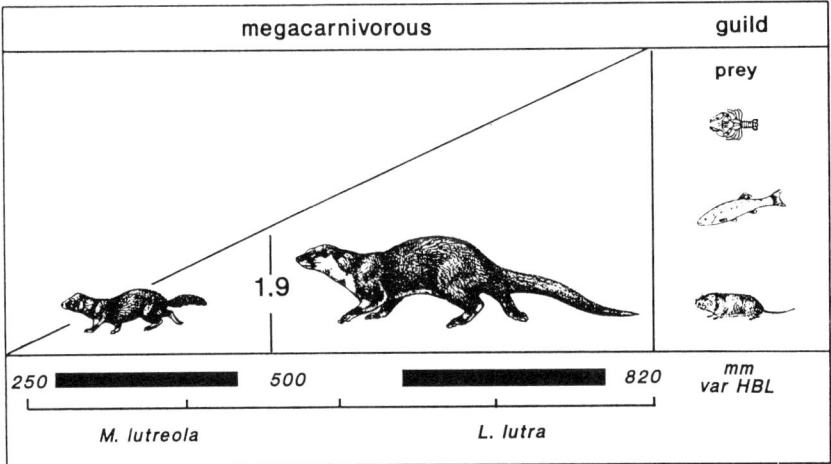

Figure 10.4b The niche occupation of the guild of megacarnivorous semiaquatic mammals on the banks of European running waters

out of Miller's shrew (*Neomys anomalus*) by the water shrew (*Neomys fodiens*) (Spitzenberger, 1990).

The availability of food is the determining factor as to which core regions of the banks are occupied by semiaquatic mammals. This is especially noticeable in relation to the dispersion of the species and in particular the carnivorous species within the river continuum (Schröpfer and Stubbe, 1992). In such river sections one finds the highest degree of population density of the

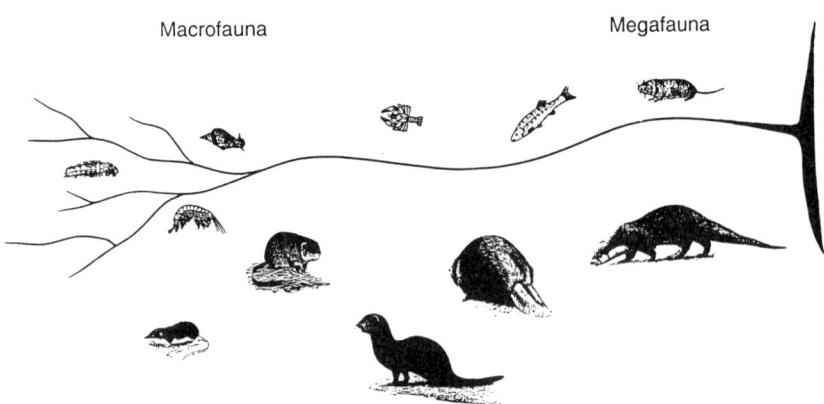

Figure 10.5 The main distribution of semiaquatic mammals within a running water continuum in Europe. Species: *N. fodiens*, *O. zibethicus*, *M. lutreola*, *C. fiber*, *L. lutra*; Prey animal taxa: macrofauna: Trichoptera, Amphipoda, Gastropoda, Potamobiidae; megafauna: Pisces, Rodentia

mammal species concerned (Figure 10.5) (compare with Harris, 1984; Vannote *et al.*, 1980).

An ecotone possesses a structure within its species community, it possesses a coenosis. It does not accommodate a random species accumulation of the ecosystems by which it is limited. It is not simply an edge with some species. Quite obviously typical ecotone occupants can easily exploit the resources of the two adjacent ecosystems; they are amphicoenous, they behave amphicoenously. They constitute a characteristic community in an ecotone.

THE ECOTONE SITUATION FOR MAMMALS

When evaluating our cultivated landscape with its many boundaries, hedgerows and scrub, one must differentiate between the obligate ecotone species with amphicoenous adaptation, attaining their species-typical social and population structures within the ecotone, and the facultative ecotone inhabitants, appearing sporadically and infrequently in the ecotone. They use the ecotone possibly for interpatch movement, thereby maintaining the genetic continuum of the metapopulation. There are probably very few terrestrial mammal species with an ecotone preference which can produce sub-populations in ecotones.

That few mammal species can be called typical ecotone inhabitants is probably due to the fact that only a few natural ecotones exist, so that only a small number of species and even fewer mammal communities were able to adapt themselves to the particular factors of an ecotone habitat. In the cultivated landscape few ecotones can be found and these are marginal areas of anthropogenic origin. This is why the biodiversity of the mammalian fauna is not increased by ecotone inhabitants.

One also has to take into account that species with similar demands will meet in the ecotone so that violent interferences may occur and the ecotone may become a fighting zone. The minimum sizes of ecotone fragments are also quite often unknown; where not the absolute size of the ecotone will be important but rather its shape, its productivity, and the distribution and quality of its resources.

It is difficult and often questionable to classify species as 'ecotone species' without discussing which biocoenosis they belong to. Thus, Estes (1991) calls the Impala antelope (*Aepyceros melampus*) of South Africa 'an edge (ecotone) species' because this antelope prefers 'light woodland with little undergrowth and grassland of low to medium height'. While Owen-Smith (1989) deals with the resource partitioning among African savanna ungulates and classifies this mixed feeding ruminant as belonging to the fine-leafed Acacia woodlands.

To determine how effectively a mammal species can exploit the ecotone, their population structure and their social behaviour have to be known. Thus the semiaquatic mammals have a strict intraspecific spatial differentiation with a high intolerance between individuals or family groups (e.g. water shrew: Voesener and Van Bemmel, 1984 ; otter: Mason and McDonald, 1986; beaver: Richard, 1970) which is certainly due to the relatively high quality linearly arranged resources.

According to the mosaic cycle concept (Remmert, 1991), it could be possible that some mammal species which settle in the succession areas of an ecosystem, where the structure of the coenosis resembles the structure of an ecotone, are also able to colonize newly created ecotones (e.g. hedges). These mammal species (e.g. *Apodemus sylvaticus* in Europe: Bauchau and Le Boulence, 1991; *Peromyscus leucopus* in North America: Wegner and Merriam, 1979) enjoy travelling (Lidicker and Stenseth, 1992), and can even reach isolated habitats rather quickly. Such colonization should, however, lead to a community of species; i.e. not one species, but a species spectrum of a coenosis should become established. Only then can one call it an ecotone habitat.

The ripicolous semiaquatic mammals offer a good opportunity to study the characteristics of vertebrates living in ecotones, which is a prerequisite for ecotone management (Dawson and Haslam, 1983; Mason *et al.*, 1984; Brooker, 1983) because we do not need edges with many species but ecotones with multi-niche communities.

ACKNOWLEDGEMENTS

For the translation I thank Ursula McCallum. Our knowledge about the behaviour of the semiaquatic mammals results from research projects supported by the Ministry of Science and Culture of Lower Saxony/Germany.

Most of the mammal drawings are copies of the pictures of various publications.

REFERENCES

Bauchau, V. and Le Boulence, E. (1991). Biologie de populations de rongeurs forestiers dans un espace hétérogène. In Le Berre, M., Le Cuelte, L. and Chabaud, C.R. (eds.) *Le Rongeur et l'Espace,* pp. 275–83. Paris

Brooker, M. (ed.) (1983). *Conservation of Wildlife in River Corridors.* Welsh Water Authority, Brecon

Dawson, F.H. and Haslam, S.M. (1983). The management of river vegetation with particular reference to shading effects of marginal vegetation. *Landscape Planning,* **10,** 147–69

Di Castri, F., Hansen, A.J. and Holland, M.M. (eds.) (1988). A new look at ecotones: emerging international projects on landscape boundaries. *Biology International.* IUBS Special Issue, Paris, **17,** 1–163

Estes, R.D. (1991). *The Behavior Guide to African Mammals.* University of California Press, Berkeley, Los Angeles, London

Hansen, A.J., Risser, P.G. and Di Castri, F. (1992). Epilogue: biodiversity and ecological flows across ecotones. In Hansen, A.J. and Di Castri, F. (eds.) *Landscape boundaries: consequences for biotic diversity and ecological flows,* pp. 423–38. Ecological Studies 92, Springer Verlag, New York

Harris, L.D. (1984). *The Fragmented Forest. Island Biogeography Theory and the Preservation of Biotic Diversity.* University of Chicago Press, Chicago

Hutchinson, G.E. (1959). Homage to Santa Rosalia, or why are there so many kinds of animals. *American Naturalist,* **93,** 145–59

Juckwer, E.A. (1990). *Galemys pyrenaicus* (Geoffroy 1811) – Pyrenäen-Desman. In Niethammer, J. and Krapp, F. (eds.) *Handbuch der Säugetiere Europas,* pp. 79–92. Academische Verlagsgesellschaft, Wiesbaden, Bd. 3/1

Lidicker, J.R. and Stenseth, N.C. (1992). To disperse or not to disperse: who does it and why? In Stenseth, N.C. and Lidicker, W.Z. (eds.) *Animal Dispersal – Small Mammals as a Model,* pp. 21–36. Chapman and Hall, London

Mason, C.F. and McDonald, S.M. (1986). *Otters – Ecology and Conservation.* Cambridge University Press, Cambridge

Mason, C.F., McDonald, S.M. and Hussey, A. (1984). Structure, management and conservation value of the riparian woody plant community. *Biological Conservation,* **29,** 201–16

McGinley, M.A. and Whitham, T.G. (1985). Central place foraging by beavers (*Castor canadensis*): a test of foraging predictions and the impact of selective feeding on the growth form of cottonwoods (*Populus fremontii*). *Oecologia,* **66,** 558–62

Owen-Smith, N.O. (1989). Morphological factors and their consequences for resource partitioning among African savanna ungulates: a simulation modelling approach. In Morris, D.W., Abramsky, Z., Fox, B.J. and Willig, M.R. (eds.) *Patterns in the Structure of Mammalian Communities,* pp. 155–65. Texas Tech University Press, Lubbock, Texas

Pagel, H.U. (1989). Untersuchungen zum Produktionsvermögen der Futtergehölze in Biberterritorien. *Arch. Nat. Schutz Landsch. Forsch., Berlin*, **1**, 29–44

Remmert, H. (1991). *The Mosaic-Cycle-Concept of Ecosystems*. Ecological Studies 85, Springer Verlag, New York

Richard, P.B. (1970). Territorialisme et aggressivité chez le castor (*C. fiver* L.). *Ecologie et Ethologie*, **1**, 97–105

Ruthardt, M. (1990). Ein öko-ethologischer Ansatz zur Erklärung der semiaquatischen Lebensweise der Wasserspitzmaus (*Neomys fodiens* Pennant 1771). Dissertation University of Osnabrück

Schröpfer, R. (1985). Symposium über semiaquatische Säugetiere und ihre Lebensräume. *Z. Angew. Zool.*, **72**, 355–67

Schröpfer, R. (1990). The Structure of European Small Mammal Communities. *Zoolgische Jahrbücher. Abteilung für Systematik, Ökologie und Geographie, Jena*, **117**, 355–67

Schröpfer, R. and Paliocha, E. (1989). Zur historischen und rezenten Bestandesänderung der Nerze *Mustela lutreola* L. 1761 und *Mustela vison* Schreber 1777 in Europa – eine Hypothesendiskussion. In Stubbe, M. (ed.) *Populationsökologie Marderartiger Säugetiere*, pp. 303–19. Wissenschaftliche Beiträge. Universität Halle 1989/37

Schröpfer, R. and Stubbe, M. (1992). The species diversity of the European semiaquatic mammals in the continuum of the river system. II. International Symposium on Semiaquatic Mammals, Osnabrück FRG, *Semiaquatische Säugetiere* (1992), pp. 9–14.

Stubbe, M. (ed.) (1992). *Populationsökologie Marderartiger Säugetiere*, pp. 303–19. Wissenschaftliche Beiträge. Universität Halle

Spitzenberger, F. (1990). *Neomys anomalus* Cabrera, 1907 – Sumpfspitzmaus. In Niethammer, J. and Krapp, F. (eds.) *Handbuch der Säugetiere Europas*, pp. 317–22. Academische Verlagsgesellschaft, Wiesbaden, Bd. 3/1

Vannote, R.L., Minshall, G.W., Cummins, K.W., Sedell, J.R. and Cushing, C.E. (1980). The river continuum concept. *Canadian Journal of Fisheries and Aquatic Sciences*, **37**, 130–7

Voesener, L.A.C.J. and Van Bemmel, A.C. (1984). Intra- and Interspecific competition in the water shrew in The Netherlands. *Acta Theriologica*, **29**, 297–301

Wegner, J.F. and Merriam, C. (1979). Movements by birds and small mammals between a wood and adjoining farmland habitats. *Journal of Applied Ecology*, **16**, 349–57

CHAPTER 11

ECOTONAL BIODIVERSITY AND SUSTAINABILITY IN UNIQUE TROPICAL LANDSCAPES

Heath J. Carney

INTRODUCTION

This chapter logically follows the previous chapters in that it deals broadly with all major types of biodiversity in tropical land–inland water ecotones. In recent years the special needs and challenges of conservation efforts in tropical developing countries have become increasingly appreciated (e.g. Soulé, 1991; Ehrlich and Wilson, 1991; Abate, 1992). This is critical given that most global biodiversity is thought to reside in these regions (Wilson, 1988). Soulé (1991) notes that habitat loss, habitat fragmentation and over-exploitation are more urgent considerations in poorer (southern) countries. Industrial pollution and climate change, on the other hand, are more important concerns in richer (northern) countries. He recognizes the need for a flexible pluralistic approach in poorer countries which adapts quickly to changing sociopolitical conditions. Ehrlich and Wilson (1991) emphasize the much greater urgency and difficulty of work in the tropics: special efforts are needed in these regions to integrate conservation with sustainable development. For these reasons, a number of special methods and programmes, including RAP (Rapid Assessment Program – Abate, 1992) and the Smithsonian BIOLAT Program (Soulé and Kohm, 1989), have been developed for tropical studies.

Paralleling these developments, considerable attention has been devoted to the problems of human poverty and environmental degradation in the tropics. Geographically, the greatest focus in Latin America, and perhaps the world, has been the Amazon. This has been symbolized by the UNCED (United Nations Conference on the Environment and Development), or 'Earth Summit,' during June, 1992 in Brazil. In this region, there are clear and urgent needs for better farming practices on fragile soils, more sustainable forestry and preservation of the tremendous biodiversity. Two conceptual tools have developed to address these needs: sustainable development and biodiversity. However, these tools were developed somewhat independently by different groups. Thus there is a greater need for integration both intellectually and in planning, policy and implementation. At a political level this is indicated by the lack of consensus on the biodiversity treaty at UNCED.

Besides the Amazon, perhaps the most notable aquatic ecosystem in Latin America is Lake Titicaca, located in the central Andes of Bolivia and Peru. Superficially Lake Titicaca and the surrounding landscape appear quite different from the Amazon. The Lake Titicaca basin is relatively cool, there are few trees, and it is clearly dominated by human activities. Beyond these apparent differences, however, there exist more profound and general similarities. For both regions there is a critical need to understand, and in the long term to optimize, the interacting dynamics of biodiversity and human development. This is generalizable to the entire developing world, including Africa and Asia, and underscores the urgent need to integrate sustainable development with biodiversity.

A number of features make the Lake Titicaca (Bolivia/Peru) watershed unique, and so of global as well as regional interest. They include the immense size of the lake (largest by volume in Latin America, 827 km^3), a cool yet otherwise tropical climate, a diverse and relatively isolated biota, extensive wetlands, and a long, rich human history. Thus this landscape is a very compelling case for research, conservation and development. Lake Titicaca is at 16° south of the equator, 3810 meters above sea level, in the Andean altiplano (plateau) of Peru and Bolivia (Figure 11.1). The Lake Titicaca drainage system consists of mountains that rise to over 6,000 m, smaller ranges of foothills nearer the lake, and a broad plain with a diversity of land uses. Recently, a field investigation has started there as part of the MAB (Man and the Biosphere)/UNESCO Land–Inland Water Ecotone Programme (Naiman et al., 1989; Naiman and Décamps, 1990). The many international limnological programmes and studies of this system have been summarized most recently by DeJoux and Iltis (1992).

This chapter provides an overview of concepts and methodologies which have been developed recently for survey and sampling of tropical ecotones. It takes into account the particular needs and challenges in the tropics. These include, first, the urgent need for rapid assessment of the largely undescribed and otherwise unknown species. Second, the ecotones in tropical regions are generally human dominated or heavily impacted, so research efforts need to be combined with sustainable development. Most of the global human population, growing relatively fast, and an even greater proportion of the world's poor, resides in the tropics. In this paper, two systems are emphasized as examples: tropical rain forests and the Lake Titicaca watershed. Since they are quite different in many respects, they provide an idea of variability within the tropics and thus provide a better idea of the truly general and profound issues and problems of tropical developing regions. From this perspective, an integrated approach to ecotonal biodiversity and sustainability is developed in this chapter.

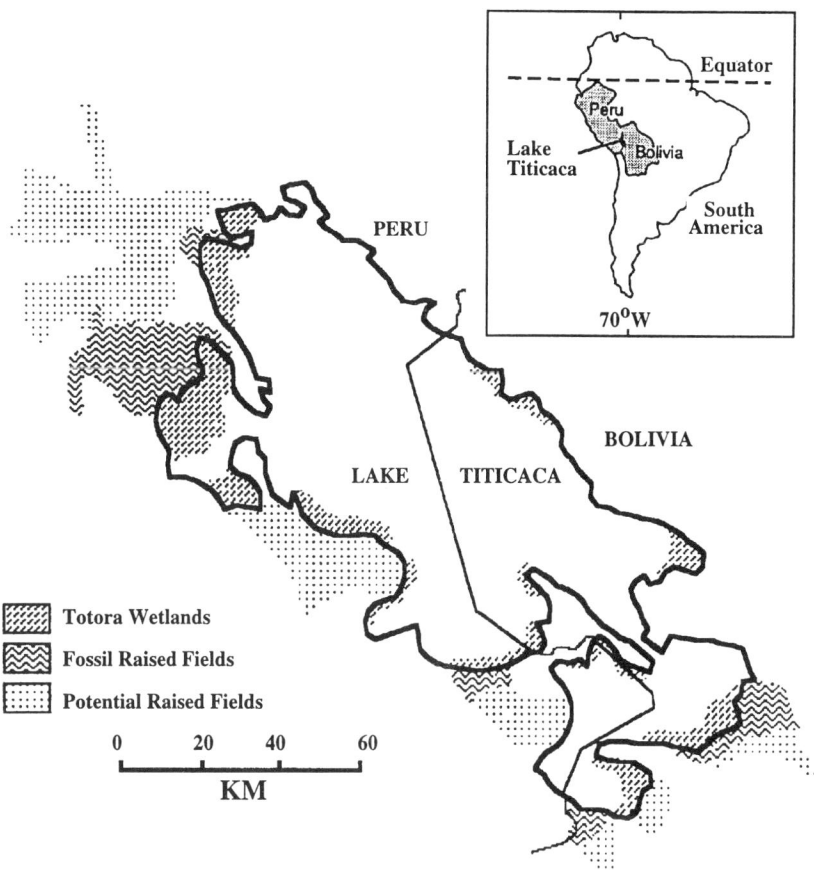

Figure 11.1 General map of the Lake Titicaca region which indicates the close proximity of wetlands to fossil, present and potential raised fields in lakeside plains. Recent rehabilitation of raised fields has been primarily in locations of major fossil fields. For additional information about both wetlands and raised fields, including more detailed maps, see DeJoux and Iltis (1992) and Kolata (1991)

SURVEY, SAMPLING AND MONITORING OF TROPICAL ECOTONES

Soulé (1991) has emphasized that conservation research and implementation must be 'tailored' to the particular circumstances of a given location. He lists as particularly important for tropical regions the problems of: (1) habitat loss, habitat fragmentation and overexploitation at the gene, population and species levels; (2) introduction of exotic species at all major levels (genes, populations, species, communities, and ecosystems); and (3) pollution at the ecosystem level. For a given location there should be an effort to identify the major impacts on biodiversity, and then design survey, sampling and monitoring accordingly.

The following major challenges and opportunities should be kept in mind for tropical developing regions. First, many to most species in land–water ecotones are as yet undescribed, and they are affected by changing, in some areas increasing, human impacts. This calls for certain rapid assessment techniques (RAP – Abate, 1992) including surveys, mapping and sampling. Second, where it is logistically possible, permanent plots for research and monitoring should be established (see BIOLAT, Appendix B in Soulé and Kohm, 1989). Permanent plots are established for long-term research and monitoring following these rapid assessment surveys. Sites should be chosen according to conservation criteria and practicality for long-term work, so it is best to have locations as convenient and logical as possible taking into account both scientific and social factors. The Smithsonian/MAB BIOLAT programme (Soulé and Kohm, 1989; Appendix B) has developed a protocol for such work in tropical rain forests especially. Its major objectives are (1) establish permanent plots, (2) inventory all species (plants, animals, microorganisms) within those plots, (3) provide checklists of species found, (4) monitor changes in populations. This programme has started in a network of areas in the western and northern Amazon basin, and expansion is planned throughout the Neotropics. Third, there should be integration with complementary projects in sustainable land use, cultural and ecological history, and landscape analysis which can help make the conservation research broader and more meaningful.

The Lake Titicaca ecotones are an excellent example for the application of these rapid assessment and permanent plot methodologies. The Lake Titicaca landscape contains both the conventional biodiversity which has been of traditional academic concern and additional elements which merit consideration of a broader perspective and approach to biodiversity for this region. As the largest and perhaps oldest (probable Miocene origin; 10–20 million years ago) lake in Latin America, it has two features which make its biodiversity a priority for conservation and restoration research (Wilson, 1988, p. 8): it is geologically ancient and relatively isolated biogeographically, and it has extensive wetlands. In this very ancient lake in a unique high altitude tropical environment, a very distinctive flora and fauna have evolved which include many probable endemics (Parenti, 1984; Carney et al., 1987; DeJoux and Iltis, 1992). 'Probable' should be emphasized because basic systematic and biogeographic studies are still very much in progress, even for some of the more abundant species.

One of the better known examples, which we have learned about only recently, are the native fish species of the genus *Orestias*. There are 43 described species in the central Andes (Parenti, 1984). More than half are

Table 11.1 Compilation of Lake Titicaca biodiversity which is presently known. These data are from various chapters in DeJoux and Iltis (1992). The two columns at the right summarize the ecotonal (shore, littoral, wetland and nearshore areas) and total species for each major group. They indicate that for all these groups, most species inhabit the land–water ecotones. Note that certain important groups such as bacteria and protozoans are not listed here because they have not yet been studied. The numbers provided are the number of taxa determined at the most detailed level of identification available (generally species, subspecies or genus)

	Ecotonal	Total
(I) Algae:	136	259
Cyanophyta (26), Chlorophyta (112), Euglenophyta (9), Pyrrophyta (7), Chrysophyta – including diatoms – (105)		
(II) Macrophytes:	23	23
Charophyta (11), higher plants (12)		
(III) Zooplankton:	33	45
Cladocera (31), Copepoda (7), Rotifera (7)		
(IV) Benthic fauna (invertebrates):	107	118
Bryozoa (2), Coelenterata (1), Porifera (1), Oligochaeta (15), Tricladida (1), Hirudinea (4), Mollusca (30), Ostracoda (11), Amphipoda (11), Hydroacarida (10), Insects: Odonata (2), Hemiptera (3), Coleoptera (5), Diptera (15), Tricoptera (7)		
(V) Fishes:	20	25
Native species – *Orestias* sp. – (23), Introduced species (2)		
(VI) Associated communities:	69	69
Batrachians – frogs and toads – (19), Snakes (1), Avifauna (birds–49)		
Grand totals:	388	539

endemic to the closed Titicaca basin, and 23 are known only from Lake Titicaca. Many more remain undescribed. Until recently fish diversity was considered unusually low for the size of the lake (Richerson *et al.*, 1977) because of this incomplete systematic work. In addition, the diversity of these fishes, invertebrates, aquatic macrophytes and algae is concentrated in the specific regions of the lake we are studying: littoral wetlands of the smaller basin, Lago Huiñamarca, and in other parts of the lake (Table 11.1). The Huiñamarca basin has a mean depth of 7 m, complex shoreline (development index = 2.57) and gradual slope (0.35%), so wetlands are a dominant feature. Almost 90% of the basin is less than 20 m deep, and about 56% of the bottom is covered with vegetation. For Lake Titicaca as a whole, about 2849 km^2 (33% of the total area) is less than 20 m deep.

DeJoux and Iltis (1992) provide the most recent summary of the species found to date in Lake Titicaca (Table 11.1). This table makes two very important points which are generalizable to a great diversity perhaps to most, tropical land–water ecotones. First, many to most species are undescribed and otherwise virtually unknown. Whole groups such as bacteria and protozoans are not yet documented, and even the most diverse and well-studied, such as fishes, benthic invertebrates and algae, are incompletely known. Second, the biodiversity in the watershed is certainly concentrated in the land–water ecotones since this is the primary habitat of the most diverse groups. This table also demonstrates the use of what can be termed a 'food web' approach to biodiversity inventory and study. Species are listed according to functional food web roles as well as systematically. Gaps in identifying important organisms cannot be filled until there is a good understanding of all species in the major groups which contribute to species interactions and ecosystem processes. Then through experimentation and monitoring a greater effort can be made to determine important species interactions and significant external perturbations which most affect total food web biodiversity.

Table 11.2 summarizes the known aquatic ecotonal biodiversity for the Amazon River according to the monograph edited by Sioli (1984). A large fraction of the Amazon basin (over 7 million km^2) can be considered an immense land–water ecotone. For several months each year the *varzea* floodplain (60,000 km^2) is flooded. Water-levels can fluctuate up to 20 m between seasons in parts of Amazonia. Currents are also substantial: 0.5 m s^{-1} at low water to 3.0 m s^{-1} during flooding. Given these strong seasonal land–water fluctuations and coupling, all the organisms listed in Table 11.2 are categorized as ecotonal. The total listed is surely a substantial underestimate. For example, both the algal and benthic invertebrate groups probably consist of well over a thousand species each. This is indicated by the authors who compiled this information. It is also suggested by the very high diversity of zooplankton and fish. The number of species in these two groups is generally substantially lower than the number of algal and benthic invertebrate species when all groups are studied thoroughly.

Major features which Amazonia has in common with the Lake Titicaca basin are the long-term occupation and extensive use of ecotonal resources by humans. While there is a popular notion that much of the Amazon is pristine tropical rain forest, in fact humans have lived throughout Amazonia for over ten thousand years (Meggers, 1984; Roosevelt *et al.*, 1991). Their settlements and resource uses have been concentrated in productive ecotonal areas. Indigenous human populations have reached much higher densities (14.6 people per km^2) near the *varzea* than in the higher *terra firma* (0.2 people per km^2) for thousands of years. This is because the *varzea* soils are much more fertile and replenished annually by floods, and the ecotonal biota is very abundant and useful. These factors continue to be important to the present, so it is imperative to assure sustainable use of land and ecotonal resources in this region.

The biodiversity of the Amazon and Lake Titicaca are summarized and compared with other major tropical and temperate systems in Table 11.3. The data provided should all be considered quite tentative and subject to substantial

Table 11.2 Summary of Amazon River biodiversity which is presently known according to various chapters in Sioli (1984). (+) to the right of a number indicates that there are many more species in addition to those documented

	Total (= ecotonal)
(I) **Algae:**	653(+)
Cyanophyta (11), Chlorophyta (319+), Euglenophyta (22), Chrysophyta (including diatoms – (231+), Other (70+)	
(II) **Macrophytes:**	55
(III) **Zooplankton:**	313(+)
Cladocera (25+), Copepoda (38), Rotifera (250)	
(IV) **Benthic fauna** (invertebrates):	?(++)
(V) **Fishes:**	1200
(all major freshwater groups except Cyprinoids)	
(VI) **Associated communities:**	43
Batrachians (9), Reptilia (27), Mammalia (7)	
Grand total:	2264(++)

revision following more thorough studies. Groups such as bacteria, protozoa, and associated communities (amphibians, reptiles, mammals) are not included in the table because they are not yet documented for many of these systems. The total biodiversity listed is probably less than half what actually exists in all cases. Thus there are probably at least a thousand ecotonal species in even moderately sized lakes and rivers, and there are many thousands in the largest systems. Despite the precautions which must be applied in interpreting these data, some generalizations can be made. First, it is clear that both ecotonal and total biodiversity are greater in the larger systems. Second, algae and benthic invertebrates are generally by far the most diverse groups. Variations in relative diversity of the major groups between systems is probably largely due to differences in the amount of sampling and taxonomic study devoted to each of these groups. For example, in the Amazon, fishes have been examined most thoroughly, while at Baikal, the benthic invertebrates have received much attention, and at Tahoe the phytoplankton have been studied most intensively. Difficulties in making comparisons are understandable since work on individual systems is still in progress. Most of the monographs referred to here were published during the past decade and their biological inventories remain quite incomplete. Even for well-studied systems such as Lake Tahoe, no comprehensive species lists have been published. The data compiled in Table 11.3 indicate the type of information which needs to be collected and published more completely for each system and more uniformly for the entire range of organisms. After this is accomplished, comparisons in the future with such information should prove quite interesting and informative.

The majority of the ecotonal biodiversity has very clear spatial distributions in relation to vegetation in the littoral zones. Thus it is important to sample along ecotonal transects which include all major biotic communities

Table 11.3 Biodiversity of major biological groups in some large tropical and temperate freshwater ecosystems. Data are provided here for the major groups of organisms for which comparative information is available. (+) indicates that the number to the left is certainly an underestimate. References: [1]DeJoux and Iltis (1992), [2]Sioli (1984), [3]Coulter (1991), [4]Kalk et al. (1979), [5]Kozhov (1963), [6]unpublished data, [7]Horie (1984)

Site	Algae	Macro-phytes	Zoo-plankton	Benthic invertebrates	Fishes	Total
TROPICAL						
L. Titicaca[1] (Area = 8,563 km^2, volume = 827 km^3)						
Ecotonal	136	23	33	107	20	319
Total	259	23	45	118	25	470
Amazon R.[2]						
Ecotonal	653(+)	55	313(+)	?(+)	1200	2221(+)
L. Tanganyika[3] (32,000 km^2, 17,827 km^3)						
Ecotonal	472	81	68	423	214	1258
Total	759	81	163	490	287	1780
L. Chilwa[4] (1,040 km^2)						
Total	87	42	13	200	35	377
TEMPERATE						
L. Baikal[5] (31,500 km^2, 22,995 km^3)						
Ecotonal	403	17	15	1173	33	1641
Total	566	23	88	1218	50	1945
L. Tahoe[6] (500 km^2, 124 km^3)						
Ecotonal	294	7	17	108	11	437
Total	575	20	34	131	15	775
L. Biwa[7] (688 km^2, 28 km^3)						
Ecotonal	18	71	76	205	46	416
Total	94	71	105	223	63	556

(Figure 11.2). Within the littoral zone of Lake Titicaca, for example, several zones mark the transition between land and open water. A transect from land to open lake-water goes through zones of intermittently inundated land, floating-leaf plants at the edge of the lake, submerged macrophytes (primarily *Myriophyllum* and *Elodea* – zone A), the totora (*Scirpus californicus* (C.A. Meyer) Steudel spp. *Tatora* (Kunth) T. Koyama – Koyama, 1963 or *Schoenoplectus tatora* Kunth (Palla). – Collot et al., 1983) beds which also contain *Potamogeton* and *Chara* (zone B), and then to deeper, open water with beds on the bottom dominated by *Chara* in shallow areas, and by *Potamogeton* at greater depth (zone C). The complexity of the littoral is related to its slope. Shallowly sloping areas contain broad expanses of each vegetation zone, often interdigitated. Steeply sloping areas of very simple ecotones may have only a narrow band of totora, only the *Myriophyllum* zone, or even simple rocky faces. These land–inland water ecotones, then, are actually one to several zones, each of which must be understood in order to

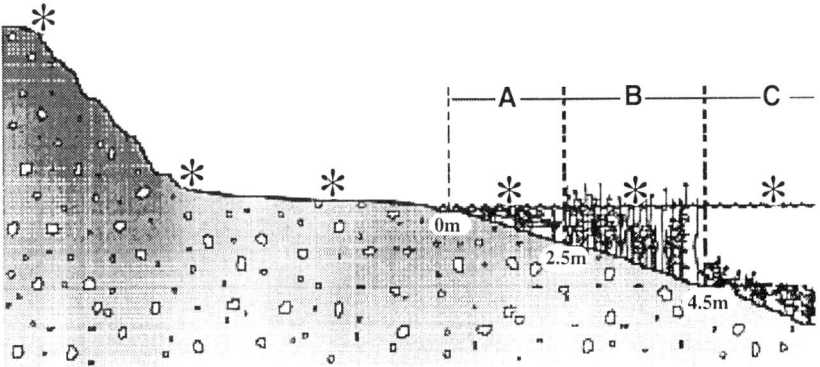

Figure 11.2 Transect sampling points and nearshore vegetation zones. Sampling points are indicated with asterisks, and the vegetation zones (A, B, C) are described in the text. For more detailed information about the vegetation zones see Collot *et al.* (1983)

characterize the system. All the other major groups of organisms have very clearly defined spatial distributions which correlate highly with these vegetation zones.

It is also important to consider and manage ecotonal biodiversity at the broader landscape level. Two additional examples make it clear that the unique biodiversity of this region extends well beyond the above wetland species. We have recently discovered a microscopic diatom, *Cyclotella andina*, which is quite abundant in the plankton and sediments (Theriot *et al.*, 1985). It is important and interesting in many ways. With a morphology intermediate between two major genera, *Cyclotella* and *Stephanodiscus*, it presents intriguing systematic and evolutionary questions. Its abundance in both the sediments and plankton will also help us understand dynamics at time scales ranging from geological to historical to recent and present day. A second example is the diversity of cultivars maintained by farmers in the lake basin. The major crop for consumption, the potato, was first cultivated in this region thousands of years ago before spreading to other parts of the world. During this time thousands of varieties have been developed and are maintained by farmers (Carney, 1980; Brush *et al.*, 1981). Thus this is a very positive example of how people in a region can help generate and maintain biodiversity which has economic and other benefits. The most productive sites for potatoes and other cultivars, both historically and presently, are the flat plains and ecotones along the lake shore (Figure 11.1). Native varieties have many important and useful properties including nutrition, yield and resistance to pathogens such as nematodes and fungi. However, many of these varieties are being replaced by relatively few introduced high-yielding hybrids. This process is termed 'genetic erosion'. Thus it is critical to include these and other crops in a regional plan for biodiversity conservation and restoration.

ECOTONAL BIODIVERSITY AND SUSTAINABILITY

While most biodiversity is in the tropics, most of the human population, and the greatest population growth, is also in these regions. Thus biodiversity work clearly needs to be integrated with sustainable development. This is clearly articulated in the US and International Sustainable Biosphere Initiatives (Lubchenco et al., 1991 and Huntley et al., 1991, respectively). The first initiative has as one of its three major objectives, 'the ecology and conservation of biological diversity'. The second initiative also has as one of its three major research priorities, 'diversity and sustainability'. Both these initiatives have a broad functional view of diversity which spans from the gene to the biosphere and thus includes as intermediate scales populations, species, communities, ecosystems, landscapes and regions.

The Lake Titicaca region is again an excellent example of how these concerns must be combined. The rural inhabitants of the Lake Titicaca watershed, with an average family income of less than $70 US per month, are among the poorest in Latin America. They do not risk starvation as some groups in sub-Saharan Africa and parts of Asia. Still, they remain at or near subsistence level because of a complex of factors including cultural isolation from the national economy and society, and scarce and/or degraded natural resources for activities such as farming. The major subsistence and economic activities of the regional human population include farming, grazing, littoral and nearshore fisheries, and tourism. All these activities are critically linked to the health and wise use of land–water ecotones. For fisheries, this is obvious. Grazing by cattle and sheep is also intensive near the lake shore. Macrophytes are often harvested for these animals, and for other uses. Large flat ecotonal areas are also best for the native raised field agriculture, as discussed below. Finally, the lake and its past and present coastal cultures are major tourist attractions. These factors certainly need to be considered in planning for the best possible use and conservation of biodiversity and other natural resources in this region.

Perhaps the most important and compelling recent change in land use in the Lake Titicaca watershed has been the expanding rehabilitation of pre-Columbian raised field agriculture in both Peru and Bolivia (Kolata, 1991). Tiwanaku, the major pre-Inca culture of this region, developed on a foundation of raised field agriculture near the lake (Figure 11.1). Raised field activity peaked 1300–1000 BP and then declined abruptly 900–800 BP (Ortloff and Kolata, 1993). It is now expanding rapidly because of clearly demonstrated economic benefits of substantially higher yields and protection from frosts (Kolata, 1991). In addition, local communities have organized enthusiastically to construct and maintain these fields. We have also found that there can be environmental benefits of high nutrient and sediment retention and recycling as described below. Both fossil and rehabilitated raised fields are in close proximity to wetlands containing high aquatic biodiversity (Figure 11.1) which makes it additionally important to consider economic and environmental aspects together.

Nutrient fluxes from land use can have very strong impacts on both the amount and kind of ecotonal biodiversity. They have the most direct effect on algae and higher plants which provide habitat for all the other ecotonal biota (Carney et al., 1993b). Nitrate is of special interest since it can indicate land use impacts which have important consequences for water quality (e.g. Addiscott, 1988). In raised field transects nitrate values decreased dramatically from 3,907 µg l^{-1} N at the inflow to raised fields to 49 µg l^{-1} N as water flowed through the raised field canals filled with algae and macrophytes (Carney et al., 1993a). This probably reflects biological uptake in the raised field canals and below in these long complex ecotones. It might also indicate denitrification. Soluble reactive phosphorus (P-SRP) also declined dramatically as the water passed through the raised fields. The inflow concentration averaged 485 µg l^{-1} P, while the outflow concentration declined to 24 µg l^{-1} P. In transects of similar complex ecotones without raised fields nutrient concentrations also decreased, but not as much (from inflow 2,800 µg l^{-1} N-NO$_3$ and 237 µg l^{-1} P-SRP to outflow 273 µg l^{-1} N-NO$_3$ and 61 µg l^{-1} P-SRP). In transects of shorter and simpler ecotones without raised fields, nutrient concentrations never declined so much *en route* to the lake. They were generally relatively constant or even increased as water flowed to the lake. If we add nitrate to TKN for total N (TN), we find that TN:TP ratios were much higher in raised field transects (mean 14.19, $n = 2$) than in other transects at the same time (mean 7.25, $n = 4$). The experimental bioassays have demonstrated that the aquatic vegetation in the canals was limited by both N and P, and so could lower concentrations of these nutrients. In summary, these nutrient results indicate that raised fields and other land uses have important impacts on ecosystem processes affecting ecotonal biodiversity.

A transect near Tiwanaku demonstrates that raised fields can also filter out suspended sediments and thus dramatically increase water clarity. At the inflow to these fields we have noted very high sediment concentrations (28.3 turbidity units). In the raised fields the concentrations declined substantially to 5 turbidity units, and at the outflow to the lake they fell to as low as 2 turbidity units. We have not observed such dramatic reductions in turbidity in any other transects. In these other transects, turbidity is relatively constant within a transect or may increase as water flows from the hills toward the lake.

Finally, we are explicitly considering human resource use, impacts on natural biota, and human attitudes towards these resources. As in many other, perhaps most, tropical regions there has been substantial human activity in the Lake Titicaca basin for thousands of years (over 3,500 years – see above). This has certainly included the diverse littoral lake ecotones, which have long supported a combination of agriculture, grazing, fisheries and other activities. As in the Amazon, high biodiversity has remained during this long period of relatively low-impact subsistence oriented activity. Thus, for most of the lake basin a programme of sustainable development including conservation objectives seems much more feasible than trying to find and set aside pristine reserve areas. Thus our research group includes anthropologists as well as ecologists and agronomists. An effort must also be made to understand and communicate with local lake shore communities because they are very close-knit and largely autonomous. For example, fishing territories are defined and defended by individuals and communities with virtually no control by regional and national governments (Levieil and Orlove, 1990). Fortunately, we have been able to establish excellent communication and working relationships with the communities in our study region through collaboration with Bolivian colleagues. We are in contact with both development agencies and conservation groups to help plan sustainable development and a MAB Biosphere Reserve at Lake Titicaca.

Landscape sustainability. Forman (1991) has emphasized that the landscape is a most critical spatial scale for sustainable development. This is quite true for land use and water quality in the Lake Titicaca basin and so is discussed briefly here. This scale corresponds to the larger watershed level ecotones which, 'include all land forms and land use patterns in the catchment' (Hillbricht-Ilkowska *et al.*, 1989). It is most important to consider the dynamics and conservation of biodiversity of the smaller littoral ecotone in the context of human activities at this larger scale. Generally, the Lake Titicaca watershed landscape includes these important elements: (1) hills with relatively poor soils, agricultural fields and sparse vegetation (grasses, some shrubs and very few trees); (2) the lake-side plain with richer soils and grasses used for grazing; (3) the lake which has extensive wetlands as described above; (4) a broad range of inflowing waters (from small streams which flow only during the rainy season to large rivers which flow throughout the year and can accumulate many materials). With respect to water flow and nutrient fluxes, retention of nutrients and other materials before they reach the lower parts of large rivers and the lake is an important need. Currently, there are several important nonpoint sources. Erosion from slopes is accelerated by grazing and agriculture. Humans and all major livestock (cattle, sheep, pigs and llamas) use inflowing streams and rivers heavily. The major activities are travel along the edges, bathing and other cleaning, grazing, drinking and defecation. These materials are flushed downstream at the beginning of the wet season especially. Largely because of these activities there is little to no riparian vegetation which could reduce these material fluxes. Thus there is a need to mitigate the impacts of these activities and improve the riparian vegetation corridors.

Another important need is to determine the optimal location for raised fields within the Titicaca basin. Several major factors to consider are efficient nutrient and water use for high productivity, impacts on basinwide hydrology, and integration with other resource-use activities. Presently, raised fields seem ideally located at the base of the *cerro* hills and the upper part of the *pampa* plain. These conditions are favourable for productivity and allow these systems to intercept nutrients and other materials from nonpoint sources. These can be recycled and enhance productivity, as described above. Currently, most activities upstream are low technology and near subsistence level, so the major materials flowing into canals are nutrients, sediments and organic matter. Thus, the major concern with the canal recycling is infection from fecal bacteria and other water-borne diseases. If and when there is greater use of toxic chemicals the recycling practices will have to be re-evaluated.

Hydrological manipulations should also be considered carefully. Mean annual rainfall is fairly high in this region, yet so is potential evaporation. Implications at the scale of individual fields are discussed above. Consideration should also be given to areas water is diverted from, especially as Lake Titicaca is already slightly saline (about 1 ppt, compared to 35 ppt for sea water). The water has a slight but clear salty taste, and freshwater organisms become osmotically sensitive at this level. Long-term salinity trends are still uncertain since the hydrological data and analyses are still inadequate. Richerson *et al.* (1977) have roughly calculated that there has been a slight increase in recent decades, but this needs more detailed work. What is clear is that evaporation from the lake is a major loss of water which must be compensated by adequate freshwater inflows if salinity is to remain relatively low. Since two other major activities, fisheries and macrophyte harvest, may be dramatically affected by salinity changes, this should be considered in freshwater diversion for land use. The littoral and nearshore areas also harbour a very substantial and unique biodiversity of algae, higher plants, invertebrates, fish and other animals (Table 11.1). Since species richness is higher in these areas, and these are transition zones, hydrological and salinity changes could have their greatest impact in these ecotonal areas.

Raised field rehabilitation also needs to be integrated with other major resource uses. Grazing has been the predominant activity on *pampa* lands, so this has presented the most direct conflict thus far. One way to resolve this is to determine the best combination of these activities within the *pampa*. For example, raised field agriculture may be best in upper parts near groundwater and surface water flows. A guiding principle can be to maximize environmental benefits of wetlands (nutrient and other retention) while also providing economic benefits. This corresponds to the more general agroecological principle of creating agricultural systems which are functional analogs of natural systems and thus mitigate environmental degradation. Grazing can complement agriculture in other lower areas. Interactions with the two major nearshore uses, macrophyte harvest and fisheries, also need consideration and continuing study. Spatially-explicit analyses, modelling and planning are needed. GIS (Geographical Information Systems)-based studies have started under the direction of M. Binford.

In summary, this example of the Lake Titicaca landscape indicates several ways in which sustainable land use at a larger scale and ecotonal biodiversity at a smaller scale must be integrated. This can be generalized to many other regions. First, retention of nutrient and sediment fluxes from land uses is beneficial for both land uses, such as agriculture, and water quality including the ecotonal biota. Second, it is also important to consider hydrological manipulations as carefully as possible in light of these two major concerns. Finally, spatial relations are quite important. The location of a given type of land use, and how it is combined with other uses, can and should be determined by optimizing the most critical economic and environmental criteria. All these and other human effects should be considered and incorporated into plans for management and conservation of ecotonal biodiversity.

CONCLUSIONS

This chapter provides an overview of recent techniques and concepts which have been developed to meet the urgent demand for work on ecotonal biodiversity and sustainability in tropical landscapes. Recently two initiatives have been developed to make the work of academic ecologists more relevant to pressing global environmental problems including loss of biodiversity and more sustainable development. The United States Sustainable Biosphere Initiative (Lubchenco *et al.*, 1991) has as its major foci Global Change, Biological Diversity, and Sustainable Ecological Systems. The International Sustainable Biosphere Project (Huntley *et al.*, 1991) places an emphasis on Diversity and Sustainability, Sustainability in a Changing Atmosphere, and Human Dimensions of Sustainability. These initiatives indicate the critical need to integrate biodiversity work with sustainable development. Watersheds such as Amazonia and the Lake Titicaca basin are excellent case studies for the need for, and start of, work called for by these initiatives. Tropical wetlands and other parts of the land–water ecotones are sites of both the greatest biodiversity and important subsistence activities of local human inhabitants for thousands of years. Thus for both conservation and sustained development in these areas it will be most important to work closely with these inhabitants in multidisciplinary programmes.

ACKNOWLEDGEMENTS

Funding for this paper has been provided primarily by the US MAB (Man and the Biosphere)/UNESCO programme (Grant No. 1753-000566) – US State Department and NSF Grant Nos. 91-03680 and 92-08122 to HJC. Jane Lam and Edith Cook assisted with preparation of this paper.

REFERENCES

Abate, T. (1992). Environmental rapid-assessment programs have appeal and critics. *BioScience*, **42**, 486–9

Addiscott, T. (1988). Farmers, fertilisers and the nitrate flood. *New Scientist*, **120**, 50–4

Brush, S.B., Carney, H.J. and Huaman, Z. (1981). Dynamics of Andean potato agriculture. *Economic Botany*, **35**, 70–88

Carney, H.J. (1980). Diversity, distribution and peasant selection of indigenous potato varieties in the Mantaro Valley, Peru: a biocultural evolutionary process. *International Potato Center Social Science Working Paper Series* 1980–3

Carney, H.J., Binford, M.W., Kolata, A., Marin, R. and Goldman, C.R. (1993a). Nutrient and sediment retention in Andean raised field agriculture. *Nature*, **364**, 131–3

Carney, H.J., Binford, M.W., Marin, R. and Goldman, C.R. (1993b). Nitrogen and phosphorus dynamics and retention in ecotones of Lake Titicaca, Bolivia/Peru. *Hydrobiologia*, **251**, 39–47

Carney, H.J., Richerson, P.J. and Eloranta, P. (1987). Lake Titicaca (Peru/Bolivia) phytoplankton: species composition and structural comparison with other tropical and temperate lakes. *Arch. Hydrobiologia*, **110**, 365–85

Collot, D., Koriyama, F. and Garcia, E. (1983). Repartitions, biomasses et productions des macrophytes du lac Titicaca. *Rev. Hydrobiol. Trop.*, **16**, 241–61

Coulter, G.W. (ed.) (1991). *Lake Tanganyika and its Life*. Oxford University Press, London

DeJoux, C. and Iltis, A. (eds.) (1992). *Lake Titicaca: a Synthesis of Limnological Knowledge*. Monographiae Biologicae Vol. 68, Kluwer Academic Publishers, Dordrecht

Ehrlich, P.R. and Wilson, E.O. (1991). Biodiversity studies: science and policy. *Science*, **235**, 758–61

Forman, R.T.T. (1991). Ecologically sustainable landscapes: the role of spatial configuration. In Zonneveld, I.S. and Forman, R.T.T. (eds.) *Changing Landscapes: an Ecological Perspective*, pp. 261–78. Springer Verlag, New York

Hillbricht-Ilkowska, A., Carney, H.J., Décamps, H., Hunsaker, C.T., Johnston, C., Klimes, L., Mulhauser, H., Nishino, M., Rambouskova, H., Salo, J. and Whigham, D. (1989). Role of ecotones in monitoring stability and change of landscape patches. In Naiman, R., Décamps, H. and Fournier, F. (eds.) Role of Land–Inland Water Ecotones in Landscape Management and Restoration. *MAB Digest 4*, pp. 65–68. UNESCO, Paris

Horie, S. (ed.) (1984). *Lake Biwa*. Monographiae Biologicae Vol. 54, Kluwer Academic Publishers, Dordrecht

Huntley, B.J., Ezcurra, E., Fuentes, E.R., Fujii, K., Grubb, P.J., Haber, W., Harger, J.R.E., Holland, M.M., Levin, S.A., Lubchenco, J., Mooney, H.A., Neronov, V., Noble, I., Pulliam, H.R., Ramakrishnan, P.S., Risser, P.G., Sala, O., Sarukhan, J. and Sombroek, W.G. (1991). A sustainable biosphere: the global imperative (The International Sustainable Biosphere Initiative). *Ecology International*, **20**, 1–14

Kalk, M., McLachlan, A.J. and Howard-Williams, C. (eds.) (1979). *Lake Chilwa: Studies of Change in a Tropical Ecoystem*. Monographiae Biologicae Vol. 35, Kluwer Academic Publishers, Dordrecht

Kolata, A. (1991). The technology and organization of agricultural production in the Tiwanaku state. *Latin American Antiquity*, **2**, 99–125

Koyama, T. (1963). The genus *Scirpus* Linn., critical species of the section Pterolepis. *Canadian Journal of Botany*, **41**, 1107–31

Kozhov, M. (ed.) (1963). *Lake Baikal and Its Life. Monographiae Biologicae* Vol. 11, Dr W. Junk, Publishers, The Hague

Levieil, D.P. and Orlove, B. (1990). Local control of aquatic resources: community and ecology in Lake Titicaca, Peru. *American Anthropologist*, **92**, 18–38

Lubchenco, J., Olson, A.M., Brubaker, L.B., Carpenter, S.R., Holland, M.M., Hubbel, S.P., Levin, S.A., MacMahon, J.A., Matson, P.A., Melillo, J.M., Mooney, H.A., Peterson, C.H., Pulliam, H.R., Real, L.A., Regal, P.J. and Risser, P.G. (1991). The Sustainable Biosphere Initiative: an ecological research agenda. *Ecology*, **72**, 371–412

Meggers, B.J. (1984). The indigenous peoples of Amazonia, their cultures, land use patterns and effects on the landscape and biota. In Sioli, H. (ed.) *The Amazon: Limnology and Landscape Ecology of a Mighty Tropical River and its Basin. Monographiae Biologicae* Vol. 56, pp. 627–48. Kluwer Academic Publishers, Dordrecht

Naiman, R.J. and Décamps, H. (eds.) (1990). *The Ecology and Management of Aquatic–Terrestrial Ecotones*. MAB Book Series 4. UNESCO/Parthenon, Paris/Carnforth

Naiman, R.J., Décamps, H. and Fournier, F. (eds.) (1989). The role of land–inland water ecotones in landscape management and restoration: a proposal for collaborative research. *MAB Digest 4*. UNESCO, Paris

Ortloff, C.R. and Kolata, A.L. (1993). Climate and collapse: agroecological perspectives on the decline of the Tiwanaku state. *Journal of Archeological Science*, **20**, 195–221

Parenti, L.R. (1984). Biogeography of the Andean killifish genus *Orestias*. In Echelle, A. and Kornfield, I. (eds.) *Evolution of Fish Species Flocks*. University of Maine Press, Orono, ME

Richerson, P.J., Widmer, C. and Kittel, T. (1977). *The Limnology of Lake Titicaca (Peru–Bolivia), a Large, High Altitude Tropical Lake*. Institute of Ecology, Publication 14. University of California, Davis, CA

Roosevelt, A.C., Housley, R.A., Imazio da Silveira, M., Maranca, S. and Johnson, R. (1991). Eighth millennium pottery from a prehistoric shell midden in the Brazilian Amazon. *Science*, **254**, 1621–4

Sioli, H. (ed.) (1984). *The Amazon: Limnology and Landscape Ecology of a Mighty Tropical River and its Basin. Monographiae Biologicae* Vol. 56, Kluwer Academic Publishers, Dordrecht

Soulé, M.E. (1991). Conservation: tactics for a constant crisis. *Science*, **253**, 744–50

Soulé, M.E. and Kohm, K.A. (eds.) (1989). *Research Priorities for Conservation Biology*. Island Press, Washington, DC

Theriot, E.D., Carney, H.J. and Richerson, P.J. (1985). Morphology, ecology and systematics of *Cyclotella andina* sp. nov. (Bacillariophyceae) from Lake Titicaca, Peru–Bolivia. *Phycologia*, **24**, 381–7

Wilson, E.O. (ed.) (1988). *Biodiversity*. National Academy Press, Washington, DC

CHAPTER 12

SCIENTIFIC BASIS FOR CONSERVING DIVERSITY ALONG RIVER MARGINS

Geoff Petts

INTRODUCTION

Biodiversity is defined as the property of groups or classes of living entities to be varied (Solbrig, 1991). It is manifest at every level of the biological hierarchy, from molecules to ecosystems, and is the outcome of many contributing forces (Diamond, 1988). However, the importance of biodiversity for the proper functioning of ecosystems is uncertain. There are two views (Solbrig, 1991): (i) that species composition is largely random within certain constraints imposed by fluctuating and variable environmental conditions, and (ii) that communities are well organized with 'limited' membership – only a fraction of the species that could be present actually forming a community at any one time. Differential survival, reproduction and death between individuals within a population has been related to environmental limits on resource availability (Townsend and Calow, 1981). This is seen particularly at the regional scale where the primary external controls (climate and geology) determine the biogeographic setting. At the local level, the complex spatial variation of communities at any point in time is explained not only by interactions between species and the physical environment, but also by randomness in nature and by interactions between species, such as competition, predation, parasitism and symbiosis. The relative abundance of species at a site will vary during a year, and from year to year, in relation to short-term (stochastic) environmental variations, complicated by the existence of biodiversity which allows some individuals to perform above average within certain environmental limits.

THE CHARACTER OF RIVER MARGINS

Within river margin ecosystems, the main environmental factor influencing biodiversity is flooding which has a continuum of effects from providing resources (moisture, silt, nutrients etc.) to causing major disturbance (channel erosion, avulsion, and sediment deposition). However, the fluvial dynamics of rivers have been markedly altered by human activity (e.g. Petts, 1984). Today, river margin ecosystems throughout Europe and much of North America are severely degraded in comparison to their characteristics 200 years ago. The anthropocentric view of 'improving' river margin systems has been a fundamental objective of river regulation by creating impoundments and channelized rivers. Thus, Ellett (1853: pp. 303–304) wrote:

'The banks of the Ohio and Mississippi may yet, in the course of a very few years, be cultivated and adorned down to the water's edge....Grounds, which are now frequently inundated and valueless, will be tilled and subdued.'

Throughout Europe, despite early floodplain deforestation from about 2500 BP (Wiltshire and Moore, 1983), along many rivers the margins remained as seasonally-flooded forests until the mid-eighteenth century. Major impacts on river margin systems occurred between 1750 and 1850 (Petts *et al.*, 1989; Petts, 1990a) when river regulation fixed the location of the river channel, preventing disturbance by fluvial erosion and flooding. The course of the regulated sector of the River Trent, UK, for example, is largely the same today as it was 200 years ago (Large and Petts *in press*). Furthermore, large areas of floodplain wetland have been drained. In Hungary, the regulation of the River Tisza, beginning in 1845, drained 12.5 million ha of floodplain marsh. The elimination of annual flooding and the lowering of groundwater levels have caused ecological change usually to a less diverse system (Bravard *et al.*, 1986; Amoros *et al.*, 1987).

Today, river margins of the temperate zone comprise fragments of natural and semi-natural habitats within an agricultural landscape, dominated by arable fields and improved pasture. An example of a 'modern' river margin, a cross-section through the River Trent floodplain, is given in Figure 12.1a. The floodplain section investigated has an area of about $0.1 km^2$ and experiences a major inundation once every 2 years on average. Floodplain topography is dominated by two former channels, both being abandoned more than 300 years ago. The number of species varies across the floodplain, in part reflecting the topographic setting. This is supported by the plant diversity and evenness indices which have highest values associated with topographic lows, and especially the margins of the present and former channels (Large *et al.*, 1994).

A hierarchy of habitat patches has been defined within the floodplain, varying in size from 0.1 ha to 10 ha. Patches were distinguished by both their floral and faunal (Coleoptera) composition. Vegetation patches were defined as those that showed little or no obvious variation in the relative abundance, physiognomy or spatial distribution of the most abundant species present (Large *et al.*, 1994).

Seventeen patches were identified and these were combined into five primary units including 126 species: 82 species in the woodland unit; 66 in the riparian unit; 52 in the seasonal wetland unit; 47 in the pasture unit; and 17 in the margins of arable fields where ruderals dominate. The definition of the units was confirmed by faunistic studies (Greenwood *et al.*, 1991) using

Scientific basis for conserving diversity

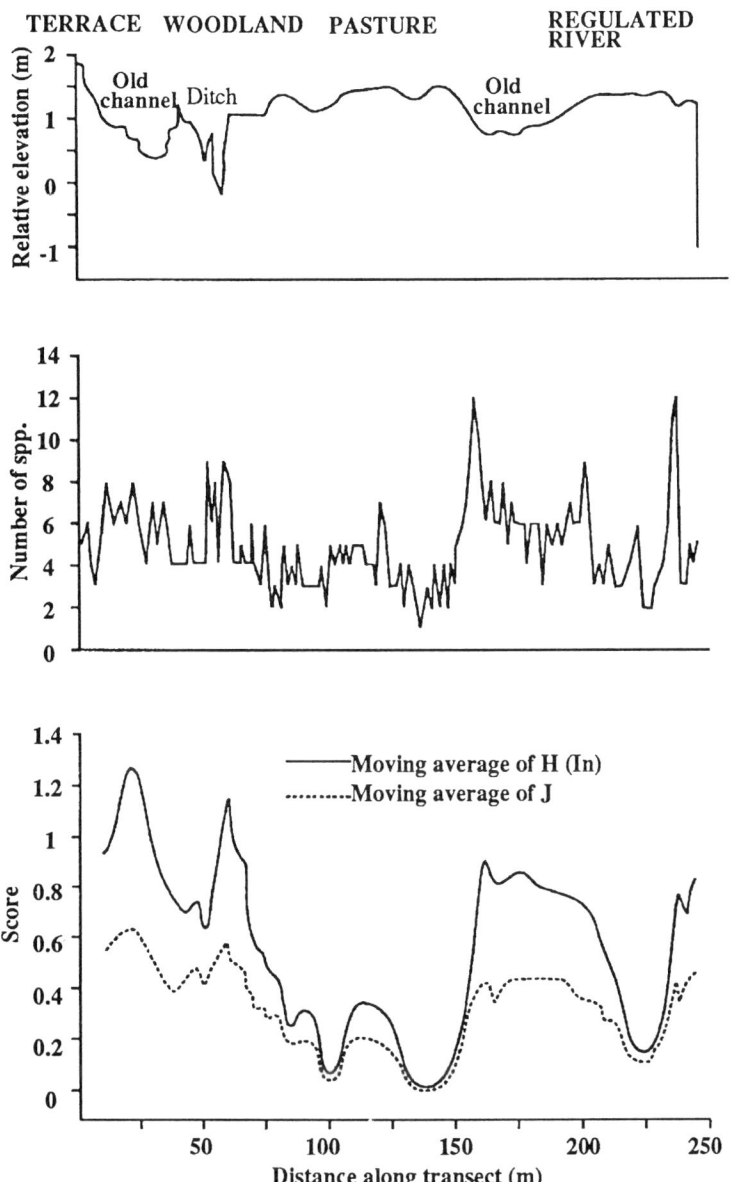

Figure 12.1a The floodplain flora and fauna in a regulated sector of the River Trent, UK at Gunthorpe. A transect across the floodplain showing (i) the variation of topography in relation to (ii) the number of vegetation species recorded in 1 m² quadrats and (iii) associated species diversity (H) and evenness (J) indices (Based on Large *et al.*, 1994)

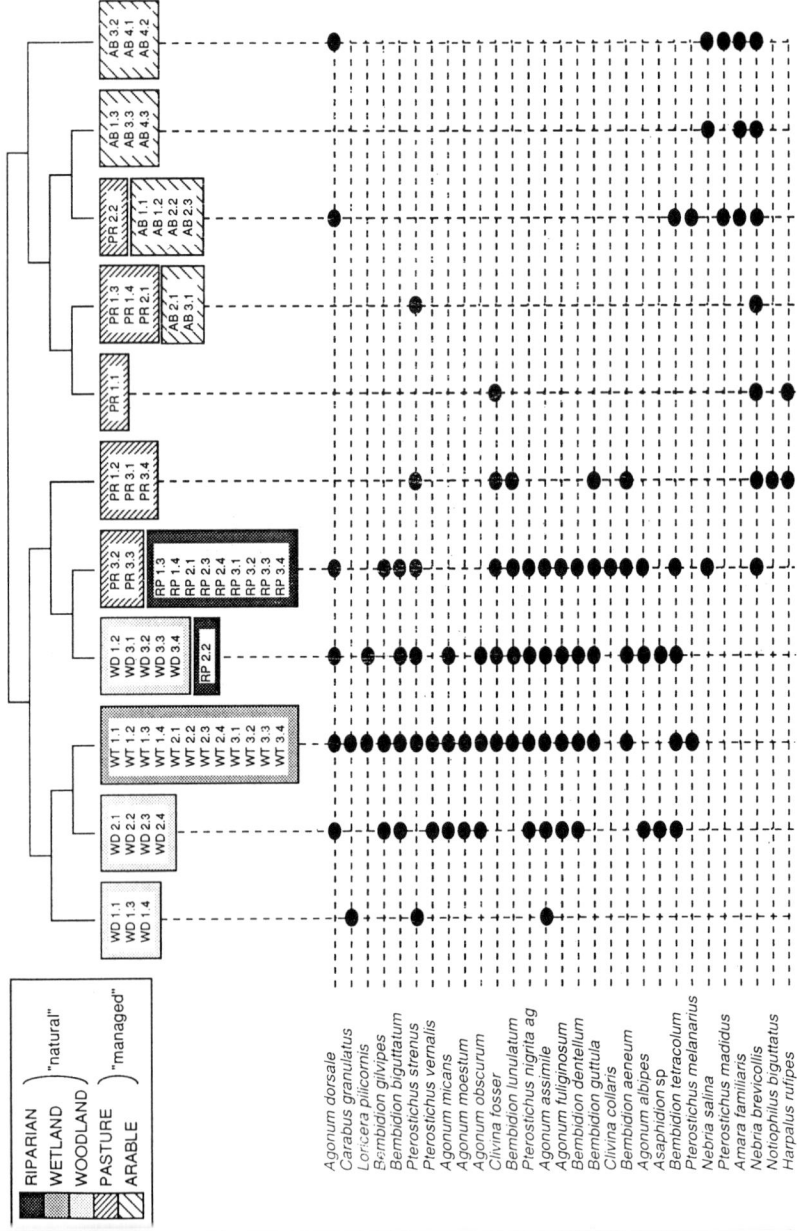

Figure 12.1b The floodplain fauna in a regulated sector of the River Trent, UK at Gunthorpe. Classification of patches from beetle (Coleoptera: Carabidae) assemblages using TWINSPAN (Hill, 1979). Units were sampled at three or four sites each comprising four pit-fall traps. Missing sites indicate that no carabids were collected. Only the most common species are shown, illustrating the main difference between sites within the dataset. Data and analysis provided by M.T. Greenwood and M.A. Bickerton

beetles (Coleoptera: Carbidae and Staphylinidae). 79 species were recorded from 60 pit-fall traps collected after 2 weeks in May, 1990. Classification of the traps (Figure 12.1b) clearly differentiated three main groups: pasture and arable; riparian sites, including riparian woodland, having relatively sandy soils; and floodplain wetland, including wet woodland, with organic, silty soils.

RATIONALE FOR MANAGING RIVER MARGINS

The rationale for managing river margins relates to their values for biological conservation. The World Conservation Strategy (IUCN, 1980) embodies the primary objectives:

- to maintain essential ecological processes and life-support systems;
- to preserve genetic diversity; and
- to ensure sustainable utilization of species and ecosystems.

The strategy focuses on the long-term benefits of biological conservation to humans, emphasizing that nature conservation does not imply the neglect of issues affecting human welfare.

Biodiversity is assumed to have an important role in the functioning of ecosystems. The most common assumptions identified by Solbrig (1991) are:

- diversity thresholds exist below which the system collapses;
- diversity permits biological systems at all levels of organization to cope with environmental stress and to recover from disturbance; and
- diversity maintains adaptive potentials.

The assumption that specific mixes of species are necessary for the proper functioning of ecosystems is fundamental to arguments for conservation. The management of river margins involves the consideration of three main options (Boon, 1992). First, for rivers that are essentially pristine, there is an overwhelming case for *preservation*; the challenge is to distinguish and then allow natural change, whilst protecting the river from artificial influences. In most cases, the pressures for land and water development, and the resulting problems of waste disposal, will require management to limit artificial changes within the catchment and to mitigate the impacts of human actions. Secondly, at the other extreme, are rivers or sectors that have become so severely degraded that in the short and medium terms the only management option is to accept *dereliction*. Acceptance of the need to define some rivers as derelict, at least in the short term, may be necessary to direct resources not only to high quality rivers and sectors deserving protection but also to those that have a fair chance of being improved by the third option, *restoration*.

THE CASE FOR RESTORATION

There is a strong case for restoring river margin ecosystems. First, river corridors are particularly important in biological conservation:

- they have high biological diversity;
- they have high biological productivity;
- they contain refuge habitats;
- they include refugia from the pre-industrial period; and
- they are sources for species dispersal.

River margins are the most species-rich components of landscape. Nilson *et al.* (1991) report 131 species of vascular plants per 200 m of river margin along the Vindel River in Sweden. The particular value of river corridors is illustrated by the study of Knopf *et al.* (1988) who show that although riparian habitat occupies less than 1% of the western North American landscape, it provides habitat for more species of bird than all other habitats combined. In arid central Arizona, the importance of the river margin has been shown to be not only the high bird density, but also its function as a distribution source, contributing 23–33% of the birds found in the adjacent desert washes and 7–15% of the birds in the adjacent desert upland (Szaro and Rinne, 1988).

Conservation status of river margin patches may be assessed using four indices (Petts *et al.*, 1992):

- *species diversity*, (e.g. Shannon–Wiener Index);
- *rarity*, based on the frequency of samples containing a species;
- *typicalness*, based on the analysis of clusters derived from a detrended correspondence analysis; and
- *specialism*, based on the relative frequency of species having a high frequency of occurrence within a specific unit to the frequency of the species in all units within the site.

With reference to the River Trent, these indices were applied using beetle (Coleoptera: Staphylinidae) assemblages (Table 12.1). Fifty-five species were recorded including four species that are new records for Nottinghamshire, including one Red Data Book species. The study highlighted the importance of the few wetland patches: having a high degree of specialism and including four nationally notable species, and a strongly typical fauna. Within the site, the woodland had a relatively large number of low frequency species, giving it the highest rarity index. Furthermore, the wetland unit was clearly differentiated from the other patches, not least the riparian unit, with which it

Table 12.1 Conservation assessment of major floodplain patches within the regulated, Gunthorpe, sector of the River Trent, UK (After Petts et al., 1992)

	Riparian	Wetland	Woodland	Pasture	Arable
Vegetation					
No. of species	66.0	52.0	82.0	47.0	17.0
Staphylinidae					
No. of species	31.0	21.0	26.0	10.0	6.0
Diversity	2.7	2.2	2.8	2.0	1.7
Rarity index	68.0	70.0	97.0	66.0	23.0
Typicalness	0.9	2.5	0.4	0.7	0.3
Specialism	3.1	5.0	2.7	0.3	0.0

Table 12.2 Rare Diptera associated with spring and wetland habitats within the headwater sectors of the Rivers Babingley and Wissey, Norfolk, UK

Species	Status
River Babingley	
Orthonevra brevicornis (Loew)	Red Data Book 3 – nationally rare
Cheilosia pubera (Zedderstedt)	Nationally notable: 15 sites known since 1960
Neoascia meticulosa (Spitoli)	Local
River Wissey	
Odontomyia argentata (Fabricius)	Red Data Book 2 – nationally vulnerable: 6 sites known since 1960
Orthonevra geniculata (Meigen)	Red Data Book 3 – nationally rare
Pelidnoptera nigripennis (Fabricius)	Red Data Book 3 – nationally rare

may have been expected to have similarities. Similar results were obtained by analysing the carabid fauna (see Figure 12.1b).

Sites of high conservation value along river margins are not only associated with rivers but also with headwater streams, especially around spring heads and at the foot of hillslopes along the valley floor where groundwater-fed 'wet' areas remain adjacent to the stream. Typically such sites have areas of much less than one hectare but contain important rarities. Examples of rare Diptera associated with such sites in north Norfolk, UK, are given in Table 12.2. The results of all studies indicate that river corridors for conservation must extend beyond the riparian zone, especially to include wetland units such as springs and seepage areas, and old cutoff channels in different successional stages.

A second dimension to the rationale for restoration is that river corridors have a range of socio-economic values related to their commercial and recreational potential (e.g. fishing, hunting, boating) and their high visual quality (Petts, 1990b). However, wild, biologically-rich reaches are considered unhealthy and hazardous, whereas parkland-style landscapes are often perceived to be of high value. Green and Tunstall (1992) showed that river corridors can be particularly attractive for casual recreation, receiving more local visits and drawing visitors from a larger area, than does the average small park. The public are attracted to river corridors that are (i) unpolluted; (ii) quiet, rich in flora and fauna, and form attractive landscapes; and (iii) have basic facilities such as toilets and paths, reflecting important public concerns for the safety of children and general public health.

In the UK, rivers have particularly high value as game and coarse fisheries. For example, declared catches of Sea Trout in England and Wales between 1983 and 1986 had an estimated minimum value of £55 million (Elliot, 1989). Public perception of a healthy environment is important. For example, the return of salmon (*Salmo salar*) to the River Taff in Wales is considered to have considerable importance, regardless of any fishery value, because salmon are symbols of clean water (Mawle, 1991). Following a long history of severe degradation caused by gross pollution, a breeding salmon population in the Taff would be a clear demonstration of improved quality. The restoration of biodiversity along river margins would be a dramatic statement of new initiatives in nature conservation and environmental enhancement for recreation and amenity.

THE SCIENTIFIC BASIS OF RESTORING RIVER MARGIN ECOSYSTEMS

The functional ecology of river margin ecosystems is the subject of much current research (e.g. Naiman *et al.*, 1989). As recognized by Solbrig (1991, p.66) a major gap in our knowledge of biodiversity concerns the effect of habitat fragmentation on species ranges and the probability of extinction and speciation. The impact of rapid fragmentation of river margin habitats may mean that present-day fragments of natural patches are rather young in terms of the relaxation time over which they respond to fragmentation (Kent, 1987). During a relaxation period, the species composition of fragmented patches will be transient states; patches holding too many species or too few species for their size. This has parallels with work on the responses of aquatic biota in fluvial systems (Petts, 1987). However, island biogeography theory (MacArthur and Wilson, 1963, 1967) remains to be properly validated (Kent, 1987) not least within river corridors where flooding creates a high degree of directional (downstream) connectivity between spatially distant patches. The effect of geographical isolation on populations and species, and the susceptibility of fragmented habitats to invasion, are important questions of particular relevance to the restoration of river margin ecosystems.

An integrated approach

Progress in the development of strategic studies with the objective of advancing the restoration of river margins requires a structured approach that gives due recognition to the ecotonal character of these systems and to the appropriate time-scales for research. River restoration requires models to predict the ecological impacts of human activities – including restoration measures! Such models must integrate knowledge from three areas of study and from three levels of investigation. Hydrological, geomorphological and ecological information must be fully integrated to develop applicable models of ecosystem, habitat, community and species responses. Information is required from three levels of scientific analysis.

At the first level of analysis, functional studies are concerned with understanding short-term process dynamics (e.g. carbon spiraling, trophic interactions, chemical exchanges between water and sediment, etc.) and with describing spatial relationships between individuals, species, communities and environmental variables. Such studies are important for determining the 'technological' basis for restoration, that is providing the scientific knowledge necessary to define management processes, such as:

- *Selection* of species, communities and/or habitat characteristics as targets for management;

- *Introduction* (e.g. fish stocking, re-introductions (e.g. the otter, *Lutra lutra*), introducing gravel to create bars, etc.);

- *Elimination* (e.g. culling to control population numbers, controls on invasive plants, etc.);

- *Control of key fluxes* (e.g. water levels, nutrient supply, primary production, siltation, etc.); and

- *Controlled disturbance* (e.g. river bank or woodland clearance to rejuvenate succession, artificial 'floods' to scour channels or inundate riparian wetlands).

The second and third levels of analysis involve historical and paleoenvironmental studies which seek to understand the ways that species, communities and habitats respond to human impacts and climate change, respectively; and to establish the former 'natural' characteristics of fluvial hydrosystems. These studies aim to define the temporal context within which restoration programmes must be set.

The most productive approach to reconstructing the sequence of changes experienced by fluvial hydrosystems over historic and palaeoenvironmental timescales is inductive (Thornes, 1987). Deductive methods are unavailable because of our inadequate knowledge of the complex physical, chemical and biological dynamics that determine river margin ecosystems. With the inductive approach (Figure 12.2) field observations, including evidence of former biological populations (Level 1) and fluvial sediments and landforms (Level 2), are used to make inferences about process dynamics (Level 3).

Biodiversity in land–inland water ecotones

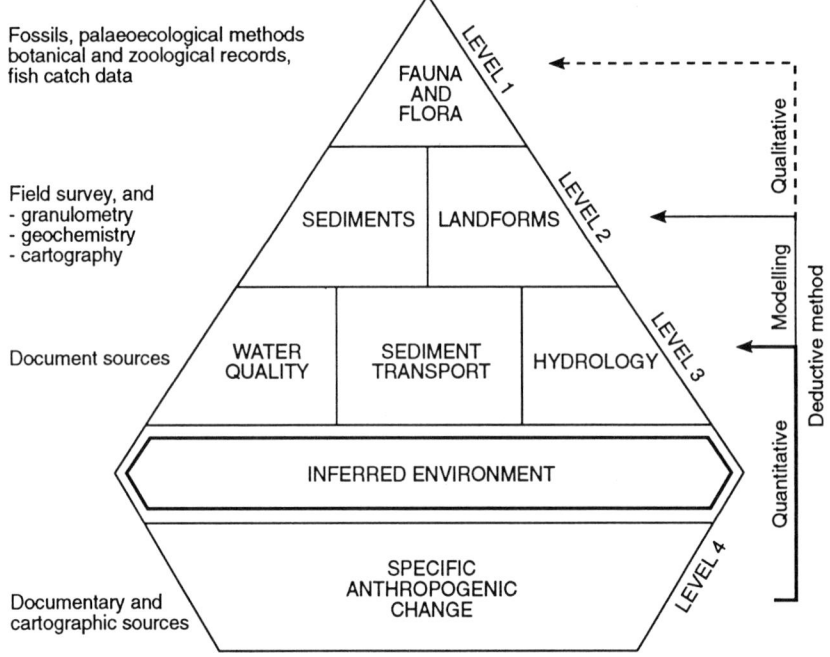

Figure 12.2 The four level inferential approach to studies of fluvial hydrosystems (From Petts, 1989)

Historical studies benefit in most cases from more or less detailed documentary evidence concerning the precise nature and timing of anthropogenic changes (Level 4). Application of the inductive method to analyse historical changes along large alluvial rivers is developed by Petts *et al.* (1989), and exemplified for the Rhône by Amoros *et al.* (1987) and for the Trent by Petts *et al.* (1992).

The fluvial hydrosystem perspective

A fluvial hydrosystem perspective (Amoros and Petts, 1993) provides a useful approach for analysing river margins by classifying rivers as a sequence of sectors (Figure 12.3). At the scale of the drainage basin, a river may be viewed as a longitudinal continuum (Vannote *et al.*, 1980) within which three primary river types can be defined (Schumm, 1977): headwater streams – the main source areas for water and sediment, middle order reaches characterized by longitudinal transfers, and lowland rivers dominated by storage and short spiral processes. Each 'type' may comprise one or a number of sectors of variable length from less than 1 km to 50 km or more. At this level of analysis variables such as altitude, distance from source and the slope of the valley floor are important. At the other extreme, pioneer patches, such as the gravel

Scientific basis for conserving diversity

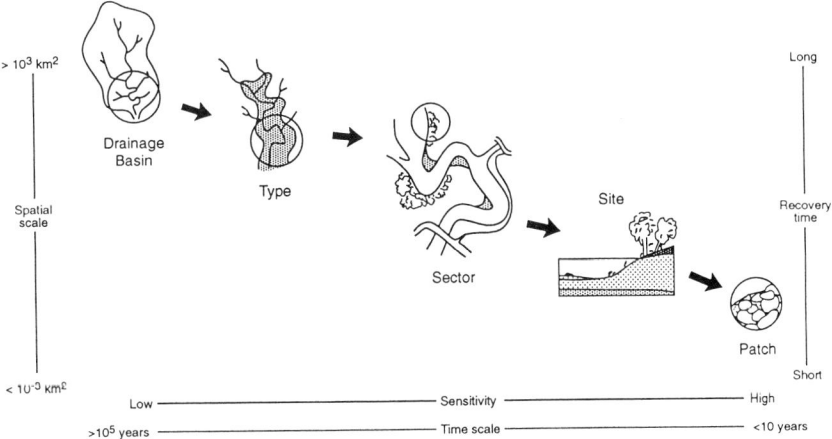

Figure 12.3 Hierarchical classification of spatial and temporal scales for analysing fluvial systems (Adapted from Frissell *et al.*, 1986)

bar, sand berm, and new cutoff channel, are highly sensitive to environmental variations, especially to changes of the magnitude and frequency of floods and of the sediment load regime (both amount and size distribution). However, they also recover relatively quickly. Each sector comprises a distinctive set of patches. Variations between individual sites within a sector, each comprising a particular subset of patches, relate to local conditions (sediment permeability, topography, groundwater flow patterns, etc.) or short-term changes, such as may occur following a major flood event.

A sector is defined by four sets of criteria, each including artificial influences:

1. *Inputs* derived from the catchment and routed through the channel network upstream. The fluvial hydrosystem is seen as a component of the drainage basin within which the primary controls are climate, geology, relief and vegetation (land use);

2. *Internal structural controls* within the sector including slope and degree of lateral confinement (by valley sides, river terraces, woodland and channel engineering structures), etc.;

3. *Internal process dynamics* within the sector such as influent or effluent flow conditions, nutrient spiraling dimensions, and morphological dynamics (aggradation and incision), etc.; and

4. *Downstream structural controls* which effectively control base level for the adjacent sector upstream, including bedrock controls, channel behaviour (aggradation and incision), and artificial controls such as weirs and bridges.

Biodiversity in land–inland water ecotones

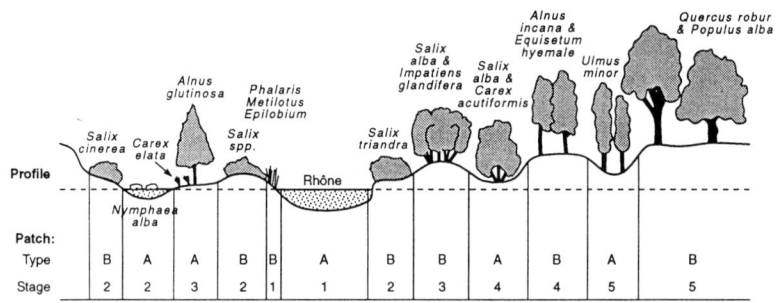

Figure 12.4 Generalized profile across the Rhône valley, France, showing the units typical of the river margin system within one sector (Based on Pautou and Décamps, 1985). A, B = Patch type (A = channel/channel fill, B = gravel/sand bar); 1–5 = successional stage

Each sector is characterized by a typical range of habitats. These are the fundamental units and their different successional stages – the biotope and biocoenosis of each elementary river landform. They include the range of sediment bar types (which may be formed of cobbles, gravels, sand or organic silt), sand levees, back swamps and cutoff channels. A cross-section through the River Rhône alluvial corridor in France (Figure 12.4) illustrates this complexity with different types of unit (cutoff channels and sedimentary deposits) in different stages of succession.

Principles for ecologically-sensitive river management

The applicability of the fluvial hydrosystem perspective in river management is justified by three important principles (Amoros and Petts, 1993). The first principle is that each sector must be viewed in the context of its catchment. Floodplain rivers may be considered as isolated from adjacent hillslopes as their hydrology, sediment supplies and water quality are dominated by inputs from the catchment upstream.

- Water quality, especially the influence of point- and diffuse-source pollution, is the major constraint for sector-scale, or site-scale, environmental enhancement;

- The allocation of water is required to maintain functioning ecosystems within the river corridor, incorporating minimum flows, an appropriate flow regime based on average monthly flows, and high flow 'events' at the appropriate time(s) of year; and

- Changes of sediment yield from headwater catchments, reflecting land use practices, and the interruption of sediment transfer along rivers by dams and weirs (many of which in Europe are several centuries old) place major constraints on the restoration of natural, dynamic systems.

The second principle is that rivers must be viewed as systems in equilibrium. Within each sector a quasi-equilibrium condition may be defined involving hydrological, geomorphological and ecological interactions. Each sector can be described by a more or less complex arrangement of aquatic, semi-aquatic and terrestrial patches. The patches will be of different type (defined primarily by morphological, sedimentological, and vegetational criteria) and different age (reflected by successional criteria for each type). The arrangement of patches within a sector changes over time in response to successional processes and disturbance (by erosion and deposition). Thus, the optimal areas for particular fauna shift to different parts of a sector in response to the build-up of sediments (e.g. Bayley, 1991). Over a timescale of 10–100 years the spatial arrangement of patches may change, but the composition of patches within each sector will remain relatively stable, about an average condition.

The third principle is that lateral exchanges play an important role in sustaining the functioning of river sectors. The role of the ecotone concept in river management is to focus attention on the transitional zone, or dynamic boundary, at the land–water interface (Petts, 1990b). Traditionally, management has focused on patches; conservation has been concerned primarily with the preservation of species within patches by defining more or less artificial boundaries around them. River margin ecosystems, however, are determined by hydrological variations and geomorphological disturbance. The ecotonal characteristics of river margins are elaborated below.

The river margin as an ecotone

It is well established that 'edges' between patches within the landscape have high habitat value, providing combinations of food resources and cover or pathways for animal movement between habitat types. Typically, edges contrast with patch interiors and support biologically diverse communities. Whilst an edge is a purely structural entity, an ecotone is defined by dynamic processes and in addition to its edge characteristics functions by regulating the flux of materials between patches. The role of the ecotone concept in managing river margins, especially for conserving and restoring biodiversity (Petts, 1990b), is to emphasize the temporal context:

- that the biological values of the ecotones are sustained by dynamic fluvial processes;
- that changes of the adjacent fluvial and terrestrial systems will affect the ecotone;
- that changes within the ecotone will affect the adjacent fluvial and terrestrial systems; and
- that protection and management of ecotonal resources require an holistic approach based upon multi-disciplinary and long-term perspectives.

The river margin is defined here as the land–water ecotone: the area of land adjacent to a river channel within which habitat patches, their composition and arrangement, are determined by fluvial processes (flooding, erosion, deposition); where biotic communities relate to the dynamic interaction of fluvial and terrestrial processes, and comprise a continuum of dependencies from terrestrial species requiring regular inundation to aquatic species requiring occasional desiccation, as well as purely aquatic and purely terrestrial taxa. The complex interactions may be simplified into two general groups: resource and disturbance.

Three examples illustrate the resource value of annual flooding. First, many species of river margin plant, such as the red gum (*Eucalyptus camaldulensis*) forests along the River Murray, Australia (Bren, 1988) and the cypress (*Taxodium distichum*) wetlands of southeastern USA (Mitsch *et al.*, 1979) are dependent upon fluvial flooding. Secondly, many species of river fish rely on the annual inundation of the river margin for reproduction and feeding. In the tropics, fish diversity and productivity are directly linked to the scale of the annual (predictable) inundation of the floodplain forest ecotone (Bailey, 1986). Despite the lack of autochthonous primary production, the Amazon river, for example, carries a high fish population that is dependent upon allochthonous forest foods as the main diet (Lowe-McConnell, 1986). Thirdly, flooding adds to the detrital and mineral nutrient pools of both the terrestrial and aquatic systems: during rising stages silts and nutrients replenish the floodplain pool whilst the drowning of mineral and organic matter added to by migratory animals and birds releases nutrients into the water; during receding flows nutrients may be returned to the river, vegetable and animal material will be stranded and these will be consumed by terrestrial animals and birds, or decompose. This 'cycle' forms the primary component of the 'flood pulse' concept (Junk *et al.*, 1989).

Disturbance changes the area occupied by a community, causes species to disappear, and creates new habitats for pioneer species, so that an equilibrium value for species diversity is an imprecise concept. Disturbance is a complex variable which can be defined in many ways such as by type, magnitude, frequency, duration, predictability, or by effectiveness in relation to system recovery processes. Biodiversity reflects the interaction of disturbances with the demographic characteristics of the system (e.g. age structure, maximum reproductive rate, maximum life span) and most communities will be influenced in some way by the historical legacy of environmental disturbances that a site has experienced over time-scales ranging from tens to thousands of years. The role of disturbance can be demonstrated by reference to three further examples. First, disturbance by erosion plays an important role in rejuvenating the channel edge and deposition provides new substrate for pioneer communities (Roux *et al.*, 1989). Secondly, major disturbance involving laterally-mobile channels and channel avulsion are important for reorganizing and rejuvenating floodplain patches (Amoros *et al.*, 1987). Thirdly, the annual inundation sustains the biological productivity and diversity of the river margin by controlling grazing by terrestrial herbivores and preventing scrub development (Davies, 1986).

The ecotonal view leads to a classification of river margins according to their characteristic process dynamics. This is illustrated in Figure 12.5. Each floodplain or floodplain patch may be associated with a typical range of water depths (Figure 12.5.i) which may be described as a depth or discharge frequency curve (Figure 12.5.ii). However, relationships are required to describe the resource value and disturbance impact of each hydrological event (Figure 12.5.iii).

The resource value of a flood for the river margin ecosystem relates to provision of moisture, nutrients, organic matter and silt. River flows below bankfull are important for maintaining groundwater levels within the floodplain. Little information is available to quantify resource value of floods for floodplains but the resource value of floodplains for river fish populations is well established and this may be used to illustrate the approach. Welcomme (1979) demonstrated highly significant linear relationships between fish catch and the flood intensity in the two previous years – the fishery in most rivers studied was based mainly on fish that are one- or two-years-old. Moreover, Welcomme suggests that the argument that larger systems favour higher diversity may be demonstrated by fish: individual fisheries of the largest floodplains having the most species.

River sectors dominated by disturbance are classically braided 'bedload' rivers within which channel migration and avulsion occur in response to frequent events. The disturbance impact also begins below the bankfull discharge because sediment transport may commence at about 0.6 bankfull. The potential magnitude of floodplain disturbance tends to increase with discharge although the effectiveness of an event will relate to the successional stage of riparian and floodplain vegetation and the developmental state of the landform comprising each floodplain patch (Figure 12.5.v).

The significance of different processes over time may be described by combining the data on flow frequency, event resource value, and event disturbance impact (Figure 12.5.iv). Low magnitude hydrological events occur frequently but have only low magnitude effects whereas extremely high magnitude events have major effects but occur so rarely that in the long term they are insignificant. Thus, the 'average' river margin is dominated by flood events of moderate magnitude and frequency (q3). Along morphologically dynamic rivers, disturbance-dominated river margins may be expected to be related to very frequent events (q1). In contrast, river margins along stable channels may be resource-dominated and adjusted to relatively infrequent events (q2). Research to investigate such relationships is required to determine the flow regime necessary to sustain river margin ecosystems along regulated rivers. Determination of resource and disturbance domains

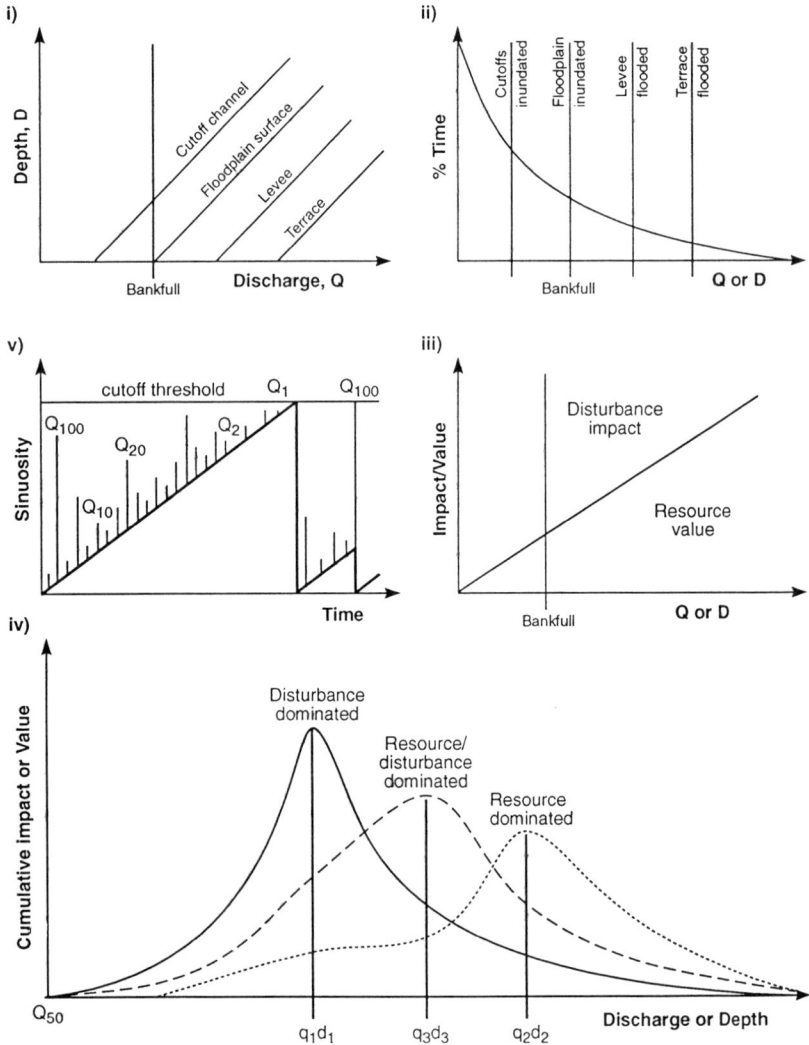

Figure 12.5 The concept of the dominant hydrological event for river margin ecotones: (i) the water depth–discharge relationship can be expressed (ii) as a frequency curve, describing the percentage of time that a particular discharge or depth occurs; (iii) each discharge or depth may be expressed in terms of its resource value and disturbance impact; (iv) combining the information in (ii) and (iii) enables the derivation of curves describing the long-term average influence of discharge events of different magnitude; and (v) illustrates the variable effectiveness of events of particular magnitude according to the nature of intrinsic thresholds and the state of the system (based on Schumm, 1977) – in this example, local channel sinuosity is seen to increase until meander cutoff occurs but a range of events may be responsible for the final cutoff and the timing of cutoff is unpredictable; here the annual flood (Q_1) and the 100-year event (Q_{100}) are seen to cause cutoff, and two 100-year floods are seen to have different impacts depending upon the state of the channel in relation to the cutoff threshold

associated with flooding may be used to define the spatial extent of the river margin ecotone. Beyond the ecotone, land may be inundated by an extreme river flood, but the flood would have no permanent effect on the ecosystem.

MANAGEMENT PERSPECTIVE

Any assessment of the options for protection and rehabilitation must be scientifically based and practicable for management. Over the past three decades scientific research on rivers has elucidated many of the complex relationships that link hydrology, geomorphology and ecology, as well as defining biotic interactions. The challenge has been to present the scientific information in such a way as to be appropriate for management. In this context, following the recommendations of the 12th Report of the Royal Commission on Environmental Pollution (1988), 'reasonably practicable' means having regard, among other things, for local conditions and circumstances, the financial implications, and the current state of technical knowledge.

A framework is required to direct scientific information to restoration objectives, recognizing the multifunctional nature of catchment management, involving: water resources, water quality, flood control and land drainage, fisheries, recreation, conservation and navigation. The approach outlined herein defines appropriate space and time scales which are both practicable for management and scientifically defensible.

Kvet *et al.* (1989) concluded that the overall aim of management should be to maintain the highest possible or attainable diversity of species, or their habitats, and of ecotone functions. A research programme on the ecosystem function of biological diversity currently being planned by the International Union of Biological Sciences (IUBS), the Scientific Committee on Problems of the Environment (SCOPE), and UNESCO (Solbrig, 1991) will provide the scientific base for restoring damaged ecosystems in the next century.

Much more scientific information is needed on the impacts of fragmentation within river margin systems on biodiversity. A particular need is to develop research on the relaxation times over which the biotic community response within relict and restored habitat fragments achieves a new equilibrium in different environmental settings. Nevertheless, immediate actions are required to prevent the continuing loss of biodiversity. The restoration of river margins can make a major contribution to this objective, recognizing that our scientific base is incomplete and the long-term success of restoration works will be uncertain. Current scientific knowledge is sufficient to allow progress in restoring river margins to protect and enhance biodiversity, providing management incorporates the monitoring and degree of flexibility necessary to allow response to both successful and unsuccessful actions. An immediate priority is to identify refugia from the pre-industrial era; to protect them, to establish their functional characteristics, and then to assess their potential as models for restoration.

ACKNOWLEDGEMENTS

I am grateful to Malcolm Greenwood, Melanie Bickerton and Andy Large for providing information on the River Trent and especially data for Figure 12.1.

REFERENCES

Amoros, C. and Petts, G.E. (eds.) (1993). *Hydrosytèmes Fluviaux*. Masson, Paris

Amoros, C., Roux, A.L., Reygrobellet, J.L., Bravard, J.P. and Pautou, G. (1987). A method for applied ecological studies of fluvial hydrosystems. *Regulated Rivers*, **1**, 17–36

Bailey, R.G. (1986). The Zaire River. In Davies, B.R and Walker, K.F. (eds.) *The Ecology of River Systems*, pp. 201–14. Dr W. Junk Publishers, Dordrecht

Bayley, P.B. (1991). The flood-pulse advantage and the restoration of river–floodplain systems. *Regulated Rivers* **6**, 75–86

Boon, P.J. (1992). Essential elements in the case for river conservation. In Boon, P.J., Calow, P. and Petts, G.E. (eds.) *River Conservation and Management*, pp. 11–34. John Wiley, Chichester

Bravard, J.P., Amoros, C. and Pautou, G. (1986). Impacts of civil engineering works on the succession of communities in a fluvial system: a methodological and predictive approach applied to a section of the Upper Rhône River. *Oikos*, **47**, 92–111

Bren, L.J. (1988). Effects of river regulation on flooding of a riparian Red Gum forest on the River Murray, Australia. *Regulated Rivers*, **2**(2), 65–78

Davies, B.R. (1986). The Zambezi River system. In Davies, B.R and Walker, K.F. (eds.) *The Ecology of River Systems*, pp. 225–68. Dr W. Junk, Publishers, Dordrecht

Diamond, J.M. (1988). Factors controlling species diversity: overview and synthesis. *Annals of the Missouri Botanical Garden*, **75**, 117–29

Ellett, C. 1853. *The Mississippi and Ohio Rivers*. Grambo and Co., Lippincott, Philadelphia

Elliot, J.M. (1989). Wild brown trout *Salmo trutta*: an important national and international resource. *Freshwater Biology*, **21**, 1–5

Frissell, C.A., Wiss, W.J., Warren, C.E. and Huxley, M.D. (1986). A hierarchical framework for stream classification: viewing streams in a watershed context. *Environmental Management*, **10**, 199–214

Green, C.H. and Tunstall, S.M. (1992). The amenity and environmental value of river corridors in Britain. In Boon, P.J., Calow, P. and Petts, G.E. (eds.) *River Conservation and Management*, pp. 423–41. John Wiley, Chichester

Greenwood, M.T., Bickerton, M.A., Castella, E., Large, A.R.G. and Petts, G.E. (1991). The use of Coleoptera (Arthropoda: Insecta) for floodplain characterization on the River Trent, UK. *Regulated Rivers*, **6**, 321–32

Hill, M.O. (1979). TWINSPAN – A FORTRAN program for arranging multivariate data in an ordered two-way table by classification of the individuals and attributes. *Ecology and Systematics*, Cornell University, Ithaca, New York

IUCN. (1980). *World Conservation Strategy*, International Union for the Conservation of Nature and Natural Resources, Gland

Junk, W.J., Bayley, P.B. and Sparks, R.E. (1989). The flood pulse concept in river–floodplain systems. Special publication of the *Canadian Journal of Fisheries and Aquatic Sciences*, **106**, 110–27

Kent, M. (1987). Island biogeography and habitat conservation. *Progress in Physical Geography*, **11**(1), 91–102

Knopf, F.L., Johnson, R.R., Rich, T., Samson, F.B. and Szaro, R.C. (1988). Conservation of riparian ecosystems in the United States. *The Wilson Bulletin*, **100**, 272–84

Kvet, J., Arunin, A., Bugenyi, B., Davis, C., Desaigues, B., Fournier, F., Holland, M., Husain, R.G.K., Jensen, A., Joly, C.A., Juge, R., Moller, H., Nachtnebel, H.P., Nontji, A., Okada, M., Petts, G., Schneider, E., Sedell, J. and Simpson, P. (1989). Management, restoration and creation of ecotones. In Naiman, R.J., Décamps, H. and Fournier, F. (eds.) Role of Land–Inland Water Ecotones in Landscape Management and Restoration: Proposals for Collaborative Research. *MAB Digest 4*, pp. 73–5. UNESCO, Paris

Large, A.R.G. and Petts, G.E. Reconstruction of the fluvial geomorphology of the pre-regulation River Trent in the 17th and 18th centuries. *Applied Geography*, in press

Large, A.R.G., Prach, K. and Wade, M. (1994). Vegetation structure of a semi-regulated floodplain: River Trent, UK. *Regulated Rivers*, **9**, 11–18

Lowe-McConnell, R.H. (1986). Fish of the Amazon system. In Davies, B.R and Walker, K.F. (eds.) *The Ecology of River Systems*, pp. 339–52. Dr W. Junk, Dordrecht

MacArthur, R.H. and Wilson, E.O. (1963). An equilibrium theory for insular zoogeography. *Evolution*, **17**, 372–87

MacArthur, R.H. and Wilson, E.O. (1967). *The Theory of Island Biogeography*. Princeton, NJ

Mawle, G. (1991). Restoration of the River Taff, Wales. In Mills, D. (ed.) *Strategies for the Rehabilitation of Salmon Rivers*, pp. 109–22. The Atlantic Salmon Trust, Institute of Fisheries Management and the Linnean Society of London, London

Mitsch, W.J., Dorge, C.L. and Wiemhoff, J.R. (1979). Ecosystem dynamics and a phosphorus budget of an alluvial cypress swamp in southern Illinois. *Ecology*, **60**, 1116–24

Naiman, R.J., Décamps, H. and Fournier, F. (eds.) (1989). Role of Land–Inland Water Ecotones in Landscape Management and Restoration: Proposals For Collaborative Research. *MAB Digest 4*, UNESCO, Paris

Nilson, C., Dynesius, M., Andersson, C. and Sylven, M. (1991). *Why Care About Wild Rivers?* Report by WWF Sweden and University of Umeå

Pautou, G. and Décamps, H. (1985). Ecological interactions between alluvial forests and hydrology of the Upper Rhône. *Archive für Hydrobiologie*, **104**, 13–37

Petts, G.E. (1984). *Impounded Rivers*. John Wiley, Chichester

Petts, G.E. (1987). Time-scales for ecological change in regulated rivers. In Craig, J. and Kemper, J.B. (eds.) *Advances in Regulated River Ecology*, pp. 257–66. Plenum, New York

Petts, G.E. (1989). Historical analysis of fluvial hydrosystems. In Petts, G.E., Moller, H. and Roux, A.L. (eds.) *Historical Change of Large Alluvial Rivers: Western Europe*, pp. 1–18. John Wiley, Chichester

Petts, G.E. (1990a). Forested river corridors: a lost resource. In Cosgrove, D.E. and Petts, G.E. (eds.) *Water, Engineering and Landscape*, pp. 12–34. Belhaven, London

Petts, G.E. (1990b). The role of ecotones in aquatic landscape management. In Naiman, R.J. and Décamps, H. (eds.) *The Ecology and Management of Aquatic–Terrestrial Ecotones*, pp. 227–60. MAB Book Series 4. UNESCO/Parthenon, Paris/Carnforth

Petts, G.E., Large, A.R.G., Greenwood, M.T. and Bickerton, M.A. (1992). Floodplain assessment for restoration and conservation: linking hydrogeomorphology and ecology. In Carling, P.A. and Petts, G.E. (eds.) *Lowland Floodplain Rivers*, pp. 217–34. John Wiley, Chichester

Petts, G.E., Moller, H. and Roux, A.L. (eds.) (1989). *Historical Change of Large Alluvial Rivers: Western Europe*. John Wiley, Chichester

Roux, A.L., Bravard, J.P., Amoros, C. and Pautou, G. (1989). Ecological changes of the French Upper Rhône River since 1750. In Petts, G.E., Moller, H. and Roux, A.L. (eds.) *Historical Change of Large Alluvial Rivers: Western Europe*, pp. 167–82. John Wiley, Chichester

Royal Commission. (1988). Best practicable environmental option. *Royal Commission on Environmental Pollution*, 12th Report, Her Majesty's Stationary Office, London

Schumm, S.A. (1977). *The Fluvial System*. John Wiley, New York

Solbrig, O. (1991). Biodiversity: Scientific Issues and Collaborative Research Proposals. *MAB Digest 9*, UNESCO, Paris

Szaro, R.C. and Rinne, J.N. (1988). *Ecosystem approach to management of southwestern riparian communities*. Transactions of the 53rd North American Wildlife and Natural Resources Conference, pp. 502–11

Thornes, J.B. (1987). Models for palaeohydrology in practice. In Gregory, K.J., Lewin, J. and Thornes, J.B. (eds.) *Palaeohydrology in Practice*, pp. 17–36. John Wiley, Chichester

Townsend, C.R. and Calow, P. (1981). *Physiological Ecology. An Evolutionary Approach to Resource Use*. Blackwell Scientific Publications, Oxford

Vannote, R.L., Minshall, G.W., Cummins, K.W., Sedell, J.R. and Cushing, C.E. (1980). The river continuum concept. *Canadian Journal of Fisheries and Aquatic Sciences*, **37**, 130–7

Welcomme, R.L. (1979). *Fisheries Ecology of Floodplain Rivers*. Longman, London

Wiltshire, P.E.J. and Moore, P.D. (1983). Palaeovegetation and palaeohydrology in upland Britain. In Gregory, K.J. (ed.) *Background to Palaeohydrology*, pp. 433–51. John Wiley, Chichester

CHAPTER 13

BIODIVERSITY: A REVIEW OF THE SCIENTIFIC ISSUES

Stephen C. Stearns

INTRODUCTION

Before discussing the scientific problems we confront in the biodiversity crisis, let me place the science in perspective. The current mass extinction is caused by the economic and reproductive behaviour of man. The problem does not lie in nature and cannot be solved by changing nature. It lies in man and can only be solved by changing man. The solution to the problem is widely recognized: fewer people who consume less.

My own personal goal is a world population of 100 million humans with a high enough standard of living to live with each other in peace. That estimate is based on a subjective estimate of number of luxury cars, jet aircraft, freeways, and the amount of intensive agriculture that will permit evolution to continue normally. That human population could be attained in about 2,500 years if each couple had two children and if there were a lot of political and economic change. By that time, most of the species on the planet will probably have disappeared, but one could then allow the survivors to continue to evolve normally.

Based on past adaptive radiations after mass extinctions, it would probably then take about 25–30 million years for evolution to repopulate the planet to pre-human levels. If that happened, those of our descendants who might be around to enjoy the situation would almost surely no longer be human.

Now let us consider what might be done during the next generation or two. We should distinguish two aspects of the biodiversity crisis. The political aspect concerns the human causes and what we can do about them. The scientific aspect concerns the factors involved in the origin and maintenance of biodiversity and in extinction at all levels of the biological hierarchy from genes through species to ecosystems (Stearns *et al.*, 1990). The kinds of research that need to be done differ depending on one's interest in the short-term political aspects – environmental activism – or the long-term scientific aspects – the truth of the matter. I discuss these in turn.

WHAT POLITICIANS NEED FROM BIODIVERSITY RESEARCH

Politicians need answers to the following questions:

How many species are there on the planet? The current best guesses range from 5 million to 100 million. This is very imprecise and politically unimpressive. One of the reasons that advocates of global warming were more successful at the United Nations Conference on the Environment in Rio de Janeiro than the advocates of biodiversity was that climate researchers can lay hard numbers on the table that document rising CO_2 levels. We do not know to within one or two orders of magnitude how many species there are on the planet. This is a scandal.

Most of the species live in humid lowland tropical forests, an area of the planet that has been poorly sampled. Estimates of the numbers of species present in tropical forests vary by an order of magnitude, and most of the variation is caused by different estimates of the number of beetles and mites that live in the forest canopy. Based on canopy collections in Panama, Erwin and Scott (1980) estimated that the number of insect species on the planet was 30 million. Later collections made in Brazil and Peru (Erwin, 1988) led him to revise this estimate upwards to 30–50 million. André et al. (1992), who include the mites, suggest that the number of species on the planet could be as high as 100 million. If this is true, then most of the species on the planet are beetles and mites living in the canopy of tropical rain forests. The beetles appear to be very locally distributed. For example, the overlap in species found in the dry season in 4 forest types more than 70 km apart at Manaus, Brazil, was only 1% of 1080 species in 61 families. At Tambopata, Peru, two plots only 50 m apart had 126 species in 7 families with an overlap of only 9%. On a single tree at Tambopata Erwin found as many species of ants as there are in the British Isles. His results motivated others to sample the insects of the tropical rain forest canopy to see if his estimates could be confirmed.

The solution to this problem is more, better systematics and taxonomy, fields that have been neglected over the last 50 years in favour of more glamorous areas like molecular biology. Systematists need to be trained at the universities, for which in many cases positions for professors will have to be created, and natural history museums will need support for curation, collection, and classification.

How fast are species going extinct? No one really knows. Most of the estimates are based on spot samples, and the entire estimation procedure is fraught with difficulty, for it is very hard to say when a species has really gone extinct.

Some recent extinctions caused by introductions of predators onto islands and into lakes have been well studied. Of the birds endemic to Hawaii before the arrival of the Polynesians, at least 68% are now extinct, about half of which disappeared before the arrival of Captain Cook (Freed et al., 1987).

The causes were hunting, habitat destruction, and introduced enemies, including the rat, the mongoose, and avian malaria. Many flightless birds were simply knocked on the head and eaten for dinner. Cowie (1992) documents extinctions of many tens of species of endemic land snails on Pacific islands following the introduction of a predatory snail for biological control. Witte *et al.* (1992) discuss the extinction of approximately 200 of the 300 endemic cichlid fish of Lake Victoria, caused in part by the introduction of a predatory fish, the Nile perch, on which a new fishery is based. It seems likely that fishery will be short-lived and will collapse. These two introductions make clear what catastrophes can result from well-meaning people with no notion of ecological consequences.

On the assumption that most of the extinctions of mammals and birds since the last ice age have been caused by human colonization of continents and islands where the fauna had no previous experience of humans, Diamond (1989) estimates that 50% of the mammal species and 20% of the bird species of the planet have been driven to extinction as the direct or indirect result of human activity during the Holocene. Diamond (1992) summarizes recent evidence indicating that the unique endemic fauna of Mediterranean islands, including a number of species of dwarf elephants, hippopotamuses, and deer, was exterminated by human hunter-gatherers about 8500 BP. Mohsin and Ambol (1983) report that an extensive attempt to find every fish species in Malaysia described in the 19th century found that 40–60% were either extinct or very rare. It would be hard to deny that many species have gone extinct as a direct or indirect result of human activity.

Two kinds of studies are needed to estimate the rate of extinction. One is the repetition of the best historical collections. This cannot be a rigorous procedure, for collecting methods were sloppy in the nineteenth century and often collection sites were poorly documented. The other is the documentation of current extinctions in disappearing habitats. That means getting to the habitats before they disappear, making good collections, then monitoring the decline and disappearance of species as the habitats vanish. This is a rather depressing sort of study that most biologists would rather forget about, but it will deliver important information with major political consequences.

What difference does it make to human self-interest? The answer to this question will determine the level of sustained political commitment. It is not a question for the natural sciences, except to the extent that they can demonstrate probable consequences for human health and well-being resulting from the extinction crisis. I doubt that there are many such consequences. Far too many wild threats of ecosystem collapse and the disappearance of oxygen from the atmosphere have been made. Humans are tough and adaptable, as tough as rats and cockroaches; they will be among the last survivors on a decimated planet. Some agronomists have calculated that humans depend directly on no more than about 100 species of plants, animals, fungi, and bacteria, and could probably survive indefinitely, in a virtually hydroponic world, on just this number. Perhaps they are right.

The question of human self-interest is a question for religion, philosophy, and politics. If it were not for the historical evidence of negative side-effects of mass religious movements, I could enthusiastically advocate the creation of a new religion which would expand the Christian dictum from 'Love thy brother' to 'Love thy fellow living being,' for it seems that religion has been the major medium through which the bulk of mankind has learned its ethics. However, it is hard to envisage a plausible new religion that could survive as more than a temporary fad in the skeptical post-industrial, post-scientific world.

My own opinion is that through gross negligence, without being able to imagine, understand, or intend what we are doing, the human race is causing a mass extinction of millions of species that have never done us any harm. Only plate tectonics and meteorite impacts have had effects of equivalent magnitude. So gigantic a perturbation of evolution and so massive an extinction is criminal and tragic. It will have consequences that will outlast our presence as a species on the planet.

WHERE BASIC RESEARCH IS NEEDED

In the broadest perspective, we need basic knowledge from the entire research programme of ecology and evolution to deal with the extinction crisis. However, certain parts of it are more relevant than others. The following emphases reflect my bias as an evolutionary ecologist interested in processes that maintain diversity.

Connecting genetic variation and ecological processes: within species and among species within communities

Does genetic variation stabilize ecological interactions? Here we have almost no information. Much work has been done on the role of genetic variation in the host as a defence against co-evolving parasites, but the general role of genetic variation in stabilizing or destabilizing competitive, predator–prey, and host–parasite interactions remains little appreciated and poorly investigated. Still less is known about whether communities consisting of genetically diverse populations are more resilient to perturbations than communities consisting of genetically impoverished populations.

Do ecological interactions evolve rapidly? We know that ecologically important traits can evolve rapidly in single species exposed to strong selection pressures on moderately large populations (reviewed in Stearns, 1992). Little is known about the rate at which ecological interactions evolve. We need to know that rate if we want to predict whether natural populations will evolve rapidly enough in response to human perturbation to persist.

What use is information on molecular variation? Recent technical breakthroughs in molecular biology are generating a flood of information on nucleic acid sequences and on various fragment polymorphisms. To get additional funding for such research, some are willing to claim that such data are essential for understanding natural biodiversity. However, molecular variation is only relevant here if it makes a difference to fitness or can be used in systematics. Its application to systematics, which is basically a matter of measurement and careful interpretation, will be more easily exploited than its use as a tool to understand the sources of variation in fitness, which requires extensive experimentation.

Putting the questions sharply. To help motivate research on basic issues connected to biodiversity, let me state a few provocative null hypotheses: Genetic diversity has no effect

- on extinction probability,
- on population dynamics,
- on interactions between two or more species, or
- on ecosystem function.

Refuting or failing to refute any of those hypotheses in a research programme that combined elegant theory with convincing experimental tests would be a fundamental contribution to the biodiversity problem (Solbrig, 1992).

Why are some communities and ecosystems more diverse than others?

To this basic question we have no satisfying general answer. Every simple answer immediately encounters a host of exceptions. The gammarid amphipods of Lake Baikal suggest that ancient communities are diverse, but the haplochromid cichlids of Lake Victoria show that young communities can be just as diverse. Humid tropical forests are diverse, but the number of tree species at La Selva in Costa Rica greatly exceeds the number of tree species at Taï in the Côte d'Ivoire, suggesting that historical and regional effects are important enough (Ricklefs, 1987) to make a general theory of community diversity anything but easy to formulate. Community ecology should be the queen of the ecological sciences and needs rejuvenation and fresh approaches. The most progress has been made by plant ecologists (e.g. Tilman, 1988). Hutchinson (1959) has long been respected for asking why there are so many kinds of animals; since his paper was published we have learned that there is quite probably no simple, general answer and that the only convincing answers will be partial, relevant to some situations and organisms but not others.

How many species are there and what can they tell us?

Systematics and taxonomy are not just stamp-collecting; they are disciplines that have been intellectually hot for twenty years. The breakthroughs in phylogenetic systematics and other tree-building methods have revolutionized our views of traditional classifications. While we can now produce trees of relationship with unprecedented rigour, we also have learned that we must view all such results with unprecedented humility, for each such result is now seen as a tentative hypothesis for which a number of equally good alternatives are often available.

Coupled with the substantial logical advances in tree-building methods, the new sources of data provided by molecular systematics will help to solve long-standing problems, such as the origin of the angiosperms and the large-scale relationships of the animal phyla. They will also help us to determine into how many genetically separate units the living world is divided, i.e. they will help to tell us how many species there are and how far the species concept can be applied at all.

The benefits of the new systematics will extend far beyond a well arranged catalogue of life on earth. A great deal can be learned from the comparative method, which has recently been widely applied in behavioural and evolutionary ecology (Harvey and Pagel, 1991). All applications of the comparative method depend absolutely on having a sound phylogeny on which to base the comparisons. Thus systematics has both an immediate, applied aim – to tell us how many species there are – and a long-term, basic benefit – to make all applications of the comparative method more reliable. These are convincing reasons to support more positions for systematists at universities, where they can train the next generation whose work we can already clearly foresee, and to support the curation, collection, and classification activities of the natural history museums of the world.

To suggest the sort of basic question about biodiversity that could be solved with systematics, I state two null hypotheses that can only be tested with the comparative method applied to a sound phylogeny:

- *Environmental change has no effect on speciation or extinction rates.*

- *Characteristics of lineages have no effect on speciation or extinction rates.*

CONCLUSIONS

We need the applied answers quickly. How many species are there on the planet and how fast are they going extinct? Answers to those questions will generate support for getting answers to the basic questions concerning the origin and maintenance of biodiversity. A well-balanced national or international research programme on biodiversity will support applied and basic projects in parallel (Figure 13.1).

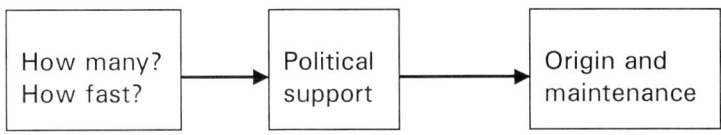

Figure 13.1 A well-balanced national research programme on biodiversity should seek quick answers to some important applied questions in order to generate continuing political support for the basic research that will get at the truth of the matter

REFERENCES

André, H.M., Lebrun, P. and Noti, M.-I. (1992). Biodiversity in Africa: a plea for more data. *Journal of African Zoology*, **106**, 3–15

Cowie, R.H. (1992). Evolution and extinction of Partulidae, endemic Pacific island land snails. *Philosophical Transactions of the Royal Society of London B*, **335**, 167–91

Diamond, J. (1989). The present, past and future of human-caused extinctions. *Philosophical Transactions of the Royal Society of London B*, **325**, 469–77

Diamond, J. (1992). Twilight of the pygmy hippos. *Nature*, **359**, 15

Erwin, T.L. (1988). The tropical forest canopy: the heart of biotic diversity. In Wilson, E.O. (ed.) *Biodiversity,* pp. 123–9. National Academy of Science, Washington, DC

Erwin, T.L. and Scott, J.C. (1980). Seasonal and size patterns, tropic structure and richness of Coleoptera in the tropical arboreal ecosystem: The fauna of the tree *Luehea seemannii* Triana and Planch in the Canal Zone of Panama. *Coleoptera Bulletin*, **34**, 305–22

Freed, L.A., Conant, S. and Fleischer, R.C. (1987). Evolutionary ecology and radiation of Hawaiian passerine birds. *Tree*, **2**, 196–203

Harvey, P.H. and Pagel, M. (1991). *The Comparative Method.* Oxford University Press, Oxford

Hutchinson, G.E. (1959). Homage to Santa Rosalia, or why are there so many kinds of animals? *American Naturalist*, **95**, 137–45

Mohsin, A.K.M. and Ambol, M.A. (1983). *Freshwater Fishes of Peninsular Malaysia.* University Pertanian Malaysia Press, Kuala Lumpur

Ricklefs, R.E. (1987). Community diversity: relative roles of local and regional processes. *Science*, **235**, 167–71

Solbrig, O. (ed.) (1992). *From Genes to Ecosystems: a Research Agenda for Biodiversity.* IUBS/SCOPE/UNESCO, Paris

Stearns, S.C. (1992). *The Evolution of Life Histories.* Oxford University Press, Oxford

Stearns, S.C., Baur, B., van Noordwijk, A.J. and Schmid-Hempel, P. (1990). Report to the Swiss Science Council on Biodiversity: a Critical Review of the US OTA Report with Recommendations for Research. Technologien zur Erhaltung der Biologischen Vielfalt, Drei Gutachten über eine OTA-Studie. Schweizerischer Wissenschaftsrat (Hrg.), Bern

Tilman, D. (1988). *Plant Strategies and the Dynamics and Structure of Plant Communities.* Princeton University Press, Princeton, NJ

Witte, F., Goldschmidt, T., Wanink, J., van Oijen, M., Goudswaard, K., Witte-Maas, E. and Bouten, N. (1992). The destruction of an endemic species flock: quantitative data on the decline of the haplochromine cichlids of Lake Victoria. *Environmental Biology of Fish*, **34**, 1–28

CHAPTER 14

LAND–INLAND WATER ECOTONES AS TRANSITIONAL SYSTEMS OF PARTICULARLY HIGH BIODIVERSITY: TOWARDS A SYNTHESIS

Jean-Bernard Lachavanne and Raphaëlle Juge

INTRODUCTION

Increasingly threatened through the growing, multi-faceted pressure exerted by man on natural ecosystems, biodiversity has become a major subject of concern at the scale of the biosphere in the past decade. This concern is rooted in the fear that a generalized reduction of biodiversity may lead to a loss of operability and stability of ecosystems (Solbrig, 1991a) and, finally, to a loss of resources which might prove vital for man in the future.

Because of several processes at work within the ecotone itself and because of its proximity and functional ties to the adjacent ecological systems, biodiversity is relatively high in ecotones (di Castri *et al.*, 1988; Risser, 1990). Consequently, ecotones and in particular land–inland water ecotones are of major importance for maintaining biodiversity.

This chapter attempts, in a condensed form, to make a synthesis of the principal established scientific facts about biodiversity patterns and regulating factors in land–water ecotones. Biodiversity is the most integrated expression of ecological systems since it is the result of varied phenomena (physical, chemical, biochemical, genetic, physiological, ecological) which operate at different paces, over scales ranging from 10^{-8} to 10^7 m (i.e. from microhabitat to the whole biosphere) and temporally from 10^{-7} to 10^8 years (Minshall, 1988). The attempt made here is obviously far from exhaustive.

THE ECOTONE CONCEPT AS APPLIED TO THE TRANSITIONAL ZONES BETWEEN TERRESTRIAL AND AQUATIC ECOSYSTEMS: THE NECESSITY OF AN AGREEMENT

Since it was introduced by Livingstone (1903) and Clements (1905), the concept of ecotone has been understood and applied in the literature in different ways (Jenik, 1992). This variety is found, to a certain extent, in the chapters of this book.

Today land–inland water ecotones are recognized as ecological systems of specific and variable abiotic and biotic characteristics influenced by dynamics at work on time scales ranging from geological through historical to recent and present day (di Castri and Younès, 1990; Naiman and Décamps, 1991; Holland et al., 1991). But according to Pieczyńska and Zalewski (Chapter 3) in most cases they are studied and described without determination of their spatial limits. Moreover, when they are specified, the limits of the ecotone as proposed by the authors do not always coincide, reflecting different conceptions (ecosystem, spatial entity, ecological function) or different spatial approaches. It therefore seems that the term ecotone may be used by the various authors to describe different realities, which makes it extremely difficult and hazardous to conduct any comparative study of the phenomena and processes at work in the zones of contact between terrestrial and aquatic ecosystems. This difficulty is further exacerbated by various terminologies used in studies of similar types of ecotones.

The lack of consensus in the definition and usage of the term ecotone stems from the fact that two fundamental notions with partially overlapping meanings – that of stress (referring to the borderline conditions prevailing at the boundaries of the tolerance spectrum of the populations, species and communities) and that of transition zone between two ecosystems at the scale of the landscape – interfere and make this term rather ambiguous (Van der Maarel, 1990; Jenik, 1992; Lachavanne, Chapter 1).

The definition proposed by the UNESCO–MAB/SCOPE Technical Consultation (Holland, 1988) contains these two notions but does not help to ensure that this term is used more precisely. Its very general character is regarded as an advantage by some authors (Holland, 1988; di Castri et al., 1988; Naiman et al., 1989), who stress the possibility of applying this concept, in a holistic approach, to any transition zone, interfaces, edges or boundaries, whatever the level of biotic and ecosystemic integration under consideration: populations, species, communities, ecosystems. So ecotone can be used to describe a wide variety of spatial and temporal scales. But such a conception is regarded as a major drawback by others (Van der Maarel, 1990; Backéus, 1993), who criticize it for causing confusion, particularly in connection with the concept of ecocline introduced by Clements (1916) and, above all, because it designates a multiple reality. One can also criticize the lack of operability resulting from such a general conception of the term ecotone. Indeed, it is not possible, based on such a conception, to answer questions as fundamental as 'What is an ecotone? Where are ecotones?' These questions have no simple answer other than 'It depends' (Gosz, 1991). Thus the identification of a particular ecotone depends on the question being asked or on the problem to be solved (Naiman and Décamps, 1991).

Several authors (Risser, 1985; Hansen et al., 1988; Naiman et al., 1988; Rusek, 1992) have underlined this need to clarify the concept of ecotone but no proposal based on a consensus has been made to date.

In order to satisfy this need, Lachavanne and Juge (in preparation) have suggested adopting a four level approach to the application of the ecotone concept to transition zones between terrestrial and aquatic ecosystems based on a Water–Soil–Vegetation (WSV) approach. They propose to consider: (1) first degree ecotones which cover the whole transitional zone and which are influenced by the complex spatio-temporal interrelation between land and water, mainly an ecotone *sensu lato*; (2) second degree ecotones which represent areas with different hydrological regimes (substrate never, occasionally or permanently flooded); (3) third degree ecotones which represent areas with different vegetated formations; and (4) fourth degree ecotones, with different plant groups/associations (see also Lachavanne, Chapter 1; Juge and Lachavanne, Chapter 5). These four different levels of approach to land–water ecotones correspond to different habitat structures and modes of operation which condition the distribution and diversity of plant and animal species.

This proposal offers the advantage of providing a theoretical framework that is indispensable for conducting comparative studies of existing ecotones in situations of different geomorphological and hydrological contexts and different latitude and elevation. It should facilitate a comparison of the results of the studies undertaken in the context of the UNESCO–MAB international collaborative research programme on the 'Role of land–inland water ecotones in landscape management and restoration'.

Finally, by designating reference frameworks that are easily identifiable, thanks to hydrophilic and hygrophilic plant groups, the WSV approach lends better operability to the concept of ecotone in the transition zones between continental terrestrial and aquatic ecosystems. It allows the scientist to address the problems of biodiversity in land–water ecotones at the different relevant functional scales and allows the manager to apply – both more easily and more effectively – the management guidelines drawn up scientifically for the conservation or restoration of biodiversity.

BIODIVERSITY ANALYSED IN AN ECOTONAL CONTEXT

Although biodiversity in ecotones had aroused the interest of many ecologists in the past – particularly following the emergence of the edge effect concept (Leopold, 1933) and the revelation of the role of corridor played in migration by ecosystem edge zones or as unique habitats for certain species (Hansen *et al.*, 1988) – it must be acknowledged that the information collected to date is still very general, descriptive and very fragmentary, which makes it difficult to develop a theory of biodiversity in ecotones. This is attributable mainly to the lack of knowledge about and – until the early 1980s – the failure to recognize the importance of transition zones for the conservation of biodiversity.

Thanks to the initiative of the Scientific Committee on Problems of the Environment (SCOPE) and UNESCO's Man and Biosphere programme

(MAB) introduced in 1988 (di Castri *et al.*, 1988), a co-ordinated set of international programmes on ecotones was launched (Naiman *et al.*, 1989) in which biodiversity occupies a place of choice. A re-evaluation and conceptualization of theory concerning biodiversity and ecotones appeared to be necessary to better explain the patterns observed in nature (di Castri and Hansen, 1992).

An understanding of the biodiversity phenomenon depends above all on our ability to apprehend the complexity of the systems and to effect a synthesis of processes operating at all and between all levels of the biosystems (cell, organism, population, community, ecosystem, landscape). One reasonable route for apprehending the complexity that underlies the expression of biodiversity at the scale of the ecological systems is to proceed in stages. The taxon-type approach used in this book is one such route. It offers certain advantages such as, for example, specifying the relevant scales to be taken into consideration in order to analyse the diversity of the different groups. It also involves some disadvantages, due particularly to the sectorial approach to biodiversity, which does not result in an overview of biodiversity in land–water ecotones. However, the numerous relations identified in the context of an analysis of the factors regulating the various biotic components considered here allow bridges to be built and provide interesting routes for a more integrated view of biodiversity in land–water ecotones (e.g. functions of plant diversity, Pieczyńska and Zalewski, Chapter 3; Juge and Lachavanne, Chapter 5; Joly and Morand, Chapter 7).

This volume treats only biodiversity in land–inland water ecotones. But if there are patterns of biodiversity rather specific to transitional zones due to the very special environmental conditions that characterize them (gradient, spatio-temporal heterogeneity), it appears that identified regulatory factors in ecotones could be more broadly meaningful for other ecosystems. Thus despite their apparent restricted scope, these proceedings from the symposium organized during the SIL (Societas Internationalis Limnologiae) congress held at Barcelona in 1992 provide a contribution to the broader subject of 'Biodiversity' and express a diversity of opinions and perspectives. The main issues do not lead to a real theory of biodiversity in land–inland water ecotones, but they do confirm the interests and the perspectives in developing research on biodiversity in ecotones.

The biological characteristics of land–water ecotones appear to be closely determined by forcing functions (i.e. geomorphology, hydrology, climate, edaphic factors, disturbances) operating on spatial and temporal scales that may be different for each major type of ecotone (Naiman and Décamps, 1990). The patterns of species diversity in ecotonal areas are complex, influenced by several factors, including grain size of environmental gradients, life history and demographic characteristics of organisms, and the structure (spatial organization) of the ecotone and adjacent patches (di Castri and

Hansen, 1992). There are clearly various complexities that must be considered to account for biodiversity patterns in ecotones.

This book documents the specific patterns and regulating factors common to the principal biotic components in land–water ecotones and attempts to furnish an overview of them in order to identify some useful conclusions for management. It provides several important items of information that shed light on the relationship between biodiversity and the land–water ecotone, but only partially meets this need. This is due to the fact that most of the data on which the line of argument is based are derived from studies aimed at answering other questions than those intended to elucidate the biodiversity phenomenon.

A reading of the chapters in this book will nevertheless allow the following main conclusions to be drawn:

Ecotones between terrestrial and aquatic ecosystems are generally characterized by high biodiversity

The large body of information collected in the chapters of this book confirms that, subject to certain conditions that have yet to be specified and quantified (qualitative and quantitative characteristics of the ecotone, regime of perturbations), land–inland water ecotones (*sensu lato*) are generally characterized by high species richness and high density of organisms. Numerous examples reported in this book by Juge and Lachavanne (plants), by Giudicelli and Bournaud (invertebrates) or by Dobrowolski (birds) illustrate this particular property of land–water ecotones.

Such a high diversity relies on the presence of heterogeneous patch mosaics of different type, quantity and quality providing combinations of vital resources for plants and animals. High biodiversity stems both from ecotone differentiation (degree of heterogeneity) and from morpho-ecological differences in species. Due to this, species utilize ecotones as breeding and feeding habitats and as corridors for migration (Dobrowolski, Chapter 9).

Biological diversity of land–water ecotones depends, among other things, on ecological valency of the species and habitat diversity. Most of the animal species present in land–water ecotonal zones exhibit high adaptability in relation to both breeding and feeding (Dobrowolski, *op. cit.*; Schröpfer, Chapter 10). This ecological plasticity results in overlapping of animal communities in these habitats. Other species find in the ecotonal zone very specific and differentiated habitats necessary for their survival.

The plant and sedentary or mobile animal species that live in land–water ecotones (see examples given by Juge and Lachavanne, Chapter 5; Joly and Morand, Chapter 7; Zalewski, Chapter 8; Dobrowolski, Chapter 9; Schröpfer, Chapter 10) can be grouped into different categories depending on the degree of dependency on the ecotone. Considering ecotone in the *sensu lato*, it is possible to distinguish schematically the species that are totally reliant on the ecotonal conditions from those that are only partially dependent on the conditions provided by the ecotone and from those that are present fortuitously or occasionally (place passed through, for example).

The differential participation (degree of presence, function) in space and time of these different categories of species (seasonal, annual flows) in the ecotone's flora and fauna is the cause of the famous edge effect (Leopold, 1933). It confers on biodiversity in the transitional land–water zones a highly dynamic character that must be taken into account in all research into biodiversity patterns and regulating factors. However, the relative participation of these different categories of species in the biodiversity of the ecotone and the conditions of the edge effect must be further specified and quantified.

At the scale of the ecotone *sensu lato* the results presented generally confirm the relationship between the degree of complexity of the ecological system (in relation to highly differentiated vegetation) and the degree of biodiversity. Indeed it is demonstrated that lake shore zones or river margins of high habitat complexity provide favourable sites for many plant and animal species.

On the other hand, at the scale of plant communities, the enrichment in species along the ecological successions provided for by the theory (Odum, 1969; Margalef, 1969) is not linear, as certain communities dominated by invading species can exhibit less diversity than younger communities of the succession series (Juge and Lachavanne, Chapter 5). Thus although an increase in plant diversity throughout the succession from aquatic to terrestrial series is seen in overall terms, the overall trend observed exhibits points of inflexion (see also Margalef, Chapter 2).

Finally, let us note that several authors describe here examples that support the validity of Connell's (1978) model, according to which the highest biodiversity is achieved in communities at intermediate stages of perturbation: e.g. effects of mesotrophic conditions on diversity of macrophyte or fish communities (Juge and Lachavanne, *op. cit.*; Zalewski, Chapter 8) and the role of the degree of spatio-temporal variability experienced by each site within a functional alluvial unit on the composition of the amphibian communities (Joly and Morand, Chapter 7).

Biodiversity in land–inland water ecotones also depends on characteristics of related ecosystems

Diversity and richness of plant and animal communities in the land–water ecotone are determined not only by internal specific environmental characteristics but also, in a proportion that remains to be determined, by the neighbouring aquatic and terrestrial ecosystems.

For species partly or totally linked with the aquatic habitat to survive (amphibians, birds, mammals), the type, the quality and extent of the related terrestrial environment is of great importance. For example, Dobrowolski (Chapter 9) reports that the highest diversity of birds occurs in the ecotone water–shrubs or water–forest whereas the ecotone water–open area is less rich in birds. An important aspect of this relationship is the connectivity of complementary suitable habitats (Joly and Morand, *op. cit.*) and the notion of networks of suitable habitats. Furthermore, while the biodiversity of ecotones depends to a certain extent on the characteristics of the related ecosystems or on much more distant ecosystems (migratory species), the converse is also true. A degraded ecotone no longer provides the resources (in quality and quantity) required by the species of the related ecosystems linked to it.

Neilson *et al.* (1992) and Hansen *et al.* (1992a, 1992b) suggest that the life history traits of organisms dictate a specific type configuration and scale of environmental conditions within which fitness is positive: thus community dynamics are largely a function of the contiguity between the environmental envelopes of species within the community, and the nature and scale of landscape patterning. The status of a population within a patterned environment is a function of the extent to which suitable conditions for survival and reproduction are present through space and time.

Thus investigations ought to concentrate on the ecotone not as an isolated and distinct entity but as an integral component of the landscape. So the understanding and the maintenance of ecotonal biodiversity require a landscape perspective and have to be viewed in the context of the multi-scale surrounding patches (multi-faceted approach).

Hydrology, a prominent regulatory factor of biodiversity in land–inland water ecotones

The influence of the hydrological factor on the structure, functioning and evolution of ecotonal zones between continental terrestrial and aquatic ecosystems (wetlands) has been the subject of a very abundant literature, particularly in connection with plant production and diversity (Heinselman, 1970; Beaumont, 1975; Hasler, 1975; Armstrong, 1975, 1978; Johnson and Bell, 1976; Good *et al.*, 1978; Gosselink and Turner, 1978; Bedinger, 1979; Carter *et al.*, 1979; Cowardin *et al.*, 1979; McKnight *et al.*, 1981; Gopal *et al.*, 1982; Etherington, 1983; Ingram, 1983; Brooks *et al.*, 1985; Mitsch and Gosselink, 1986; Dykyjová and Kvet, 1978; Gibert *et al.*, 1990).

It is accepted today that the hydrological regime within which a land–inland water ecotone operates is the most fundamental determinant of its space–time behaviour (Wetzel *et al.*, 1989) and consequently of its biodiversity. Indeed, the environment of land–inland water ecotones is influenced by four types of natural disturbances driven by hydrological factors: floods, drought, erosion and sedimentation. The hydrological regime has a continuum of effects ranging from provision of resources (moisture, silt, nutrients, etc.) to causing major disturbances (erosion, avulsion and sediment deposition) when extreme variability in hydrology is registered from season-to-season and year-to-year (Petts, Chapter 12). Water flow rates and patterns determine the transport and distribution of nutrients and carbon inside the ecotone. The circulation of water in sediments partially determines their oxidation–reduction potentials which, in turn, affect rates of nutrient mineralization and gaseous fluxes, detritus and carbon processing, and sediment weathering rates (Wetzel et al., 1989; Aragno and Ulehlova, Chapter 4).

The interaction of the processes driven by spatial and temporal variability in hydrodynamics creates environmental complexity and heterogeneity within ecotonal areas providing highly differentiated habitats for plant and animal species. Depending on the scale considered and the strength of forces implicated, the degree of disturbance by more or less extreme hydrological events affects plant diversity both positively (maintaining successive stages of different age) and negatively (breaking or uprooting plants and removing fine sediments, Haslam, 1978). It influences animal diversity through its effects on suitable sites for breeding and feeding (surface area, patchy distribution, fragmentation/connectivity; see Merriam and Wegner, 1992). In summary, hydrological disturbance changes the area occupied by a community, causes species to disappear and creates new habitats for pioneer species, so that an equilibrium value for species diversity in land–inland water systems is an imprecise concept (Petts, *op. cit.*).

Higher plants as a major determinant factor for biodiversity

Diversity of habitat has long been recognized as a major contribution to the pattern of diversity of species, since fundamentally different habitats promote the colonization of a system by different species (MacArthur, 1972; Cody, 1975; Jacobs, 1975). Within certain spatial and functional limits, the more heterogeneous and complex the physical environment, the more complex the plant and consequently animal communities and the higher the species diversity.

Vegetation plays a central role in nutrient cycling inside an ecosystem and the quantitative role of higher plants as an energy base for the trophic networks of predators and decomposers has been the subject of numerous studies since the first work of Odum (1968). In contrast, their qualitative role (through their presence or absence) as a factor conditioning the diversity of the other taxonomic groups is still unexplored. Now higher plants constitute, on account of their specific, strategic and structural diversity (life-forms), one of the major factors in the structuring and spatio-temporal and functional complexity of ecosystems, providing a great diversity of habitats for the other plant species (microbial, microphyte) and animal species (Pieczyńska, 1976; Aragno and Ulehlova, Chapter 4; Pieczyńska and Zalewski, Chapter 3; Juge and Lachavanne, Chapter 5; Joly and Morand, Chapter 7). A rich, diversified vegetation of variable density and the accumulation of detritus help to make the biotope in land–inland water ecotones more complex and to generate a diversity of habitats for the other taxons.

Several observations reported by the various authors confirm the importance of this 'conditioner' role of vegetation biodiversity. For example, Pieczyńska and Zalewski mention that high biodiversity is generally noticed in shore zones with rather rich, differentiated and forested vegetation; a decrease of animal species richness (both of permanent inhabitants and seasonal visitors) usually being observed with a decrease in plant cover. Similar observations are reported by Joly and Morand (special role of patchiness, connectivity/fragmentation for amphibians), and by Dobrowolski (influence of type of vegetation prevailing in the ecotone: water–forest/water–open area for birds). It should be noted, however, that vegetation cover does not constitute a guarantee of optimal development of biodiversity. Indeed, dense cover may prevent colonization of the system by species developing preferentially or exclusively on freshly uncovered habitats. This is the case, for example, of the Carabidae (Giudicelli and Bournaud, Chapter 6) or of some species of fish (Salmonidae, whose egg laying sites must be free from vegetation (Zalewski, Chapter 8)).

An analysis of the influence of the diversity of higher plants on the biodiversity of the other taxons, both in its nutritional and structural components, thus appears to be a promising route towards a better understanding of the biodiversity phenomenon and its spatio-temporal dynamic. Such an approach should also provide useful information about its function.

BIODIVERSITY IN LAND–INLAND WATER ECOTONES AT RISK

Land–inland water ecotones are among the ecological systems that are the most degraded and the most threatened by human activities (Good *et al.*, 1978; Lachavanne *et al.*, 1986; Mitsch and Gosselink, 1986; Loeffler, 1990; National Research Council, 1992; Kent, 1994; Mitsch, 1991; most chapters of this book) through the consequences on the configuration (impairment of physical integrity, size, etc.) and the quality (pollution) of ecotones inhabited by plant and animal species.

The chapters of this book document, by means of numerous examples, the effects of the various perturbations of anthropogenic origin that are common in the transitional zones between continental terrestrial and aquatic ecosystems. Today most river margins and lowland lake ecotones of the temperate zone comprise only fragments of natural and semi-natural habitats within an agricultural landscape, dominated by arable fields and improved pasture (Lachavanne *et al.*, 1986; Petts, *op. cit.*).

The structural and functional integrity of land–inland water ecotones that underlies biodiversity is particularly threatened in ecotones because they may be located such as to suffer the cumulative effects of the perturbations affecting the two adjacent ecosystems. Under certain conditions 'the edge effect of the perturbations' that may result from this means that biodiversity may be more threatened in ecotones than in the two ecosystems that they delimit.

On the aquatic ecosystem side, the main threats are connected with the control of the hydrological regime (altering the regime of watercourse flow rates, limiting variations in the water level of lakes), with the embankment of shores (flood control), eutrophication and (toxic) pollution of the water and with the exploitation of resources (gravel for construction) or even the over-exploitation of certain resources (e.g. fish, irrigation water, etc.).

On the terrestrial ecosystem side, the growing needs for human housing, agricultural and industrial activities and transport are the main perturbations liable to affect or destroy the structural and functional integrity of biodiversity-generating land–inland water ecotones.

These disturbances affect biodiversity in land–inland water ecotones through: (1) a decrease of the whole ecotone surface (e.g. in the river section where embankments are erected or in lakes with a controlled water table) and consequently the loss or restriction of the surface of suitable sites for breeding and feeding; (2) alteration of structural and functional characteristics of ecotones as a consequence of increased abiotic stress (e.g. resulting from river channeling, point source and non-point source pollution leading to eutrophication) and consequent reduction of ecotone complexity and habitat diversity; (3) greater fragmentation of ecotone surface, thus increasing the isolation of suitable habitats, increasing vulnerability of a population to extinction and (4) reduction of the connectivity of terrestrial environment (e.g. for amphibians, birds).

MANAGING ECOTONAL ZONES FOR PRESERVING AND RESTORING BIODIVERSITY: AN URGENT NEED

The protection of land–inland water ecotones is a priority in temperate zones because they are reservoirs of biodiversity, they are vulnerable and their structural and functional integrity are under threat from human activity. In recent years the special needs and challenges of conservation efforts in land–inland water ecotones have become increasingly appreciated (di Castri *et al.*, 1988; Naiman and Décamps, 1990; Hansen and di Castri, 1992). This is critical, given that most regional biodiversity is thought to reside in these transition systems.

The different chapters of this book furnish much relevant information that justifies this importance: (1) on the generally high diversity of flora and fauna (specific richness and diversity, variety of life-forms and strategies, spatio-temporal patterns), in fluviatile and circumlacustrine land–inland water ecotones; (2) on the principal regulating factors controlling the spatio-temporal variability of these patterns, and (3) on the threats connected with natural perturbations or those of anthropogenic origin. However, all the authors stress that much more scientific information is needed on the impacts of man's activities on the structural and functional characteristics of land–inland water ecotones that affect biodiversity. Many questions are still open but the risks run require action to be taken urgently to ensure better conservation of biodiversity in land–inland water ecotones.

Despite the numerous gaps in our knowledge of the existing conditions of particularly threatened species or communities, it is now possible, in the light of progress made in conservation biology (Soulé, 1987; Simberloff, 1988; McNeely et al., 1990; Pimm, 1991; Cockburn, 1991), to establish a number of conservation principles and to draw up guidelines for the benefit of managers, some being general in scope, others tailored to the specific features of land–inland water ecotones (e.g. Petts, Chapter 12, for conservation status of river margins). The basic recommendations for biodiversity conservation in land–inland water ecotones are as follows:

- The necessity of recognizing the land–inland water ecotone as a transitional system having special structural and functional properties that can be located in space (see method proposed by Lachavanne and Juge in preparation in Lachavanne, Chapter 1). As a bioindicator of degrees of hygrometry and hydrometry, vegetation can be utilized advantageously for the delimitation of the land–inland water ecotone and thereby facilitate recognition and individualized consideration of the transition systems by managers in the context of regional planning and land management;

- Processes determining biodiversity in land–inland water ecotones have to be described from a holistic perspective, considering different scales and levels of integration including physical, chemical, biological, ecological and human effects on both aquatic and terrestrial systems (di Castri et al., 1988). Given that the numerous links maintained by the land–inland water ecotone (ecological flows: material, energy, organisms) with the adjacent or more or less distant ecosystems (catchment basin, other regions or other continents *via* migratory species and air movements), investigations ought to concentrate on the ecotone not only as an isolated and distinct entity but as an integral component of the landscape (Risser, 1990) and the biosphere. As a consequence of interrelations between ecosystems, biodiversity conservation in land–inland water ecotones requires a multi-speed approach (depending on the species or function under consideration) and a landscape perspective;

- Since the environmental requirements of the populations and communities may be different or contradictory, it is paramount to set spatially and temporally clear conservation objectives that aim to maintain the highest possible or attainable diversity of plants and animals (Kvet et al., 1989) and to optimize biodiversity at all spatial scales (local, regional, national, global). The management measures to be promoted will in fact not be the same, depending on whether one aims at the highest diversity of species in an ecotone at the local scale or whether one seeks to give priority to protecting a particular species threatened with extinction at the regional, national or global scale. Furthermore a measure that may appear to be judicious at a given scale may prove inappropriate at another spatial scale or in the course of time. For example, systematically aiming at a regime of medium-sized perturbations to promote biodiversity in land–inland water ecotones (Connell, 1978; Ward and Stanford, 1983), a hypothesis verified in this book on macrophytes (Juge and Lachavanne) and on fish (Zalewski) in connection with water eutrophication, could lead to a decrease in regional or national diversity by eliminating the species that are characteristic of unperturbed or slightly perturbed habitats (oligotrophic species) or of highly perturbed habitats (species adapted to eutrophic-to-hypertrophic conditions);

- According to Petts (Chapter 12), any assessment of the options for protection and rehabilitation of land–inland water ecotonal zones must be scientifically based and practicable for management. A framework is required to direct scientific information to restoration objectives, recognizing the multifunctional nature of catchment management, involving: water resources, water quality, flood control and land drainage, fisheries, recreation, conservation and navigation. Biodiversity conservation research and implementation must be 'tailored' to the particular circumstance of a given location (Soulé, 1991; see also Carney, Chapter 11);

- In ecotones as in any ecosystem, the arsenal of biodiversity conservation measures must comprise in parallel:
  - a set of curative measures aimed at durably repairing the damage suffered by restoring the structural and functional integrity of ecotones wherever this is possible;
  - a set of preventive measures aimed at limiting as much as possible the adverse effects of human activities on the diversity of populations and communities. These measures concern primarily the land–inland water ecotone itself and the adjacent ecosystems but also the catchment basin (Hillbricht-Ilkowska et al., 1989) and more distant terrestrial and aquatic ecosystems (migratory species) or even at a more global scale in relation to air pollution (Soulé, 1991; Hansen and di Castri, 1992);

- The principal harmful effects of human activities on biodiversity in land–inland water ecotones are connected with nutritional and toxic pollution and with various physical alterations of the environment (drainage, facilities such as dykes and ports, etc.). These modifications of the flora and fauna's living conditions may lead to the disappearance, reduction, erosion and fragmentation of habitats.

The restoration measures should therefore be targeted primarily at:

- recreating the dynamic (especially hydrodynamic) conditions needed to durably restore the processes of differentiation of habitats within the ecotone (rejuvenation of life conditions in order to maintain the coexistence of pioneering stages with sub-climactic stages). The surfaces 'returned' to the ecotone during the restoration procedures should take account primarily of the biodiversity conservation requirements while also taking account of the other useful functions of the ecotone (e.g. filter effect and anti-pollution buffer zone). The choice of the sectors to be restored and the characteristics of the ecotones targeted should be specified on a case-by-case basis depending on the balance of the socioeconomic and natural environment interests involved; and

- controlling pollutant discharges of domestic, industrial and agricultural origin (point and non-point sources of pollution) at the scale of the ecotone itself, of the adjacent ecosystems, of the catchment basin and, finally, at the scale of the region and even the biosphere (air pollution).

Today the main general causes of the erosion of biodiversity in land–inland water ecotones have been identified. On the other hand, the specific causes of the regression or disappearance of particular species are less clear. Future research will have to focus on evaluating the relative importance (quantification) of the deleterious factors for the different levels of biosystems (individuals, populations, communities, ecosystems) in order to be in a position, through a multi-criteria and multi-scale approach, to devise a priority action plan within the framework of previously well-defined objectives.

The efficient conservation approach should also combine the complexity of interactions which appear at the land–inland water ecotone in relation to the various forms of cumulative impact in antagonistic and synergistic ways influencing the whole community complex (Zalewski, Chapter 8). As pointed out by Sedell *et al.* (1991), policy-makers need a list of options locating the best opportunities for protecting land–inland water ecotones and for restoring degraded areas to a self-sustaining state by incorporating desirable, natural and designed features.

The restoration objectives in terms of surface of ecotonal zone 'to be recovered' ought to take into consideration both the natural zone of hydric influence before perturbations (historic data) and the constraints linked to the presence of man or the infrastructure necessary for the development of his activities. In such an approach it may be useful to draw a distinction between apparent or expressed biodiversity, which reflects the environmental conditions at a given point in time whether under the influence of perturbations or not, and potential or real diversity, which comprises all plant and animal species present in the region or in the form of resistant organs (seeds, eggs, etc.) in the site and capable of appearing in the various states of the transition system in a given geomorphological and climatic context (Margalef, Chapter 2).

CONCLUSIONS

Although the large body of information collected in this book is relatively heterogeneous and essentially qualitative in nature, it sheds much light on biodiversity in land–inland water ecotones and demonstrates that ecotones are key landscape elements for biodiversity conservation as a whole. They contribute to a better understanding of biodiversity patterns and regulating factors at the scale of ecosystems, showing the degree of their interdependence at the different spatial and temporal scales. The information documents how in temperate regions the ecotones are generally human-dominated or heavily impacted, so research efforts need to be combined with sustainable development.

In recent years freshwater ecologists have made great strides in understanding the structure, function and vulnerability of lakes, ponds and rivers. After having centred research on homogeneous ecological systems, the parallel development of new research on the interface zones between ecological systems thus appears to be of major interest and very promising.

Based on the volume of empirical evidence emerging from case studies, the authors demonstrate the manifold interactions between the different biotic groups (especially between animal groups and vegetation) and the importance of terrestrial–aquatic interactions in determining biological diversity.

It has been shown that an understanding of the biodiversity phenomenon depends very clearly on our ability to comprehend the complexity of ecological systems and the relations that link them at different temporal and spatial scales. It requires the development of a new type of research which no longer aims only to reduce as much as possible the number of explanatory variables but, on the contrary, to conserve and take into account the diversity of all the processes and mechanisms at work in their spatio-temporal context.

This book also shows the difficult path that lies ahead if we are to understand the biodiversity phenomenon (particularly evaluation of the human impacts) and to be able to manage it intelligently (identification and application of the most appropriate protection measures). It clearly reveals the critical need for knowledge in order to understand and, in the long term, to try and optimize, the interaction dynamics between human development and biodiversity.

The basic characteristics and the questions raised in this book justify continuing research into ecotones along two main lines of investigation: the structure and functioning of ecotones at different scales of perception, and the effects of human activities on this functioning and thus on ecotone management. The answers that will be provided to the numerous pending questions will enable us in future to gain a better understanding not only of the determinants of biodiversity but also – and this is one of the major objectives of the 'DIVERSITAS' programme launched by the International Union of Biological Sciences (IUBS), the Scientific Committee on Problems of the Environment (SCOPE) and UNESCO (Solbrig, 1991b) – to understand better the functions that it performs at the scale of ecological systems.

Finally, as responsible scientists, we cannot conclude without launching an urgent appeal:

- to scientists, to develop research on biodiversity in land–inland water ecotones and not to hesitate to come down from their ivory towers and get more involved in solving the concrete problems that exist and that require urgent solutions in collaboration with the administrative, socioeconomic and political bodies;

- to regional planning managers, to take into account the specific characteristics and vulnerability of land–inland water ecotones;

- to education officials (undergraduate, school, university), to develop education and training in the (natural and socioeconomic) environmental sciences and increase the funds necessary for studying biodiversity (ecology and taxonomy, for example);

- to managers of the media, to develop the necessary means – tailored to the cultural context – of creating an awareness in the broadest possible public of the issues connected with the erosion of biodiversity;

- to politicians, to grant biodiversity conservation a central place in their many concerns and to finance the research programmes necessary for an understanding of ecological systems and their biodiversity, so that judicious and sustainable management of resources can be guaranteed;

- finally, to each individual, to take account of biodiversity in his or her decisions and behaviour patterns.

REFERENCES

Aragno, M. and Ulehlova, B. (1996). Microbial diversity and functions in land–inland water ecotones. In Lachavanne, J.-B. and Juge, R. (eds.) *Biodiversity in Land–Inland Water Ecotones*, pp. 81–108. UNESCO/Parthenon, Paris/Carnforth

Armstrong, W. (1975). Waterlogged soils. In Etherington, J.R. (ed.) *Environment and Plant Ecology*, pp. 181–218. John Wiley, London

Armstrong, W. (1978). Root aeration in wetland conditions. In Hook, D.D. and Crawford, R.M.M. (eds.) *Plant Life in Anaerobic Environment*, pp. 269–97. Ann Arbor Science Publishers, Ann Arbor, MI

Backéus, I. (1993). Ecotone *versus* ecocline: vegetation zonation and dynamics around a small reservoir in Tanzania. *Journal of Biogeography*, **20**, 209–18

Beaumont, P. (1975). Hydrology. In Whitton, B. (ed.) *River Ecology*, pp. 1–38. Blackwell Scientific Publications, Oxford

Bedinger, M.S. (1979). Relation between forest species and flooding. In Greeson, P.E., Clark, J.R. and Clark, J.E. (eds.) *Wetland Values and Functions: the State of Our Understanding*, pp. 427–35. Proceedings of the national symposium on wetlands, (1979). American Water Resources Association, Minneapolis, MN

Brooks, R.P, Samuel, D.E. and Hill, J.B. (eds.) (1985). *Wetlands and water management on mined lands*. Proceedings of a Conference Oct. 23–24, 1985, The Pennsylvania State University, University Park, PA

Carter, V., Bedinger, M.S., Novitzki, R.P. and Wilen, W.O. (1979). Water resources and wetlands. In Greeson, P.E., Clark, J.R. and Clark, J.E. (eds.) *Wetland Functions and Values: the State of our Understanding*, pp. 344–76. Proceedings of the national symposium on wetlands, (1979). American Water Resources Association, Minneapolis, MN

Clements, F.E. (1905). *Research Methods in Ecology*. University Publishing Company, Lincoln, NE

Clements, F.E. (1916). *Plant Succession: An Analysis of the Development of Vegetation*. Publication 242, Carnegie Institution of Washington, Washington, DC

Cockburn, A. (1991). *An Introduction to Evolutionary Ecology*. Blackwell Scientific Publications, Oxford

Cody, M.L. (1975). Towards a theory of continental species diversities: bird distributions over Mediterranean habitat gradients. In Cody, M.L. and Diamond, J.M. (eds.) *Ecology and Evolution of Communities*, pp. 214–57. Belknap Press of Harvard University Press, Cambridge, MA

Connell, J.H. (1978). Diversity in tropical rain forests and coral reefs. *Science*, **199**, 1302–10

Cowardin, L.M., Carter, V., Golet, F.C. and LaRoe, E.T. (1979). *Classification of Wetlands and Deep Water Habitats of the United States*. US Fish and Wildlife Service, Pub. FWS/OBS-79/31, Washington, DC

Di Castri, F. and Hansen, A.J. (1992). The environment and development crises as determinants of landscape dynamics. In Hansen, A.J. and di Castri, F. (eds.) *Landscape boundaries: consequences for biotic diversity and ecological flows*, pp. 3–18. Ecological Studies 92, Springer Verlag, New York

Di Castri, F., Hansen, A.J. and Holland, M.M. (eds.) (1988). A New Look at Ecotones: Emerging International Projects on Landscape Boundaries. *Biology International*, Special Issue **17**, 1–163

Di Castri, F. and Younès, T. (1990). Ecosystem Function of Biological Diversity. *Biology International*, Special Issue, **22**, 1–20

Dobrowolski, K.A. (1996). Bird diversity in ecotonal habitats. In Lachavanne, J.-B. and Juge, R. (eds.) *Biodiversity in Land–Inland Water Ecotones*, pp. 205–21. UNESCO/Parthenon, Paris/Carnforth

Dykyjová, D. and Kvet, J. (eds.) (1978). *Pond Littoral Ecosystems. Structure and Functioning*. Ecological Studies 28, Springer Verlag, New York

Etherington, J.R. (1983). *Wetland Ecology*. Edward Arnold, London

Gibert, J., Dole-Olivier, M.J., Marmonier, P. and Vervier, P. (1990). Surface water–ground water ecotones. In Naiman, R.J. and Décamps, H. (eds.) *The Ecology and Management of Aquatic–Terrestrial Ecotones,* pp. 199–225. MAB Book Series 4. UNESCO/Parthenon, Paris/Carnforth

Giudicelli, J. and Bournaud, M. (1996). Invertebrate biodiversity in land–inland water ecotonal habitats. In Lachavanne, J.-B. and Juge, R. (eds.) *Biodiversity in Land–Inland Water Ecotones,* pp. 143–160. UNESCO/Parthenon, Paris/Carnforth

Good, R.E., Whigham, D.F. and Simpson, R.L. (eds.) (1978). *Freshwater Wetlands: Ecological Processes and Management Potential.* Academic Press, New York

Gopal, B., Turner, R.E., Wetzel, R.G. and Whigham, D.F. (eds.) (1982). *Wetlands: Ecology and Management.* National Institute of Ecology, International Science Publishers, Jaipur

Gosselink, J.G. and Turner, R.E. (1978). The role of hydrology in freshwater wetland ecosystems. In Good, R.E., Whigham, D.F. and Simpson, R.L. (eds.) *Freshwater Wetlands: Ecological Processes and Management Potential,* pp. 63–78. Academic Press, New York

Gosz, J.R. (1991). Fundamental ecological characteristics of landscape boundaries. In Holland, M.M., Risser, P.G. and Naiman, R.J. (eds.) *Ecotones: the Role of Landscape Boundaries in the Management and Restoration of Changing Environments,* pp. 8–31. Chapman and Hall, New York and London

Hansen, A.J. and di Castri, F. (eds.) (1992). *Landscape boundaries: consequences for the biotic diversity and ecological flows.* Ecological Studies 92, Springer Verlag, New York

Hansen, A.J., di Castri, F. and Naiman, R.J. (1988). Ecotones: what and why? In di Castri, F., Hansen, A.J. and Holland, M.M. (eds.) A New Look at Ecotones: Emerging International Projects on Landscape Boundaries. *Biology International,* Special Issue, **17**, 9–46

Hansen, A.J., Risser, P.G. and di Castri, F. (1992a). Epilogue: biodiversity and ecological flows across ecotones. In Hansen, A.J. and di Castri, F. (eds.) *Landscape boundaries: consequences for biotic diversity and ecological flows,* pp. 423–38. Ecological Studies 92, Springer Verlag, New York

Hansen, A., Urban, D.L. and Marks, B. (1992b). Avian community dynamics: the interplay of landscape trajectories and species life histories. In Hansen, A.J. and di Castri, F. (eds.) *Landscape boundaries: consequences for biotic diversity and ecological flows,* pp. 170–95. Ecological Studies 92, Springer Verlag, New York

Haslam, S.M. (1978). *River Plants. The Macrophytic Vegetation of Watercourses.* Cambridge University Press, Cambridge

Hasler, A.D. (ed.) (1975). *Coupling of Land and Water Systems.* Springer Verlag, Berlin

Heinselman, M.L. (1970). Landscape evolution, peatland types and the environment in the Lake Agassiz Peatland Natural Area, MN. *Ecological Monographs,* **40**, 235–61

Hillbricht-Ilkowska, A., Carney, H., Décamps, H., Hunsaker, C.T., Johnston, C., Klimes, L., Muhlhauser, H., Nishino, M., Rambouskova, H., Salo, J. and Whigham, D. (1989). Role of ecotones in monitoring changes or stability of landscape patches. In Naiman, R.J., Décamps, H. and Fournier, F. (eds.) Role of Land–Inland Water

Ecotones in Landscape Management and Restoration: a Proposal for Collaborative Research, pp. 65–68. *MAB Digest No 4*, UNESCO, Paris

Holland, M.M. (compiler). (1988). SCOPE/MAB technical consultations on landscape boundaries. Report of a SCOPE/MAB workshop on ecotones. In di Castri, F., Hansen, A.J. and Holland, M.M. (eds.) A New Look at Ecotones: Emerging International Projects on Landscape Boundaries. *Biology International*, Special Issue, **17**, 47–106

Holland, M.M., Risser, P.G. and Naiman, R.J. (eds.) (1991). *Ecotones: The Role of Landscape Boundaries in the Management and Restoration of Changing Environments*. Chapman and Hall, New York and London

Ingram, H.A.P. (1983). Hydrology. Vol 4A, Mires: swamp, bog, fen and moor, general studies. In Gore, A.J.P. (ed.) *Ecosystems of the World*, pp. 67–158. Elsevier, Amsterdam

Jacobs, J. (1975). Diversity, stability and maturity in ecosystems influenced by human activities. In Van Dobben, W.H. and Lowe-McConnell, R.H. (eds.) *Unifying Concepts in Ecology*, pp. 187–207. Dr W. Junk, The Hague

Jenik, J. (1992). Ecotone and ecocline: two questionable concepts in ecology. *Ekologia (CSFR)* **11**(3), 243–50

Johnson, F.L. and Bell, D.T. (1976). Plant biomass and net primary production along a flood frequency gradient in the streamside forest. *Castanea*, **41**, 156–65

Joly, P. and Morand, A. (1996). Amphibian diversity and land–water ecotones. In Lachavanne, J.-B. and Juge, R. (eds.) *Biodiversity in Land–Inland Water Ecotones*, pp. 161–82. UNESCO/Parthenon, Paris/Carnforth

Juge, R. and Lachavanne, J.-P. (1996). Patterns and regulation of plant diversity in lacustrine ecotones. In Lachavanne, J.-B. and Juge, R. (eds.) *Biodiversity in Land–Inland Water Ecotones*, pp. 109–45. UNESCO/Parthenon, Paris/Carnforth

Kent, D.H. (ed.) (1994). *Applied Wetlands Science and Technology*. Lewis Publishers, London

Kvet, J., Arunin, A., Bugenyi, B., Davis, C., Desaigues, B., Fournier, F., Holland, M., Husain, R.G.K., Jensen, A., Joly, C.A., Juge, R., Moller, H., Nachtnebel, H.P., Nontji, A., Okada, M., Petts, G., Schneider, E., Sedell, J. and Simpson, P. (1989). Management, restoration and creation of ecotones. In Naiman, R.J., Décamps, H. and Fournier, F. (eds.) Role of land–inland water ecotones in landscape management and restoration: a proposal for collaborative research, pp. 73–5. *MAB Digest No 4*, UNESCO, Paris

Lachavanne, J.-P. (1996). Why study biodiversity in land–inland water ecotones? In Lachavanne, J.-B. and Juge, R. (eds.) *Biodiversity in Land–Inland Water Ecotones*, pp. 1–45. UNESCO/Parthenon, Paris/Carnforth

Lachavanne, J.-P., Juge, R. and Noetzlin, A. (1986). Evolution des macrophytes du Léman (rives genevoises 1972–1984). *Sciences de l'Eau*, **5**, 419–33

Leopold, A. (1933). *Game Management*. Charles Scribner's Sons, New York

Livingstone, D.C. (1903). The distribution of the upland societies of Kent County, Michigan. *Botanical Gazette*, **35**, 36–55

Loeffler, H. (1990). Wetland objects: human uses. In Patten, B.C. (ed.) *Wetlands and Shallow Continental Water Bodies*, pp. 17–30. SPB Academic Publishing, The Hague

MacArthur, R.H. (1972). *Geographical Ecology: Patterns in the Distribution of Species*. Harper & Row, New York

Margalef, R. (1969). Diversity and stability: a practical proposal and a model of interdependence. In *Diversity and Stability in Ecological Systems,* pp. 25–38. Brookhaven Symposium in Biology No 22

Margalef, R. (1996). Dynamics of diversity and its expression over gradients and boundaries. In Lachavanne, J.-B. and Juge, R. (eds.) *Biodiversity in Land–Inland Water Ecotones*, pp. 47–59. UNESCO/Parthenon, Paris/Carnforth

McKnight, J.S., Hook, D.D., Langdon, O.G. and Johnson, R.L. (1981). Flood tolerance and related characteristics of trees of the bottomland forests of the southern United States. In Clark, J.R. and Benforado, J. (eds.) *Wetlands of Bottomland Hardwood Forests,* pp. 29–69. Elsevier, Amsterdam, Oxford and New York

McNeely, J.A., Miller, K.R., Reid, W.V., Mittermeier, R.A. and Werner, T.B. (1990). *Concerning the World's Biological Diversity.* IUCN, WRI, CI, WWF-US, the World Bank, Gland Switzerland and Washington, DC

Merriam, G. and Wegner, J. (1992). Local extinctions, habitat fragmentation and ecotones. In Hansen, A.J. and di Castri, F. (eds.) *Landscape boundaries: consequences for biotic diversity and ecological flows,* pp. 150–69. Ecological Studies 92, Springer Verlag, New York

Minshall, G.W. (1988). Stream ecosystem theory: a global perspective. *Journal of the North American Benthological Society*, 7, 263–88

Mitsch, W.J. (1991). Ecological engineering – approaches to sustainability and biodiversity in US and China. In Constanza, R. (ed.) *Ecological Economic: The Science and Management of Sustainability*, pp. 428–48. Columbia University Press, New York

Mitsch, W.J. and Gosselink, J.G. (1986). *Wetlands.* Van Nostrand Reinhold, New York

Naiman, R.J. and Décamps, H. (eds.) (1990). *The Ecology and Management of Aquatic–Terrestrial Ecotones.* MAB Book Series 4. UNESCO/Parthenon, Paris/Carnforth

Naiman, R.J. and Décamps, H. (1991). Landscape boundaries in the management and restoration of changing environments: a summary. In Holland, M.M., Risser, P.G. and Naiman, R.J. (eds.) *Ecotones: the Role of Landscape Boundaries in the Management and Restoration of Changing Environments,* pp. 130–7. Chapman and Hall, New York and London

Naiman, R.J., Décamps, H. and Fournier, F. (eds.) (1989). Role of land–inland water ecotones in landscape management and restoration: a proposal for collaborative research. *MAB Digest 4*, UNESCO, Paris

Naiman, R.J., Holland, M.M., Décamps, H. and Risser, P.G. (1988). A new UNESCO programme: research and management of land–inland water ecotones. In di Castri, F., Hansen, A.J. and Holland, M.M. (eds.) A New Look at Ecotones: Emerging International Projects on Landscape Boundaries. *Biology International*, Special Issue, **17**, 107–36

National Research Council. (1992). *Restoration of Aquatic Ecosystems: Science, Technology and Public Policy.* National Academy Press, Washington, DC

Neilson, R.P., King, G.A., DeVelice, R.L. and Lenihan, J.M. (1992). Regional and local vegetation patterns: the responses of vegetation diversity to subcontinental air masses. In Hansen, A.J. and di Castri, F. (eds.) *Landscape boundaries: consequences for biotic diversity and ecological flows,* pp. 129–49. Ecological Studies 92, Springer Verlag, New York

Odum, E.P. (1968). Energy flow in ecosystems: a historical review. *American Zoologist,* **8**, 11–18

Odum, E.P. (1969). The strategy of ecosystem development. *Science,* **164**, 262–70

Petts, G.E. (1996). Scientific basis for conserving diversity along river margins. In Lachavanne, J.-B. and Juge, R. (eds.) *Biodiversity in Land–Inland Water Ecotones,* pp. 249–68. UNESCO/Parthenon, Paris/Carnforth

Pieczyńska, E. (ed.) (1976). *Selected Problems of Lake Littoral Ecology.* Warsaw University, Warsaw

Pieczyńska, E. and Zalewski, M. (1996). Habitat complexity in land–inland water ecotones. In Lachavanne, J.-B. and Juge, R. (eds.) *Biodiversity in Land–Inland Water Ecotones,* pp. 61–79. UNESCO/Parthenon, Paris/Carnforth

Pimm, S.L. (1991). *The Balance of Nature.* Chicago University Press, Chicago

Risser, P.G. (compiler). (1985). *Spatial and Temporal Variability of Biospheric and Geospheric Processes: Research Needed to Determine Interactions with Global Environmental Change.* The International Council of Scientific Union Press, Paris

Risser, P.G. (1990). The ecological importance of land–water ecotones. In Naiman, R.J. and Décamps, H. (eds.) *The Ecology and Management of Aquatic–Terrestrial Ecotones,* pp. 7–21. MAB Book Series 4. Parthenon Publishing, Carnforth and UNESCO, Paris

Rusek, J. (1992). Distribution and dynamics of soil organisms across ecotones. In Hansen, A.J. and di Castri, F. (eds.) *Landscape boundaries: consequences for the biotic diversity and ecological flows, pp.* 196–214. Ecological Studies 92, Springer Verlag, New York

Schröpfer, R. (1996). Mammal diversity in inland water ecotone habitats. In Lachavanne, J.-B. and Juge, R. (eds.) *Biodiversity in Land–Inland Water Ecotones,* pp. 223–31. UNESCO/Parthenon, Paris/Carnforth

Sedell, J.R., Steedman, R.J., Regier, H.A. and Gregory, S.V. (1991). Restoration of human impacted land–water ecotones. In Holland, M.M., Risser, P.G. and Naiman, R.J. (eds.) *Ecotones: The Role of Landscape Boundaries in the Management and Restoration of Changing Environments,* pp. 110–29. Chapman and Hall, New York and London

Simberloff, D. (1988). The contribution of population and community biology to conservation science. *Annual Review of Ecological Systems,* **19**, 473–511

Solbrig, O. T. (ed.) (1991a). *From Genes to Ecosystems: a Research Agenda on Biodiversity.* IUBS–SCOPE–UNESCO. Cambridge, MA

Solbrig, O. T. (1991b). Biodiversity: Scientific Issues and Collaborative Research Proposals. *MAB Digest 9,* UNESCO, Paris

Soulé, M.E. (ed.) (1987). *Viable Populations for Conservation.* Sinauer Associates, Inc. Publishers, Sunderland, MA

Soulé, M.E. (1991). Conservation: tactics for a constant crisis. *Science,* **253**, 744–50

Van der Maarel, E. (1990). Ecotones and ecoclines are different. *Journal of Vegetation Science,* **1**, 135–8

Ward, J.V. and Stanford, J.A. (1983). The intermediate-disturbance hypothesis: an explanation for biotic diversity patterns in lotic ecosystems. In Fontaine, T.D. and Bartell, S.M. (eds.) *Dynamics of Lotic Ecosystems,* pp. 347–56. Ann Arbor Science, Ann Arbor, MI

Wetzel, R., Adams, M., Almassy, A., Dinka, M., Eiseltova, M., Fustec, E., Gibert, J., Gopal, B., Leichtfried, M., Livingstone, R.J., Mathieu, J., Parma, S., Pastor, J., Pieczyńska, E., Roberts, G., Sagova, M., Thompson, K. and Uerhoog, F. (1989). Nutrient, energy and water flows through land–water ecotones. In Naiman, R.J., Décamps, H. and Fournier, F. (eds.) Role of Land–Inland Water Ecotones in Landscape Management and Restoration: a Proposal for Collaborative Research, pp. 61–4. *MAB Digest No 4*, UNESCO, Paris

Zalewski, M. (1996). Fish diversity and ecotonal habitat. In Lachavanne, J.-B. and Juge, R. (eds.) *Biodiversity in Land–Inland Water Ecotones*, pp. 183–203. UNESCO/Parthenon, Paris/Carnforth

INDEX

Page numbers in **bold** type refer to figures, and those in *italics* to tables.

abiotic factors in ecosystem regulation 184–187
acetogenesis 89–90
adenosine triphosphate (ATP), in chemotrophic metabolism 87–88, 91
aerobic conditions, chemotrophic metabolism in 90–92
agriculture
   and bird diversity 212, **213**, 218
   and conservation of genetic diversity 4
   effect on riparian ecosystems 193–194
   effects of on Lake Titicaca 242–246
   and threat to ecotones 22, 23
air breathing adaptations, in fish 186
alder woods, and bird diversity 210, 214, **216**, 217–218, **219**
algae
   in amphibian larval diet 164, 167, 168, 170
   diversity in Lake Titicaca **237**, 238
   effects of 67–68
   and macrophytes 68
   and raised field agriculture 243
Amazon River, biodiversity of 238–239, *240*
ammonia, in dinitrogen fixation 92–93
amphibians
   adaptations to ecological constraints 168–173, 175–176

diversity of 161–176
extinction probability and population size 172–173, 176
and fish predation 171–172
habitats and human pressure 173–175
larval diet 164–165, 175
tadpoles in nutrient cycle 167–168
amphicoenous adaptation
   by birds 208, 211–212, 215–219
   by mammals 228–229
amphicoenous adaptation *see also* related ecosystems
anaerobic conditions, chemotrophic metabolism in 88
anaerobic respiration, by bacteria 88–90
anoxic conditions
   and plant litter accumulation 96
   in wetlands, in regulation of plant diversity 124
anoxygenic photosynthesis, by bacteria 87
aquatic bank of rivers, characteristics of 145
Arc, hydrographic basin, and invertebrate communities 146–148
Archaea, metabolism of 85, 90

bacteria
   biomass measurement of 83–84
   diversity assessment in 84–86

functions of 86–94
  in the rhizophere 99
  in swamp conditions 96–97
  and water table changes 102
beetles
  aquatic life-style of 150
  in floodplain habitats 251, **253**, *255*
behaviour strategies, of semiaquatic mammals 225–228
Biebrza River, and bird diversity 205–208, **209**, 212–214
biodiversity
  conservation of in ecotones 19–25
  definition of 1–2
  and dynamics of diversity 47–58
  and the ecotonal concept 279–285
  and ecotonal degradation 285–286
  erosion of 2–8
  evaluation of in microbial habitats 83–86
  factors affecting 249
  and hydrology 283–284
  importance of 233–234, 241, 242, 251, 254
  of invertebrates, evaluation of 143–144
  issues on 31–32
  in Lake Titicaca *237*, 245
  needs for protection of 291
  and politics 270–272
  and related ecosystems 282–283
  scientific issues of 269–275
  study needs of 27–30
biodiversity *see also* diversity
biomass measurement, in ecotonal environments 83–84
biopolymers, in chemotrophic metabolism 87–88
biotic factors in ecosystem regulation 184–187
bird diversity 205–209
  and forest structure 210–211
  in lake ecotones 217–219
  in river ecotones 205–217
  and urban habitats 215–217
bird diversity *see also* waterfowl
boundaries, complexities of 27, 52–53
breeding *see* reproduction

breeding sites
  for amphibian reproduction 165–167
  for insect reproduction 151
  salmon, for egg-laying 285
canals, microhabitats in 154, **155**
Carabidae
  abundance of 148
  in floodplain habitats 251, **253**
carbon
  in chemotrophic metabolism 87
  and litter decomposition 96–98
  transport of across ecotones 9
  utilization of by invertebrates 53
carbonate respiration 89–90, 91
carnivorous semiaquatic mammals, niche occupation by 226–227
channelization
  and ecotone damage 71
  impact of on biodiversity 173, 183, 250
chemolithoautotrophic bacteria, in swamp conditions 101
chemolithoautotrophic oxidations 87, 90–92
chemotrophic metabolism, of bacteria 82, 87–92
climate change, importance of 184, 233
Coleoptera
  abundance of 146, 147, 148
  in floodplain habitats 251, **253**
  in river banks 152, 153
  use of aquatic conditions 151
commensal relationships, by microbial organisms 94–95
community dynamics
  fish in regulation of 189–193
  and related ecosystems 282–283
competition
  in amphibians 166
  between aquatic plants, effects of 125
conservation
  of biodiversity in ecotones 19–25
  need for 4, 233–234, 241, 242, 251, 286–287
Corsica, species richness in mountain streams 152–153
cultural methods, in assessment of microbial diversity 83–84, 86

300

current velocity, effects of 144, 145

Daphnia, in fish nutrient supply 69, 192
denitrification (nitrate respiration) 88–89
diatoms, and index of diversity 50–51
dinitrogen fixation 92–94
   in swamp conditions 99–101
Diptera
   abundance of 147
   in rarity index 255
   in river banks 152, 153, 255
   use of aquatic conditions 151
"Diversitas" programme, objectives of 31, 32
diversity
   of amphibia 161–176
   dynamics of 47–58
   of plants in lacustrine ecotones 111–115
   regulation of in plants 119–126
   roles of in plants 126–127
   of species, need for research on 270, 274
diversity *see also* biodiversity
dormancy
   bacterial 84
   of plants 112

ecocline
   concept of 13–15
   definition of 54
ecological plasticity, and life strategy in plants 113
ecological successions
   and index of diversity 50–51
   study of 28–30
ecology, and ecotones 11–12
ecosystem dynamics 68–71
ecosystems, related, and community dynamics 282–283
ecotones
   characteristics of 61–68
   concept of 11–19, 277–279
   construction of 72–73
   definition of 8–11, 54
   degradation of 285–286
   destruction of 71–72
   differentiation in 279–282

   filtering function of 72–74
   importance of 19–25
   management needs of 25
   as microbial habitats 81–83
   perturbation of, and plant diversity *122–123*, 125–126
   rehabilitation of 288
   restoration of 254–265, 289–290
   river margin characteristics 261–265
   roles of 30–32, 68–71, 72–74, 174
   size of
      and management 73
      and water-level fluctuations 61–65
   study needs of 25–27
   threats to biodiversity in 22–25, 285–286
ecotones *see also* lakeside; river
edge effect
   concept of 109–110, 282
   definition of 20–21
   and ecotonal degradation 286
   and invertebrate diversity 149–151
   study of 28
   in wetlands 117, **118**
egg size, fish, and ecosystem regulation 186
energy flux, in the rhizosphere 98–101
energy supply, for micro-organisms 82
energy supply *see also* nutrient supply
environmental changes
   and invertebrate habitats 149–150
   and species diversity 2–3, 5, 21
environmental conditions
   in ecotones 279, 280
   and invertebrate diversity 144–151
   and life strategy in plants 112–113
environmental enhancement, and restoration of river margin ecosystems 254–256
environmental perturbation
   causes of 9, 14
   study of 30
Ephemeroptera, use of aquatic conditions 151, 152
ergocline, definition of 53, 54
erosion *see* shore-line erosion
eutrophication
   and fish diversity 183, 187–189

of wetlands, effects of 125–126
exopolymer biosynthesis 94
extinction probability, and population
    size, in amphibian communities
    172–173
extinction of species
    and erosion of biodiversity 2, 5
    in fish 196–197
    need for research on 270–272, 274

farmland *see* agriculture
feedback processes, and ecosystem
    dynamics 49, 68–71
feeding habits, and niche occupation of
    semiaquatic mammals 226–227
feeding methods, and invertebrate
    groups 146
feeding resources, and amphibian
    adaptations 168, 170–171
feeding sites, of waterfowl 205, **207**
fermentative metabolism, microbial 88,
    102
ferric oxide, in anaerobic respiration **89**,
    90
fighting zone, and ecotone size for
    mammals 229
filtering function
    of ecotones 11, 72–75
    of plants 126
    in raised field agriculture 243
fish
    and amphibian populations 165
    in community dynamics regulation
        189–193
    diversity
        and agricultural catchments
            193–194
        and ecosystem regulation 184–187
        and human pressure 183–184
        in Lake Titicaca 237
        stocking effects 194–195
    and ecotone damage 71
    and flooding in river margin
        ecosystems 262–264
    introduced species, dangers of 271
    morphological changes in 195
    as predators of amphibian larvae
        171–172

productivity of 187, 188
    in water quality biomanipulation
        68–69
fishing
    on Lake Titicaca 242, 244
    and loss of biodiversity 195, 197–198
    and river margin ecosystems 256
    and water quality biomanipulation 69
flooding
    and amphibians 166–167
    and biodiversity 249
    value of 262–264
    and vegetation type 18–19
floodplains
    and agriculture 193–194
    biodiversity of 154, 155–156
    invertebrate habitats on 150
    River Trent, biodiversity of 250, **252**,
        253
fluvial hydrosystem perspective
    257–260
forest *see also* trees
forest structure, and bird diversity
    210–211
form diversity, of semiaquatic
    mammals 224
fossil fields, by Lake Titicaca **235**, 242
functional diversity, in microbes,
    assessment of 86
functional studies, on river margin
    ecosystems 257
fungi, metabolism of 87, 91, 94, 95

genetic diversity
    loss of in fish 196–197
    need for research on importance of
        272
grazing, on aquatic plants, effects of 125
greenhouse gases, and global
    temperature 184

habitats
    for amphibians 162–163, 169–170,
        173–175
    and behaviour strategies of
        semiaquatic mammals 225–228
    and bird diversity 205–209, **210**,
        215–217

canals as 154, **155**
diversity of 10, 19–20, 22, 61–75
and environmental disturbance 2–3,
    5, 169–170, 173–175, 262–264
and fish diversity 187, 192, **193**, 194
in floodplains 250–251, **252**, **253**
and human pressure 173–175
and invertebrate diversity 146–149,
    156
loss, importance of 233
macrophytes in provision of 66, 69
and mammal diversity 223–229
microbial 81–86
plants in creation of 109–110,
    117–118, 123, 126–127, 284–285
of river banks 144–145
role of ecotones 72–73
heavy metals, depollution of 104
herbivorous semiaquatic mammals,
    niche occupation by 226
High Atlas mountains, hydrographic
    basins in, and invertebrate
    communities 146–149
human pressure
    in Amazonia 238
    and amphibian habitats 173–175
    control of effects of 288–289
    and ecotonal degradation 22–23,
        71–72, 285–286
    and environmental degradation
        183–184
    and eutrophication 187–188
    and impact on biodiversity 2–3, 5–6,
        249–250, 269, 271–272
    on Lake Titicaca 240–241, 242–246
    in plant diversity regulation 119,
        122–123
    in river margin ecosystems 255–256
    studies on 30, 257
    and sustainable development needs
        233–234, 242–246
Hydrachnellidae, in river banks 152–153
hydrogen-oxidizing bacteria 92, 99–101
hydrogenase activity 93
hydrological conditions
    and biodiversity 283–284
    disturbance
        and amphibians 166–167

in river margin ecosystems
    262–264
and ecotone type 18–19, 30
and human pressure 286
importance of 8–9, 14
in Lake Titicaca 245
and plant diversity 116–117, **118**,
    120–125
hydrophytes, growth forms of 113, **114**

index of diversity 48–49, 50, 51
index of rarity, of river margin
    ecosystems 254–255
inductive approach, to studies of fluvial
    hydrosystems 257–258
industrial development, and threat to
    ecotones 23
interference competition, in amphibians
    166
invertebrates
    and amphibian predation 164–165
    biodiversity of 143–156
        in Lake Titicaca **237**, 238, 240–241
    life-cycle, requirements of 150–151
    morphological reductions in
        psammobiont species 152
    plants in species diversity of 127
iron-oxidizing bacteria 92

'knallgas' bacteria, in swamp
    conditions 101

Lake Titicaca, features of 234, **235**,
    236–240, *240*, **241**, 242–246
lakeside ecotones
    bird diversity in 217–219
    characteristics of 61–65
    plant diversity in 109–128
    in pools 63–64
land–inland water ecotones *see* ecotones
landscape sustainability, on Lake
    Titicaca 244–246
life strategy, diversity of in plants 112–113
life-cycle
    of amphibians, and habitat 161–162,
        168–170
    of invertebrates, and habitat 150–152,
        155–156, 165–167

light
  availability of 64
  in regulation of plant diversity 123
  as energy source
    for micro-organisms 82
  and nitrogen fixation 93
limes convergens, definition of 52
limes divergens, definition of 52, 54
lithotrophy, process of 91
littoral zones, and vegetation types 115–119
Luknajno Lake, bird diversity in 217–218

MAB (Man and the Biosphere) Programme, aims of v–vi
macrophytes
  and amphibia 164
  diversity in Lake Titicaca **237**
  effects of 65–67, 68
  and raised field agriculture 243
mammal diversity 223–229
Man and the Biosphere Programme, aims of v–vi
management
  need for 286–290
  of river margin ecosystems 265
marshes
  drainage of, effects on biodiversity 250
  trees in 116–117
medicine, and conservation of genetic diversity 4
methane (swamp gas), bacterial production of 96–98
methanogenesis 89–90
  and water table changes 102
microbial biodiversity, assessment of 84–86
microbial ecology, importance of 95–102
microbial habitats
  biodiversity in 83–86
  ecotones as 81–83
microecotones, and microbial ecology 81
migration
  by amphibians 161, 166, 172–175
  by birds 205, 211, 215
  by fish, disturbance of 189–192, 194
  and ecotones 20–22
Mississippi river, invertebrate diversity in 151, **152**
molecular biology, in assessment of taxonomical diversity 85–86
molecular variation, need for research on importance of 273, 274
morphological adaptations, of semiaquatic mammals 224–225
Murray river, invertebrate diversity in 151

Narew River, and bird diversity 212, **213**, 214
nitrate respiration 88–89
nitrates, and raised field agriculture 243
nitrifying bacteria 91
nitrogen *see also* denitrification; dinitrogen
nitrogen fixation, in the rhizosphere 99
nitrogenase activity, test for 93
nutrient cycling
  across ecotones 9, 10
  amphibian tadpoles in 167–168
  by Lake Titicaca 244–246
  dynamics of 65–66, 68–71
  and ecotone construction 72–73, 75
  and fish 184, 188
  importance of 63
  nitrogen in amphibian life-cycle 167–168
  plant detritus in 126
  in raised field agriculture 243
  in streams 155–156
  and vegetation 284–285
  and water quality 68–69
nutrient supply
  and fish diversity 185–186, 187
  and invertebrate habitats 150
  in river margin ecosystems 261–264
nutrient supply *see also* energy supply

Oligochaetae
  abundance of 146, 147, **148**
  in river banks 153
organic matter, in streams 155–156

Index

organic pollutants, depollution of 104
organo-heterotrophic mineralization 90
Ostracada, larvae of, in river banks 152
oxbow lakes
  and amphibians 163
  and bird diversity 212
  and fish diversity 194
oxygen
  in bacterial photosynthesis 87
  in microbial metabolism 82–83
oxygen consumption
  in depollution 103, 104
  in submerged conditions 96
oxygen levels, and fish diversity 185–186
oxygen levels *see also* eutrophication
oxygen requirements, and water table changes 102
oxygenation, of sediments, by rooted plants 126–127

parasitic algae, in interference competition in amphibians 166
parasitic relationships, by microbial organisms 95
parental protection, by fish 186
patches, in ecotones 9, 15
perch, in elimination of *Daphnia* 69
phospholipid fatty acids (PLFA), in biomass determination 84–85
phosphorus, and raised field agriculture 243
photosynthesis
  by bacteria 87
  by micro-organisms 82
  and carbonate precipitation 66
phytoplankton
  in fish nutrient supply 187, **190–191**, 192
  and index of diversity 50–51
  and macrophytes 66, 68
  in nutrient cycling 68–69
  spatial differences in diversity in 53, **54**, *55*, *56*, 57
pike, predatory nature of 188, 192
plankton *see* phytoplankton
plant litter
  and aerobic degradation 82

  anaerobic decay of 88, 90
  decomposition of, and carbon flux 96–98
  effects of 61–62, 65–67
  and invertebrate feeding 147, 150
  in nutrient cycle 126
plants
  diversity
    in lacustrine ecotones 109–128
    regulation of 119–126
    roles of 126–127
  importance of 127–128
  structural diversity in 113
  succession in 117–119
  taxonomic diversity of 112, **115**, 117–119
Plecoptera, larvae of, in river banks 152
Poland
  Lake Mikolajskie 62, 63
  lake sediment, composition of 63
  Lake Sniardwy 61
politicians, and biodiversity research 270–272, **275**
pollution
  and amphibians 174
  control of 11, 72–73, 194, 288–289
  and ecotonal degradation 286
  and ecotone damage 71
  and erosion of biodiversity 5
  and fish 183, 194, 195
  importance of 233
  in Lake Titicaca 244, 245
  micro-organisms in removal of 103–104
  and morphological changes in fish 195
  vegetation in removal of 72–73
pollution *see also* industrial development
polymer *see* exopolymer
ponds, and amphibians 163, 165–166
population dynamics, and species diversity 47, 48
population size, of fish 196
population structure, of semiaquatic mammals 229
potato, genetic diversity of in Lake Titicaca basin 241
predation
  by amphibians on invertebrates 164–165, 168, 171

by fish 188, 192
by microbial organisms 94–95
by newts and salamanders 164

raised fields, by Lake Titicaca **235**, 242–246
rarity index, of river margin ecosystems 254–255
reed beds, and bird diversity 211–212, **213**, 217–219
rehabilitation, assessment of 288
related ecosystems, and community dynamics 282–283
related ecosystems *see also* amphicoenous adaptation
reproduction
  of amphibians, flexibility in 163, 168–170
  of birds
    and habitat 212, 214, 218
    waterfowl 205, **207**, **208**
  of fish, and habitat **193**, 194
  and life strategy in plants 113
research, on biodiversity, need for 270–275, 287, 290
reservoirs, eutrophication of 189
respiratory metabolism
  systems of 88–90
  and water table changes 102
restoration
  measures for 289–290
  of river margin ecosystems 254–265
Rhine river, aquatic banks of 154, **155**
rhizophere, environment in 98–101
Rhône river
  and amphibians 162, 165–166
  and invertebrate communities 144, 145, 146, 151
  river margin ecosystems of 260
rice paddy, nitrogen fixation in 93
river margin ecosystems
  biodiversity in 54–58
  bird diversity in 205–217
  characteristics of 61–65, 249–256, 261–265
  corridors, for vertebrates 20–21
  faunal diversity in 151–156
  human pressure on 255–256

invertebrate biodiversity in 144–149
management of 260–261, 265
patches in 258–261
restoration of 254–265
roach, functional adaptations of 189
roots, roles of in microbial metabolism 98–101
Rotifera, in the Murray river 151

salinity, in Lake Titicaca 245
salmon
  artificial stocking of 195
  egg size **186**
  egg-laying sites of 285
  and restoration of river ecosystems 256
scientific basis, of restoring river margin ecosystems 256–265
scientific issues, of biodiversity 269–275
sediment
  and amphibians 167
  and biodiversity 283–284
  as biodiversity stores 51, 57
  composition of 63
  importance of 9–10
  and invertebrate habitats 149–150
  oxygen diffusion in 82
  and plant communities 119
  quality of in aquatic banks 145, 147
  and raised field agriculture 243, 246
  and river management 259, 260
semiaquatic mammals, diversity of 223–229
shore-line development, and threat to ecotones 22–23
shore-line erosion
  and amphibians 167
  and biodiversity 283–284
  by Lake Titicaca 244
  and ecotone damage 71
  and invertebrate habitats 149–150
  and plant communities 119
  studies of 289
shore-line slope
  and ecotone size 15
  in ecotones 9, 10
  vegetation on 115–119, 240–241

Index

size diversity, of semiaquatic mammals 224–225
social behaviour, of semiaquatic mammals 229
soil type, in floodplain habitats 251
solar radiation, and damage to amphibian eggs 174
space, and dynamics of diversity 51–52
stocking, and fishery improvement 194–195
stress zones, ecotones as 11–15, 278
successional change, and diversity 50–51
sulfate-reducing bacteria 99, 101
sulfo-oxidizing bacteria 92
sulfur, bacterial use of 87, 89, **90**
sustainable development, need for 7, 233–234, 242–246
swamp gas (methane), bacterial production of 96–98
swamps
  and bird habitats 205–208, **209**, 212–215
  drainage of, and bird habitat degradation 214
  microbial ecology in 95–102
  oxygen levels in 96
symbiotic relationships, by microbial organisms 94
systematics, need for application of 270, 273

taxonomic diversity
  of invertebrates 146–149, 151–154
  in microbes, assessment of 85–86
  of plants 112, **115**, 117, **118**
taxonomy, importance of 7–8
terrestrial bank of rivers, characteristics of 145–149
time-dependence, of diversity 50–51
Tipulidae, use of aquatic conditions 151
tourism and leisure
  effects of on Lake Titicaca 242
  in river margin ecosystems 255–256
  and threat to ecotones 22–23
toxicity, of pollutants 103–104
transition zone, of ecotones 278–279
trees
  and bird diversity 210–211
  and ecotone construction 69, 73
  effects of on fish 188
  forest structure, and bird diversity 210–211
  in the littoral zone 116–117
Trent river, floodplain of 250, **252**, **253**, 255
Trichoptera, use of aquatic conditions 151
tropical ecotones, monitoring of 235–242

urban development, and threat to ecotones 22
urban habitats, and bird diversity 215–217

vegetation
  and biodiversity 284–285
  and bird diversity 211–212
  composition of
    and water-level fluctuation 66–67
    in wetlands 63
  continuum, study of 28–30
  effects of 17, 65–67
  and fish diversity **190–191**, 192, **193**, 194
  and invertebrate habitats 149–150
  on sloping shores 240–241
  and water quality 72–75
vegetation *see also* plants; trees
Volga River, and bird diversity 205, **208**

water, free-flowing in rivers 56–57
water flow, factors affecting 64, 65–66
water movement, and nutrient cycling 66
water quality
  biomanipulation of 68–69
  and fish biodiversity 193–194
  and invertebrate diversity 147, 148
  and raised field agriculture 243, 246
  in regulation of plant diversity 123
  and river management 260
  and vegetation 72–75
water table, changes in 102
water temperature
  and amphibian reproduction 163

and fish diversity 185–186
and reproductive flexibility in
   amphibians 169
of water, factors affecting 64–65
water-level changes 9–10, 15
   and Amazon River biodiversity 238
   and amphibians 162–163
   and biodiversity 283–284
   and fish diversity 68–70, 185,
      190–191, 192
   and invertebrate habitats 149–150
   and oxygen in soil 83
   and shoreline movement 61–65
   and vegetation types 29, 66–67,
      115–116, 117, **118**, 124–125

waterfowl, habitats of 205, **206**, **207**,
   209
waterfowl *see also* bird diversity
waves, action of on vegetation 116
wetlands
   characteristics of 62–63
   drainage of 22, 173, 183
   species richness in 117, **118**
wetlands *see also* marshlands; swamps
willow scrub, and bird diversity 210,
   212, 215–217, **218**, **219**
wind, influence of 64

zonation, of circumlacustrine vegetation
   115–119